SCALE ISSUES IN REMOTE SENSING

Edited by
QIHAO WENG

WILEY

Published by John Wiley & Sons, Inc., Hoboken, New Jersey
Published simultaneously in Canada

For general information on our other products and services or for technical support, please contact our Customer Care Department within the United States at (800) 762-2974, outside the United States at (317) 572-3993 or fax (317) 572-4002.

Wiley also publishes its books in a variety of electronic formats. Some content that appears in print may not be available in electronic formats. For more information about Wiley products, visit our web site at www.wiley.com.

Library of Congress Cataloging-in-Publication Data:

Scale issues in remote sensing / edited by Qihao Weng.
     pages cm
   ISBN 978-1-118-30504-1 (cloth)
  1. Remote sensing–Mathematics. 2.   Ecology—Mathematical models.
3.   Spatial ecology—Mathematical models. I. Weng, Qihao.
   G70.4.S24 2014
   621.36'78—dc23
                                          2013031571

Printed in the United States of America

10 9 8 7 6 5 4 3 2 1

CONTENTS

ACKNOWLEDGMENTS

I wish to extend my sincere thanks to all the contributors of this book for making this endeavor possible. Moreover, I offer my deepest appreciation to all the reviewers, who have taken precious time from their busy schedules to review the chapters. Finally, I am indebted to my family for their enduring love and support. It is my hope that this book will stimulate students and researchers to perform more in-depth analysis of scale issues in remote sensing and Geographic Information Science.

The reviewers of the chapters are listed here in alphabetical order: Aleksey Boyko, Alexander Buyantuyev, Anatoly Gitelson, Angelos Tzotsos, Benjamin Bechtel, Caiyun Zhang, Cedric Vega, Charles Emerson, Gang Chen, Guangxing Wang, Haibo Yao, Hannes Taubenboeck, Hongbo Su, Hong-lie Qiu, Iryna Dronova, Jianjun Ge, Lee De Cola, Prasad Thenkabail, Qi Chen, Shelley Meng, Xin Miao, Yuhong He, and Zhixiao Xie.

CONTRIBUTORS

Demetre Argialas, Remote Sensing Laboratory, National Technical University of Athens, Athens, Greece

Toby N. Carlson, Department of Meteorology, Pennsylvania State University, University Park, PA, USA

Manfred Ehlers, Institute for Geoinformatics and Remote Sensing, University of Osnabrück, Osnabrück, Germany

Fang Fang, School of Urban and Environmental Engineering, Ulsan National Institute of Science and Technology (UNIST), Ulju-gun, Ulsan, South Korea

Edward P. Glenn, Environmental Research Laboratory of the University of Arizona, Tucson, AZ, USA

Geoffrey J. Hay, Foothills Facility for Remote Sensing and GIScience, Department of Geography, University of Calgary, Calgary, Alberta, Canada

Yuhong He, Department of Geography, University of Toronto Mississauga, Mississauga, Ontario, Canada

Yang Hong, School of Civil Engineering and Environmental Science; Advanced Radar Research Center; Center for Analysis and Prediction of Storms, University of Oklahoma, Norman, OK, USA; Water Technology for Emerging Region (WaTER) Center, University of Oklahoma, Norman, OK, USA

Alfredo R. Huete, University of Technology Sydney, Sydney, New South Wales, Australia

Jungho Im, School of Urban and Environmental Engineering, Ulsan National Institute of Science and Technology (UNIST), Ulju-gun, Ulsan, South Korea; Department of Environmental Resources Engineering, State University of New York College of Environmental Science and Forestry (SUNY ESF), Syracuse, NY, USA

Konstantinos Karantzalos, Remote Sensing Laboratory, National Technical University of Athens, Athens, Greece

Sascha Klonus, Institute for Geoinformatics and Remote Sensing, University of Osnabrück, Osnabrück, Germany

Manqi Li, School of Urban and Environmental Engineering, Ulsan National Institute of Science and Technology (UNIST), Ulju-gun, Ulsan, South Korea

Bingqing Liang, Department of Geography, University of Northern Iowa, Cedar Falls, IA, USA

Hua Liu, Old Dominion University, Norfolk, VA, USA

Jeffrey C. Luvall, NASA Marshall Space Flight Center, Huntsville, AL, USA

Kiril Manevski, Department of Agro-Ecology and Environment, Aarhus University, Blichers Allé, Tjele, Denmark

Pamela L. Nagler, U.S. Geological Survey, Southwest Biological Science Center, Sonoran Desert Research Station, University of Arizona, Tucson, AZ, USA

George P. Petropoulos, Department of Geography and Earth Sciences, University of Aberystwyth, Wales, UK

Lindi J. Quackenbush, Department of Environmental Resources Engineering, State University of New York College of Environmental Science and Forestry (SUNY ESF), Syracuse, NY, USA

Dale A. Quattrochi, NASA, Marshall Space Flight Center, Huntsville, AL, USA

José L. Silvan-Cárdenas, Centro de Investigación en Geografía y Geomática "Ing. Jorge L. Tamayo" A.C., Mexico City, Mexico

Angelos Tzotsos, Remote Sensing Laboratory, National Technical University of Athens, Athens, Greece

Guangxing Wang, Geography and Environmental Resources, Southern Illinois University, Carbondale, IL, USA

Le Wang, Department of Geography University at Buffalo, State University of New York, Buffalo, NY, USA

Qihao Weng, Center for Urban and Environmental Change, Department of Earth and Environmental Systems, Indiana State University, Terre Haute, IN, USA

Maozhen Zhang, College of Environment and Resources, Zhejiang A&F University, Lin-An, ZheJiang, China

Yu Zhang, School of Civil Engineering and Environmental Science; Advanced Radar Research Center; Center for Analysis and Prediction of Storms, University of Oklahoma, Norman, OK, USA

AUTHOR BIOGRAPHY

 **Dr. Qihao Weng** is the Director of the Center for Urban and Environmental Change and a professor of geography at Indiana State University. He was a visiting NASA senior fellow (2008–2009). Dr. Weng is also a guest/adjunct professor at Peking University, Hong Kong Polytechnic University, Wuhan University, and Beijing Normal University, and a guest research scientist at Beijing Meteorological Bureau, China. He received his Ph.D. in geography from the University of Georgia in 1999. In the same year, he joined the University of Alabama as an assistant professor. Since 2001, he has been a member of the faculty in the Department of Earth and Environmental Systems at Indiana State University, where he has taught courses on remote sensing, digital image processing, remote sensing–GIS integration, GIS, and environmental modeling and has mentored 11 doctoral and 10 master students.

Dr. Weng's research focuses on remote sensing and GIS analysis of urban ecological and environmental systems, land use and land cover change, environmental modeling, urbanization impacts, and human–environment interactions. He is the author of over 150 peer-reviewed journal articles and other publications and 8 books. Dr. Weng has worked extensively with optical and thermal remote sensing data and more recently with lidar data, primarily for urban heat island study, land cover and impervious surface mapping, urban growth detection, image analysis algorithms, and the integration with socioeconomic characteristics, with financial support from U.S. funding agencies that include NSF, NASA, USGS, USAID, NOAA, National Geographic Society, and Indiana Department of Natural Resources. Dr. Weng was the recipient of the Robert E. Altenhofen Memorial

Scholarship Award by the American Society for Photogrammetry and Remote Sensing (1999), the Best Student-Authored Paper Award by the International Geographic Information Foundation (1998), and the 2010 Erdas Award for Best Scientific Paper in Remote Sensing by ASPRS (first place). At Indiana State University, he received the Theodore Dreiser Distinguished Research Award in 2006 (the university's highest research honor) and was selected as a Lilly Foundation Faculty Fellow in 2005 (one of the six recipients). In May 2008, he received a prestigious NASA senior fellowship. In April 2011, Dr. Weng was the recipient of the Outstanding Contributions Award in Remote Sensing in 2011 sponsored by American Association of Geographers (AAG) Remote Sensing Specialty Group. Dr. Weng has given over 70 invited talks (including colloquia, seminars, keynote addresses, and public speeches) and has presented over 100 papers at professional conferences (including co-presenting).

Dr. Weng is the Coordinator for GEO's SB-04, Global Urban Observation and Information Task (2012–2015). In addition, he serves as an associate editor of *ISPRS Journal of Photogrammetry and Remote Sensing* and is the series editor for both the Taylor & Francis Series in Remote Sensing Applications and the McGraw-Hill Series in GIS&T. His past service includes National Director of American Society for Photogrammetry and Remote Sensing (2007–2010), Chair of AAG China Geography Specialty Group (2010–2011), and Secretary of ISPRS Working Group VIII/1 (Human Settlement and Impact Analysis, 2004–2008), as well as a panel member of the U.S. DOE's Cool Roofs Roadmap and Strategy in 2010.

INTRODUCTION

1

CHARACTERIZING, MEASURING, ANALYZING, AND MODELING SCALE IN REMOTE SENSING: AN OVERVIEW

Qihao Weng

1.1 SCALE ISSUES IN REMOTE SENSING

Scale is a fundamental and crucial issue in remote sensing studies and image analysis. The University Consortium for Geographic Information Science (UCGIS) identified it as a main research priority area (1996). Scale influences the examination of landscape patterns in a region. The change of scale is relevant to the issues of data aggregation, information transfer, and the identification of appropriate scales for analysis (Krönert et al., 2001; Wu and Hobbs, 2002). Extrapolation of information across spatial scales is a needed research task (Turner, 1990). It is suggested that spatial characteristics could be transferred across scales under specific conditions (Allen et al., 1987). Therefore, we need to know how the information is transferred from a fine scale to a broad scale (Krönert et al., 2001). In remote sensing studies, use of data from various satellite sensors may result in different research results, since they usually have different spatial resolutions. Therefore, it is significant to examine changes in spatial configuration of any landscape pattern as a result of using different spatial resolutions of satellite imagery. Moreover, it is always necessary to find the optimal scale for a study in which the environmental processes operate. Theories, methods, and models for multiscaling are crucial to understand the heterogeneity of landscapes (Wu and Qi, 2000; Wu and Hobbs, 2002). Methods and techniques are important for the examination of spatial arrangements at a wide range of spatial scales. Regionalization

Scale Issues in Remote Sensing, First Edition. Edited by Qihao Weng.
© 2014 John Wiley & Sons, Inc. Published 2014 by John Wiley & Sons, Inc.

describes a transition from one scale to another, and upscaling or downscaling is an essential protocol in the transition (Krönert et al., 2001).

Characterized by irregularity and scale independence, fractals are recognized as a suitable method to capture the self-similarity property of the spatial structure of interest (Zhao, 2001). Self-similarity represents invariance with respect to scale. In geoscience, the property of self-similarity is often interpreted as scale independence (Clarke, 1986). However, most environmental phenomena are not pure fractals at all scales. Rather, they only exhibit a certain degree of self-similarity within limited regions and over limited ranges of scale, which is measurable by using statistics such as spatial auto-covariances. The underlying principle of fractals is to use strict or statistical self-similarity to determine the fractal dimension (FD) of an object/ surface, which is often used as an indicator of the degree of irregularity or complexity of objects. When fractals are applied to remote sensing, an image is viewed as a complex "hilly terrain surface" whose elevations are represented by the digital numbers. Consequently, FDs are readily computable and can be used to denote how complicated the "image surfaces" are. Remote sensing studies assume that spatial complexity directly results from spatial processes operating at various levels, and higher FD occurs at the scale where more processes operate. With FDs, the spatial processes that occurred at different scales are measurable and comparable. Compared to other geospatial algorithms in image analysis such as landscape metrics, fractals offer a better benefit in that they can be directly applied to raw images without the need for classification or land cover feature identification, in addition to their sound mathematic bases. Therefore, it is not surprising to see a growing number of researches utilize fractals in remote sensing image analysis (De Jong and Burrough, 1995; Emerson et al., 1999, 2005; Lam, 1990; Lam and De Cola, 1993; Myint, 2003; Qiu et al., 1999; Read and Lam, 2002; Weng, 2003). Fractal-derived texture images have also been used as additional layers in image classification (Myint, 2003).

Spatial resolution has been another focus in remote sensing studies. It is necessary to estimate the capability of remote sensing data in landscape mapping since the application of remote sensing may be limited by its spatial resolution (Aplin, 2006; Buyantuyev and Wu, 2007; Ludwig et al., 2007). Imagery with finer resolution contains greater amount of spatial information, which, in turn, enables the characterization of smaller features better. The proportion of mixed pixels is expected to increase as spatial resolution becomes coarser (Aplin, 2006). Stefanov and Netzband (2005) identified weak positive and negative correlations between the normalized vegetation index (NDVI) and landscape structure at three different resolutions (250, 500, and 1000 m) when they examined the capability of the Moderate Resolution Imaging Spectroradiometer (MODIS) NDVI data in the assessment of arid landscape characteristics in Phoenix. Asner et al. (2003) examined the significance of subpixel estimates of biophysical structure with the help of high-resolution remote sensing imagery and found a strong correlation between the senescent and unmixed green vegetation cover values in a deforested area. Agam et al. (2007) sharpened the coarse-resolution thermal imagery to finer resolution imagery based on the analysis of the relationship between vegetation index and land surface temperature. The results showed that the vegetation index–based sharpening method provided an effective way to improve the spatial resolution of thermal imagery.

Adaptive choice of spatial and categorical scales in landscape mapping was demonstrated by Ju et al. (2005). They provided a data-adaptive choice of spatial scale varying by location jointed with categorical scale by the assistance of a statistical finite mixture method. Buyantuyev and Wu (2007) systematically analyzed the effects of thematic resolution on landscape pattern analysis. Two problems need to be considered in landscape mapping: the multiplicity of classification schemes and the level of detail of a particular classification. They found that the thematic resolution had obvious effects on most of the landscape metrics, which indicated that changing thematic resolution may significantly affect the detection of landscape changes. However, an increase in spatial resolution may not lead to a better observation since objects may be oversampled and their features may vary and be confusing (Hsieh et al., 2001; Aplin and Atkinson, 2004). Although coarse resolution may include fewer features, imagery with too fine resolution for a specific purpose can be degraded in the process of image resampling (Ju et al., 2005). Remote sensing data may not be always be sufficient when specific problems were addressed at specific scales and on-ground assessment may be needed, since coarser imagery cannot provide sufficient information about the location and connectivity in specific areas (Ludwig et al., 2007).

Substantial researches have previously been conducted on scale-related issues in remote sensing studies, as discussed above. This book intends to revisit and reexamine the scale and related issues. It will also address how new frontiers in Earth observation technology since 1999—such as very high resolution, hyper-spectral, lidar sensing, and their synergy with existing technologies and advances in remote sensing imaging science such as object-oriented image analysis, data fusion, and artificial neural networks—have impacted the understanding of this basic but pivotal issue. The scale-related issues will be examined from three interrelated perspectives: in landscape properties, patterns, and processes. These examinations are preceded by a theoretical exploration of the scale issue by a group of authorities in the field of remote sensing. The concluding section prospects emerging trends in remote sensing over the next decade(s) and their relationship with scale.

1.2 CHARACTERIZING, MEASURING, ANALYZING, AND MODELING SCALE

This book consists of 5 parts and 14 chapters, in addition to this introductory chapter. Part I focuses on theoretical aspects of scale and scaling. Part II deals with the estimation and measurement of vegetation parameters and ecosystems across various spatial and temporal scales. Part III examines the effect of scaling on image segmentation and object extraction from remotely sensed imagery. Part IV exemplifies with case studies on the scale and scaling issues in land cover analysis and in land–atmosphere interactions. Finally, Part V addresses how new frontiers in Earth observation technology, such as hyperspectral and lidar sensing, have impacted the understanding of the scale issue.

Three chapters are included in Part I. In Chapter 2, Ehlers and Klonus examine data fusion results of remote sensing imagery with various spatial scales. The scales are thought to relate to the ground sampling distances (GSDs) of the respective sensors. They find that for electro-optical sensors GSD or scale ratios of 1:10 (e.g., IKONOS and SPOT-5 fusion) can still produce acceptable results if the fusion method is based on a spectral characteristic-preserving technique such as the Ehlers fusion. Using radar images as a substitute for high-resolution panchromatic data is possible, but only for scale ratios between 1:6 and 1:20 due to the limited feature recognition in radar images. In Chapter 3, Quattrochi and Luvall revisit an article published in *Landscape Ecology* in 1999 by them and examine the direct or indirect uses of thermal infrared (TIR) remote sensing data to analyze landscape biophysical characteristics to offer insights on how these data can be used more robustly for furthering the understanding and modeling of landscape ecological processes. In Chapter 4, Weng discusses some important scale-related issues in urban remote sensing. The requirements for mapping three interrelated entities or substances in the urban space (i.e., material, land cover, and land use) and their relationships are first examined. Then, the relationship between spatial resolution and the fabric of urban landscapes is assessed. Next, the operational scale/optimal scale for the studies of land surface temperature are reviewed. Finally, the issue of scale dependency of urban phenomena is discussed via reviewing two case studies, one on land surface temperature (LST) variability across multiple census levels and the other on multiscale residential population estimation modeling.

Part II also contains three chapters. Vegetation indices can be used to separate landscape components into bare soil, water, and vegetation and, if calibrated with ground data, to quantify biophysical variables such as leaf area index and fractional cover and physiological variables such as evapotranspiration and photosynthesis. In Chapter 5, Glenn, Nagler, and Huete use a case study approach to show how remotely sensed vegetation indices collected at different scales can be used in vegetation change detection studies. The primary sensor systems discussed are digital pheno-cams, Landsat and MODIS, which cover a wide range of spatial (1 cm–250 m) and temporal (15 min–16 days) resolutions/scales. Sources of error and uncertainty associated with both ground and remote sensing measurements in change studies are also discussed. In Chapter 6, Wang and Zhang combine plot data and Thematic Mapped (TM) images to map above-ground forest carbon at a 990-m pixel resolution in Lin-An, Zhejiang Province, China, by using two upscaling methods: point simple cokriging point cosimulation and point simple cokriging block cosimulation Their results suggest that both methods perform well in scaling up the spatial data as well as in revealing the propagation of input data uncertainties from a finer spatial resolution to a coarser one. The output uncertainties reflect the spatial variability of the estimation accuracy caused by the locations of the input data and the values themselves. In Chapter 7, Yuhong He intends to bridge the gap in spatial scales through estimating grassland chlorophyll contents from leaf to landscape level using a simple yet effective canopy integration method. Using data collected in a heterogeneous tall grassland located at Ontario, Canada, Yuhong's study first scales leaf level chlorophyll measurements to canopy and landscape levels and then investigates the relationships between a chlorophyll spectral index and vegetation chlorophyll contents at the leaf, canopy, and landscape scales. Significant relationships are found at all three scales, suggesting that it is feasible to accurately estimate chlorophyll contents using both ground and space remote sensing data.

In remote sensing, image segmentation has a longer history and has its roots in industrial image processing but was not used extensively in the geospatial community in the 1980s and 1990s (Blaschke, 2010). Object-oriented image analysis has been increasingly used in remote sensing applications due to the advent of high-spatial-resolution image data and the emergence of commercial software such as eCognition (Benz et al., 2004; Wang et al., 2004). In the process of creating objects, a scale determines the occurrence or absence of an object class. Thus, the issue of scale and scaling are fundamental considerations in the extraction, representation, modeling, and analyses of image objects (Hay et al., 2002; Tzotsos et al., 2011).

The three chapters in Part III focus on discussion of these issues. In Chapter 8, Hay introduces a novel geo-object-based framework that integrates hierarchy theory and linear scale space (SS) for automatically visualizing and modeling landscape scale domains over multiple scales. Specifically, this chapter describes a three-tier hierarchical methodology for automatically delineating the dominant structural components within 200 different multiscale representations of a complex agro-forested landscape. By considering scale-space events as critical domain thresholds, Hay further defines a new scale-domain topology that may improve querying and analysis of this complex multiscale scene. Finally, Hay shows how to spatially model and visualize the hierarchical structure of dominant geo-objects within a scene as "scale-domain manifolds" and suggests that they may be considered as a multiscale extension to the hierarchical scaling ladder as defined in the hierarchical patch dynamics paradigm. Chapter 9 by Tzotsos, Karantzalos, and Argialas introduces a multiscale object-oriented image analysis framework which incorporates a region-merging segmentation algorithm enhanced by advanced edge features and nonlinear scale-space filtering. Initially, edge and line features are extracted from remote sensing imagery at several scales using scale-space representations. These features are then used by the enhanced segmentation algorithm as constraints in the growth of image objects at various scales. Through iterative pairwise object merging, the final segmentation can be achieved. Image objects are then computed at various scales and passed on to a kernel-based learning machine for classification. This image classification framework was tested on very high resolution imagery acquired by various airborne and spaceborne panchromatic, multispectral, hyperspectral, and microwave sensors, and promising experimental results were achieved. Chapter 10, by Im, Quackenbush, Li, and Fang, provides a review of recent publications on object-based image analysis (OBIA) focusing on determination of optimum scales for image segmentation and the related trends. Selecting optimum scale is often challenging, since (1) there is no standardized method to identify the optimality and (2) scales in most segmentation algorithms are arbitrarily selected. The authors suggest that there should be transferable guidelines regarding segmentation scales to facilitate the generalization of OBIA in remote sensing applications, to enable efficient comparison of different OBIA approaches, and to select optimum scales for the multitude of different image components.

Part IV introduces three case studies on the scale and scaling issues in analysis of land cover, landscape metrics, and biophysical parameters. Chapter 11 by Liu and Weng assesses the effect of scaling on the relationship between landscape pattern and land surface temperature with a case study in Indianapolis, Indiana. A set of spatial resolutions were compared by using a landscape metric space. They find that the spatial resolution of 90 m is the optimal scale to study the relationship and think that it is the operational scale of the urban thermal landscape in Indianapolis. In Chapter 12, Liang and Weng provide an evaluation of the effectiveness of the triangular prism fractal algorithm for characterizing urban landscape in Indianapolis based on eight satellite images acquired by five different sensors: Landsat Multispectral Scanner, Landsat Thematic Mapper, Landsat Enhanced Thematic Mapper Plus, Advanced Spaceborne Thermal Emission and Reflection, and IKONOS. Fractal dimensions computed from the selected original, classified, and resampled images are compared and analyzed. The potential of fractal measurement in the studies of landscape pattern characterization and the scale/resolution issues are further assessed. Chapter 13 by Hong and Zhang provides important insights into the spatiotemporal scales of remotely sensed precipitation. This chapter first overviews the precipitation measurement methods—both traditional rain gauge and advanced remote sensing measurements; then develops an uncertainty analysis framework that can systematically quantify the remote sensing precipitation estimation error as a function of space, time, and intensity; and finally assesses the spatiotemporal scale-based error propagation in remote sensing precipitation estimates into hydrological prediction.

The last part of this book looks at how new frontiers in Earth observation technology have transformed our understanding of this foremost issue in remote sensing. Chapter 14 examines lidar data processing, whereas Chapter 15 explores hyperspectral remote sensing for land cover mapping. Digital terrain models (DTMs) are basic products required for a number of applications and decision making. Nowadays, high-spatial-resolution DTMs are primarily produced through airborne laser scanners (ALSs). However, the ALS does not directly deliver DTMs; rather it delivers a dense point cloud that embeds both terrain elevation and height of natural and human-made features. Hence, discrimination of above-ground objects from terrain is a basic processing step. This processing step is termed ground filtering and has proved especially difficult for large areas of varied terrain characteristics. In Chapter 14, Silvan-Cárdenas and Wang revise and extend a filtering method based on a multiscale signal decomposition termed the multiscale Hermite transform (MHT). The formal basis of the latter is presented in the context of scale-space theory, a theory for representing spatial signals. Through the unique properties of the MHT, namely local spatial rotation and scale-space shifting, the original filtering algorithm was extended to incorporate higher order coefficients in the multiscale erosion operation. Additionally, a linear interpolation was incorporated through a truncated Taylor expansion which allowed improving the ground filtering performance along sloppy terrain areas. Practical considerations in the operation of the algorithm are discussed and illustrated with examples. In Chapter 15, Petropoulos, Manevski, and Carlson assess the potential of hyperspectral remote sensing systems for improving discrimination among similar land cover classes at different scales. The chapter provides first an overview of the current state of the art in the use of field spectroradiometry in examining the spectral discrimination between different land cover targets. In this framework, techniques employed today and linked with the most important scale factors are critically reviewed and examples of recent related studies and spectral libraries are provided. Then, it focuses on the use of hyperspectral remote sensing for obtaining land use/cover mapping from space. An overview of the different satellite sensors and techniques employed is furnished, providing examples taken from recent studies. The chapter closes by highlighting the main challenges that need to be addressed in the future towards a more precise estimation of land cover from spectral information acquired from hyperspectral sensing systems at variant spatial scales.

REFERENCES

Agam, N., Kustas, W. P., Anderson, M. C., Li, F., and Neale, C. M. U. 2007. A vegetation index based technique for spatial sharpening of thermal imagery. *Remote Sensing of Environment* 107(4):545–558.

Allen, T. F. H., O'Neill, R. V., and Hoekstra, T. W. 1987. Interlevel relations in ecological research and management: Some working principles from hierarchy. *Journal of Applied Systems Analysis* 14:63–79.

Aplin P. 2006. On scales and dynamics in observing the environment. *International Journal of Remote Sensing* 27(11):2123–2140.

Aplin, P., and Atkinson, P. M. 2004. Predicting missing field boundaries to increase per-field classification accuracy. *Photogrammetric Engineering and Remote Sensing* 70:141–149.

Asner, G. P., Bustamante, M. M. C., and Townsend, A. R. 2003. Scale dependence of biophysical structure in deforested areas bordering the Tapajos National Forest, Central Amazon. *Remote Sensing of Environment* 87(4):507–520.

Benz, U. C., Hofmann, P., Willhauck, G., Lingenfelder, I., and Heynen, M. 2004. Multi-resolution, object-oriented fuzzy analysis of remote sensing data for GIS-ready information. *ISPRS Journal of Photogrammetry & Remote Sensing* 58:239–258.

Blaschke, T. 2010. Object based image analysis for remote sensing. *ISPRS Journal of Photogrammetry and Remote Sensing* 65(1):2–16.

Buyantuyev, A., and Wu, J. 2007. Effects of thematic resolution on landscape pattern analysis. *Landscape Ecology* 22(1):7–13.

Clarke, K. C. 1986. Computation of the fractal dimension of topographic surfaces using the triangular prism surface area method. *Computers and Geosciences* 12:713–722.

De Jong, S. M., and Burrough, P. A. 1995. A fractal approach to the classification of Mediterranean vegetation types in remotely sensed images. *Photogrammetric Engineering and Remote Sensing* 61:1041–1053.

Emerson, C. W., Lam, N. S. N., and Quattrochi, D. A. 1999. Multiscale fractal analysis of image texture and pattern. *Photogrammetric Engineering and Remote Sensing* 65:51–61.

Emerson, C. W., Lam, N. S. N., and Quattrochi, D. A. 2005. A comparison of local variance, fractal dimension, and Moran's I as aids to multispectral image classification. *International Journal of Remote Sensing* 26:1575–1588.

Hay, G. J., Dube, P., Bouchard, A., and Marceau, D. J. 2002. A scale-space primer for exploring and quantifying complex landscapes. *Ecological Modelling* 153(1–2):2–49.

Hsieh, P. F., Lee, L. C., and Chen, N. Y. 2001. Effect of spatial resolution on classification errors of pure and mixed pixels in remote sensing. *IEEE Transactions on Geoscience and Remote Sensing* 39(12):2657–2663.

Ju, J. C., Gopal, S., and Kolaczyk, E. D. 2005. On the choice of spatial and categorical scale in remote sensing land-cover classification. *Remote Sensing of Environment* 96(1):62–77.

Krönert, R., Steinhardt, U., and Volk, M. 2001. *Landscape Balance and Landscape Assessment*. New York: Springer.

Lam, N. S. N. 1990. Description and measurement of Landsat TM images using fractals. *Photogrammetric Engineering and Remote Sensing* 56:187–195.

Lam, N. S. N., and De Cola, L. 1993. *Fractals in Geography*. Englewood Cliffs, NJ: Prentice Hall.

Ludwig, J., Bastin, G., Wallace, J., and McVicar, T. 2007. Assessing landscape health by scaling with remote sensing: When is it not enough? *Landscape Ecology* 22(2):163–169.

Myint, S. W. 2003. Fractal approaches in texture analysis and classification of remotely sensed data: Comparisons with spatial autocorrelation techniques and simple descriptive statistics. *International Journal of Remote Sensing* 24:1925–1987.

Qiu, H.-L., Lam, N. S. N., Quattrochi, D. A., and Gamon, J. A. 1999. Fractal characterization of hyperspectral imagery. *Photogrammetric Engineering and Remote Sensing* 65:63–71.

Read, J. M., and Lam, N. S. N. 2002. Spatial methods for characterizing land cover and detecting land-cover changes for the tropics. *International Journal of Remote Sensing* 23:2457–2474.

Stefanov, W. L., and Netzband, M. 2005. Assessment of ASTER land-cover and MODIS NDVI data at multiple scales for ecological characterization of an arid urban center. *Remote Sensing of Environment* 99(1–2):31–43.

Turner, M. G. 1990. Spatial and temporal analysis of landscape patterns. *Landscape Ecology* 4(1):21–30.

Tzotsos, A., Karantzalos, K., and Argialas, D. 2011. Object-based image analysis through nonlinear scale-space filtering. *ISPRS Journal of Photogrammetry and Remote Sensing* 66:2–16.

Wang, L., Sousa, W. P., Gong, P., and Biging, G. S. 2004. Comparison of IKONOS and QuickBird images for mapping mangrove species on the Caribbean coast of panama. *Remote Sensing of Environment* 91:432–440.

Weng, Q. 2003. Fractal analysis of satellite-detected urban heat island effect. *Photogrammetric Engineering and Remote Sensing* 69:555–566.

Wu, J., and Hobbs, R. 2002. Key issues and research priorities in landscape ecology: An idiosyncratic synthesis. *Landscape Ecology* 17:355–365.

Wu, J., and Qi, Y. 2000. Dealing with scale in landscape analysis: An overview. *Geographic Information Sciences* 6(1):1–5.

Zhao, W. 2001. Multiscale Analysis for Characterization of Remotely Sensed Images. Ph.D. Dissertation, Louisiana State University.

PART I

SCALE, MEASUREMENT, MODELING, AND ANALYSIS

2

SCALE ISSUES IN MULTISENSOR IMAGE FUSION

Manfred Ehlers and Sascha Klonus

2.1 SCALE IN REMOTE SENSING

Scale is a term that is used in many scientific applications and communities. Typical well-known measurement scales are, for example, the Richter scale for the magnitude of earthquakes and the Beaufort scale for wind speed. We speak of large-scale operations if they involve large regions or many people. Cartographers use the term scale for the description of the geometric relationship between a map and real-world coordinates. In remote sensing, scale is usually associated with the latter meaning the map scale for typical applications of remote sensors. To a large degree, scale is dependent on the geometric resolution of the sensor which can be measured in ground sampling distance (GSD). The GSD is usually the same or similar to the final pixel size of the remote sensing data set. In addition to the GSD, scale is also associated with the level and quality of information that can be extracted from remotely sensed data.

Especially with the launch of the first SPOT satellite with independent panchromatic and multispectral sensors, it became evident that a combined analysis of the high-resolution panchromatic sensor and the lower resolution multispectral images would yield better results than any single image alone. Subsequently, most Earth observation satellites, such as the SPOT and Landsat series, or the very high resolution (VHR) sensors such as IKONOS, QuickBird, or GeoEye acquire image data in two different modes, a low-resolution multispectral and a high-resolution panchromatic mode. The GSD or scale ratio between the panchromatic and the multispectral image can vary between 1 : 2 and 1 : 8 with 1 : 4 the most common value. This ratio can even become smaller when data from different sensors are used, which is, for example, necessary if

Scale Issues in Remote Sensing, First Edition. Edited by Qihao Weng.
© 2014 John Wiley & Sons, Inc. Published 2014 by John Wiley & Sons, Inc.

satellite sensors with only panchromatic (e.g., WorldView-1) or only multispectral (e.g., RapidEye) information are involved. Consequently, efforts started in the late 1980s to develop methods for merging or fusing panchromatic and multispectral image data to form multispectral images of high geometric resolution. In this chapter, we will investigate to what degree fusion techniques can be used to form multispectral images of larger scale when combined with high-resolution black-and-white images.

2.2 FUSION METHODS

Similar to the term scale, the word *fusion* has different meanings for different communities. In a special issue on data fusion of the *International Journal of Geographical Information Science(IJGIS)*, Edwards and Jeansoulin (2004, p. 303) state that "data fusion is a complex process with a wide range of issues that must be addressed. In addition, data fusion exists in different forms in different scientific communities. Hence, for example, the term is used by the image community to embrace the problem of sensor fusion, where images from different sensors are combined. The term is also used by the database community for parts of the interoperability problem. The logic community uses the term for knowledge fusion."

Consequently, it comes as no surprise that several definitions for data fusion can be found in the literature. Pohl and van Genderen (1998, p. 825) proposed that "image fusion is the combination of two or more different images to form a new image by using a certain algorithm." Mangolini (1994) extended data fusion to information in general and also refers to quality. He defined data fusion as a set of methods, tools and means using data coming from various sources of different nature, in order to increase the quality (in a broad sense) of the requested information (Mangolini, 1994). Hall and Llinas (1997, p. 6) proposed that "data fusion techniques combine data from multiple sensors, and related information from associated databases." However, Wald (1999) argued that Pohl and van Genderen's definition is restricted to images. Mangolini's definition puts the accent on the methods. It contains the large diversity of tools but is restricted to these. Hall and Llinas refer to information quality in their definition but still focus on the methods.

The Australian Department of Defence defined data fusion as a "multilevel, multifaceted process dealing with the automatic detection, association, correlation, estimation, and combination of data and information from single and multiple sources" (Klein, 2004, p. 52). This definition is more general with respect to the types of information than can be combined (multilevel process) and very popular in the military community. Notwithstanding the large use of the functional model, this definition is not suitable for the concept of data fusion, since it includes its functionality as well as the processing levels. Its generalities as a definition for the concept are reduced (Wald, 1999). A search for a more suitable definition was launched by the European Association of Remote Sensing Laboratories (EARSeL) and the French Society for Electricity and Electronics (SEE, French affiliate of the Institute of Electrical and Electronics Engineers) and the following definition was adopted in January 1998: "Data fusion is a formal framework in which are expressed means and tools for the alliance of data originating from different sources. It aims at obtaining information of greater quality; the exact definition of 'greater quality' will depend upon the application" (Wald, 1999, p. 1191).

Image fusion forms a subgroup within this definition, with the objective to generate a single image from multiple image data for the extraction of information of higher quality (Pohl, 1999). Image fusion is used in many fields such as military, medical imaging, computer vision, the robotics industry, and remote sensing of the environment. The goals of the fusion process are multifold: to sharpen multispectral images, to improve geometric corrections, to provide stereo-viewing capabilities for stereo-photogrammetry, to enhance certain features not visible in either of the single data sets alone, to complement data sets for improved classification, to detect changes using multitemporal data, and to replace defective data (Pohl and van Genderen, 1998). In this article, we concentrate on the image-sharpening process (iconic fusion) and its relationship with image scale.

Many publications have focused on how to fuse high-resolution panchromatic images with lower resolution multispectral data to obtain high-resolution multispectral imagery while retaining the spectral characteristics of the multispectral data (e.g., Cliche et al., 1985; Welch and Ehlers, 1987; Carper et al., 1990; Chavez et al., 1991; Wald et al., 1997; Zhang 1999). It was evident that these methods seem to work well for many applications, especially for single-sensor, single-date fusion. Most methods, however, exhibited significant color distortions for multitemporal and multisensoral case studies (Ehlers, 2004; Zhang, 2004).

Over the last few years, a number of improved algorithms have been developed with the promise to minimize color distortion while maintaining the spatial improvement of the standard data fusion algorithms. One of these fusion techniques is Ehlers fusion, which was developed for minimizing spectral change in the pan-sharpening process (Ehlers and Klonus, 2004). In a number of comprehensive comparisons, this method has tested superior to most of the other pan-sharpening techniques (see, e.g., Ling et al., 2007a; Ehlers et al., 2010; Klonus, 2011; Yuhendra et al., 2012). For this reason, we will use Ehlers fusion as the underlying technique for the following discussions. The next section presents a short overview of this fusion technique.

2.3   EHLERS FUSION

Ehlers fusion was developed specifically for spectral characteristic-preserving image merging (Klonus and Ehlers, 2007). It is based on an intensity–hue–saturation (IHS) transform coupled with a Fourier domain filtering. The principal idea behind spectral characteristic-preserving image fusion is that the high-resolution image has to sharpen the multispectral image without adding new gray-level information to its spectral components. An ideal fusion algorithm would enhance high-frequency changes such as edges and gray-level discontinuities in an image without altering the multispectral components in homogeneous regions. To facilitate these demands, two prerequisites have to be addressed. First, color information and spatial information have to be separated. Second, the spatial information content has to be manipulated in a way that

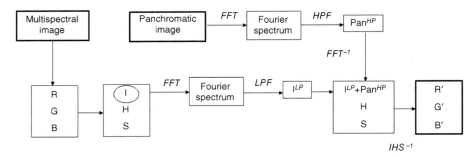

FIGURE 2.1   Concept of Ehlers fusion.

allows an adaptive enhancement of the images. This is achieved by a combination of color and Fourier transforms (Figure 2.1).

For optimal color separation, use is made of an IHS transform. This technique is extended to include more than three bands by using multiple IHS transforms until the number of bands is exhausted. If the assumption of spectral characteristic preservation holds true, there is no dependency on the selection or order of bands for the IHS transform. Subsequent Fourier transforms of the intensity component and the panchromatic image allow an adaptive filter design in the frequency domain. Using fast Fourier transform (FFT) techniques, the spatial components to be enhanced or suppressed can be directly accessed. The intensity spectrum is filtered with a low-pass (LP) filter whereas the spectrum of the high-resolution image is filtered with an inverse high-pass (HP) filter. After filtering, the images are transformed back into the spatial domain with an inverse FFT and added together to form a fused intensity component with the low-frequency information from the low-resolution multispectral image and the high-frequency information from the high-resolution panchromatic image. This new intensity component and the original hue and saturation components of the multispectral image form a new IHS image. As the last step, an inverse IHS transformation produces a fused RGB image that contains the spatial resolution of the panchromatic image and the spectral characteristics of the multispectral image. These steps can be repeated with successive three-band selections until all bands are fused with the panchromatic image. The order of bands and the inclusion of spectral bands for more than one IHS transform are not critical because of the color preservation of the procedure (Klonus and Ehlers, 2007). In all investigations, the Ehlers fusion provides a good compromise between spatial resolution enhancement and spectral characteristics preservation (Yuhendra et al., 2012), which makes it an excellent tool for the investigation of scale and fusion. Fusion techniques to sharpen multispectral images have focused primarily on the merging of panchromatic and multispectral electro-optical (EO) data with emphasis on single-sensor, single-date fusion. Ehlers fusion, however, allows multiple-sensor, multiple-date fusion and has recently been extended to fuse radar and EO image data (Ehlers et al., 2010). Consequently, we will address scale-fusion issues for multisensor, multitemporal EO image merging as well as the inclusion of radar image data as a "substitute" for panchromatic images.

FIGURE 2.2   Study site for EO image fusion (image © Google).

2.4   FUSION OF MULTISCALE ELECTRO-OPTICAL DATA

2.4.1   Data Sets and Study Site

For our investigation, we selected a study area in Germany representing a part of the village of Romrod (Figure 2.2). This area was chosen because it contained several important features for fusion quality assessment such as agricultural lands (fields and grasslands), man-made structures (houses, roads), and natural areas (forest).

It was used as a control site of the Joint Research Centre of the European Commission (JRC) in the project "Control with Remote Sensing of Area-Based Subsidies" (CwRS) (http://agrifish.jrc.it/marspac/DCM/). A series of multitemporal multispectral remote sensing images were available for this site and formed the basis for our multiscale fusion investigation (Table 2.1).

2.4.2   Multisensor Image Fusion

The multisensor images cover a time frame of almost four years and virtually all seasons and thus pose an excellent challenge for a combined scale–fusion investigation. All multispectral images were fused using the only panchromatic data set, that is, the IKONOS image from June 18, 2005. As a first part of our investigation, all multispectral images were registered to the panchromatic IKONOS image, resampled

TABLE 2.1 Multisensor Remote Sensing Data Sets for Study Site (in Chronological Order)

Satellite Sensor	Recording Date	Recording Mode	Ground Sampling Distance
Landsat 5 TM	December 21, 2004	Multispectral (six bands)	30 m
SPOT-2	April 21, 2005	Multispectral (three bands)	20 m
IKONOS	June 18, 2005	Panchromatic	1 m
		Multispectral (four bands)	4 m
SPOT-5	June 19, 2005	Multispectral (four bands)	10 m (short-wave infrared (swir): 20 m)
DMC	July 28, 2005	Multispectral (three bands)	32 m
RapidEye5	August 16, 2009	Multispectral (five bands)	6.5 m (resampled to 5 m)

to its resolution (1 m), and fused using the Ehlers fusion method. The scale ratio between the panchromatic and the multispectral data sets varied from 1 : 4 to 1 : 32. Results are shown in Figures 2.3–2.9. To demonstrate the effects of fusion on resolution and spectral characteristics, only single bands of the multispectral data sets are displayed.

In all images the fusion could improve the spatial resolution of the original multispectral image. But if the scale ratio between the high-resolution panchromatic image and the low-resolution multispectral image exceeds 1 : 10, artifacts from the

FIGURE 2.3 Panchromatic IKONOS image (left) recorded on June 18, 2005, and equivalent multispectral band 3 (red) resampled to 1 m (right).

FIGURE 2.4 Fused multispectral image band 3 using Ehlers fusion. Geometric quality of the image has significantly improved and the spectral characteristics of the multispectral image are well preserved.

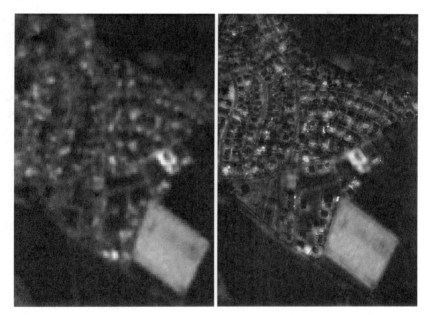

FIGURE 2.5 Band 3 (red) of RapidEye image from August 16, 2009 (resampled from 6.5 to 1 m, left) and IKONOS/RapidEye fusion result (right). Besides a much improved spatial resolution, the result shows an excellent spectral fidelity despite the four-year difference between the images.

FIGURE 2.6   Band 2 (red) of a SPOT-5 image from June 19, 2005 (resampled from 10 to 1 m, left) and IKONOS/SPOT-5 fusion result (right). Despite the scale ratio of 1 : 10, the fusion still produces a good-quality image with spatial improvement and spectral fidelity.

FIGURE 2.7   Band 2 (red) of SPOT-2 image from April 21, 2005 (resampled from 20 to 1 m, left) and IKONOS/SPOT-2 fusion result. Spatial features are mostly from the panchromatic image and due to the scale ratio of 1 : 20 it is hard to judge the spectral quality. Still, the fused image shows a lot of improvement over the original.

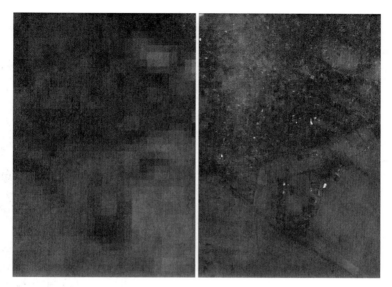

FIGURE 2.8   Band 3 (red) of Landsat TM 5 image from December 21, 2004 (resampled from 30 to 1 m, left) and IKONOS/TM fusion result (right). Again, the spatial features are mostly from the panchromatic image. Despite the scale ratio of 1 : 30 the spectral values seem to be well preserved. Image interpretation, however, is questionable.

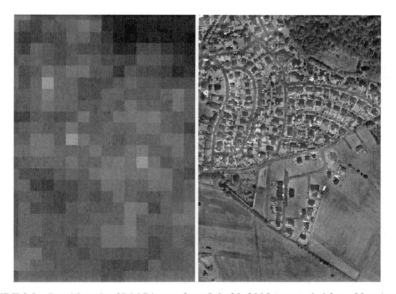

FIGURE 2.9   Band 3 (red) of DMC image from July 28, 2005 (resampled from 32 to 1 m, left) and IKONOS/DMC fusion result (right). The spatial features are from the panchromatic IKONOS image. Because of a general low contrast in the DMC image, the spectral quality cannot really be analyzed. The fused image looks more like the IKONOS image with some gray-level changes. Multispectral analysis seems impossible.

resampling method appear on the fields. The fused images seem to be blurred. For the assessment of spectral characteristic preservation one has also to consider that one pixel in a 20-m multispectral image represents 400 pixels in the 1-m panchromatic image. There is an enormous information gap which makes the images look like an IKONOS image with "some color overlay." Based on this analysis with real multisensor data, the cutoff scale ratio for image fusion seems to lie near 1 : 10.

2.4.3 Image Fusion with Variable Spatial Resolution

For a thorough analysis of the scale effects for EO image fusion, we used only the single-sensor IKONOS data set, thus alleviating all multisensor and multitemporal influences in the fusion process. In this investigation the IKONOS multispectral image was reduced to different spatial resolutions, which represent common satellite sensors used in remote sensing. These resolutions are 5, 8, 10, 20, 30, and 32 m. Again, the Ehlers fusion method was used to fuse all artificially created multiresolution IKONOS data sets with the panchromatic 1-m image. The results are shown in Figures 2.10–2.15. For better demonstration of the spectral quality the fusion results for a scale ratio of 1 : 10 and less are displayed in color with bands 2 (green), 3 (red), and 4 (near infrared).

FIGURE 2.10 Multispectral IKONOS image band 3 reduced to 5 m (left) and fusion result with panchromatic 1-m IKONOS image (right). Spatial improvement and spectral fidelity are clearly visible.

FIGURE 2.11   Band 3 of multispectral IKONOS image reduced to 8 m (left) and fusion result with panchromatic 1-m IKONOS image (right). There is no visible difference from the 5-m image from Figure 2.10.

FIGURE 2.12   Multispectral IKONOS image in false-color CIR display (band 4: red, band 3: green, band 2: blue) reduced to 10 m (left) and fusion result with panchromatic 1-m IKONOS image (right). The fusion process produces excellent pan-sharpening and spectral preservation results and is evaluable as a 1-m multispectral image. (See the color version of this figure in Color Plates section.)

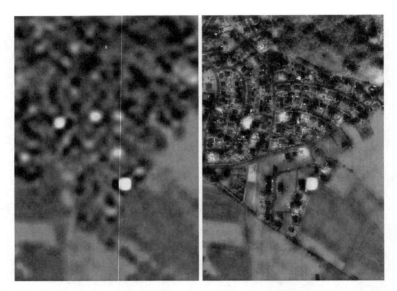

FIGURE 2.13 Multispectral IKONOS image in false-color CIR display (band 4: red, band 3: green, band 2: blue) reduced to 20 m (left) and fusion result with panchromatic 1-m IKONOS image (right). The fusion process produces a multispectral image of improved spatial resolution but some spectral changes begin to occur which would make a multispectral analysis questionable. Also the fused image begins to look like a "colorized black-and-white photo." (See the color version of this figure in Color Plates section.)

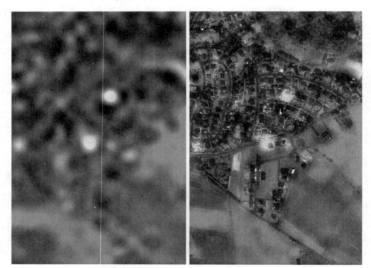

FIGURE 2.14 Multispectral IKONOS image in false-color CIR display (band 4: red, band 3: green, band 2: blue) reduced to 30 m (left) and fusion result with panchromatic 1-m IKONOS image (right). The fusion process no longer produces a single image. The effects of "color overlay" on a panchromatic image make this image very hard to interpret. (See the color version of this figure in Color Plates section.)

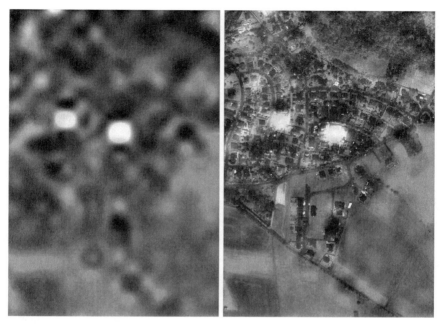

FIGURE 2.15   Multispectral IKONOS image in false-color CIR display (band 4: red, band 3: green, band 2: blue) reduced to 32 m (left) and fusion result with panchromatic 1-m IKONOS image (right). The fused image shows no significant differences from the fused 30-m image in Figure 2.14. (See the color version of this figure in Color Plates section.)

These results confirm the findings of the previous section. To prove the excellent spectral characteristic preservation, the correlation coefficients between the fused and the original bands for the multisensor and single-sensor multiscale IKONOS images are presented in Table 2.2. The coefficient values show that the spectral quality of the images is nearly the same for all different scale ratios.

TABLE 2.2   Correlation Coefficients between Original and Fused Bands

Sensor	Scale Ratio	Band 1	Band 2	Band 3	Band 4	Band 5	Band 7	Mean
IKONOS	1:4	0.90	0.90	0.93	0.95	—	—	0.92
RapidEye	1:6.5	0.94	0.98	1.00	0.99	0.99	—	0.98
SPOT-5	1:10	0.96	0.99	0.98	0.98	—	—	0.98
SPOT-2	1:20	0.89	0.97	0.99	—	—	—	0.95
Landsat 7 TM	1:30	0.97	0.98	0.99	0.99	0.99	0.96	0.94
DMC	1:32	0.99	0.95	0.91	—	—	—	0.95
IKONOS 5 m	1:5	0.90	0.91	0.93	0.95	—	—	0.92
IKONOS 8 m	1:8	0.90	0.91	0.94	0.95	—	—	0.92
IKONOS 10 m	1:10	0.90	0.92	0.96	0.96	—	—	0.93
IKONOS 20 m	1:20	0.91	0.93	0.96	0.96	—	—	0.94
IKONOS 30 m	1:30	0.91	0.92	0.95	0.96	—	—	0.93
IKONOS 32 m	1:32	0.91	0.93	0.96	0.96	—	—	0.94

FIGURE 2.16   Band 2 of the SPOT-2 image from the April 21, 2005 (left) and Ikonos/SPOT-2 fusion result which shows spectral fidelity and improved delineation for agricultural fields.

Based on the results with real multisensor and artificially created multiscale IKONOS images it can be concluded that scale ratios of up to 1 : 10 can safely be used for image fusion provided that the spectral fidelity of the fusion process can be guaranteed. Fusion with a ratio below 1 : 10 is only meaningful for a limited number of applications such as agricultural analyses. For scale ratios of 1 : 20, positive experiences with the Ehlers fusion technique are reported by project partners in the European Union agricultural monitoring program "Control with Remote Sensing." In one of the reports they wrote: "Ehlers image fusion method can get fusion images from different sensors and from different date of image acquisition. The profits of this method are located in the radiometric aspect, since this method keeps the radiometric properties of the multispectral image giving certain texture that improves the photointerpretation task. The improvement in the spatial resolution affects two factors during the computer aided photo interpretation (CAPI), namely permanent land uses which present not much soil cover are more precisely recognized and a better delineation of the field boundaries can be made, without a decrease of the radiometric quality" (JRC, 2008, pp. xx–xx; see Figure 2.16).

2.5   FUSION OF ELECTRO-OPTICAL AND RADAR DATA

Radar remote sensing offers a number of advantages over electro-optical (EO) sensing. Most notable is the independence of atmospheric and sunlight conditions.

There are, however, also disadvantages for the combination with electro-optical data: The side-looking mode makes it difficult to register radar data with nadir images; the speckle effects caused by the coherent electromagnetic wavelength make interpretation problematic; and the image acquisition process of the active sensor predominantly caused by the roughness and the orientation of the terrain provides completely different imaging effects. Despite these problems, radar remote sensing may be an alternative for panchromatic remote sensing data for image fusion if no other images are available. Consequently, iconic fusion with radar data has been practiced as early as in the 1980s (Welch and Ehlers, 1988; Harris et al., 1990; Ehlers, 1991). Preliminary results from the fusion of 1-m-resolution TerraSAR-X and 2.4-m-resolution multispectral QuickBird data for the pyramid fields of Giza, however, showed that even though there was some structural enhancement, an improvement in spatial resolution was not really visible (Klonus and Ehlers, 2008). This poses the immediate question as to what extent these two types of data are comparable in terms of terrain feature interpretation and subsequently to radar-based pan sharpening. It was also important to know what role the spatial resolution plays in both types of data and how different resolutions may affect data fusion.

Terrain feature interpretation is difficult to compare between radar and electro-optical systems, given the differences in image acquisition. While the optical sensors detect the reflected sunlight from objects, mostly in the visible and infrared wavelengths, the radar satellites produce coherent electromagnetic radiation in the microwave region that is then received as they bounce against the objects. Sensor and object orientation, distance from the object, and relative movement are important factors for radar remote sensing. Based on the preliminary results, it is evident that spatial resolution has to be differently defined for EO and radar data. Consequently, we investigated a number of representative terrain objects of different types and sizes chosen in a scene in Spain and compared in terms of visibility and identifiability in both types of images. The study area is located in the North of Spain representing the region around Santo Domingo de la Calzada. This area was also used as a control site of the JRC and provided EO remote sensing data as well as a TerraSAR-X scene (Ehlers et al., 2010). We made use of an IKONOS panchromatic image and a TerraSAR-X scene—both with a nominal spatial resolution of 1 m. The IKONOS image was acquired May 30, 2005, and theTerraSAR-X spotlight image on May 3, 2008. The TerraSAR-X image was despeckled with a proprietary algorithm developed at the University of Würzburg in Germany.

A terrain feature was considered visible when it was possible to differentiate it from the background and was considered identifiable when it could be precisely recognized on the image. Features that were highly variable in time or across season, such as agricultural fields, were not considered. Of all the features that were visible in IKONOS at 1 m ground resolution (being power lines the smallest feature), only features of size 7–10 m or larger could be seen in TerraSAR-X. This is exemplified by features such as the main road and a single tree (Figure 2.17 and Table 2.3).

In the case of the dirt road, the low visibility in TerraSAR-X is probably due to the lack of contrast in the surrounding areas due to only small differences in surface roughness. Except for the main road, none of the features were identifiable in

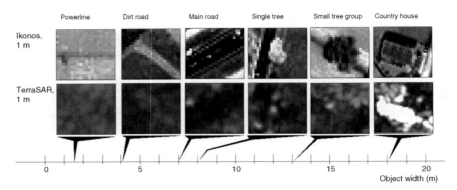

FIGURE 2.17 Comparison of exemplary features between TerraSAR-X and IKONOS (both at 1 m nominal spatial resolution).

TerraSAR-X. In contrast to country roads, asphalt or concrete roads have a surface roughness that is distinct from the surroundings. The roof in Figure 2.17 (below) cannot be clearly recognized because only the part of the roof at a certain angle has a strong reflection, whereas the other portion is completely in the radar shadow. With respect to all features, it becomes clear that image interpretability is greatly impeded in radar images (Table 2.3). These results are emphasized by the examples in Figure 2.17, which prove that pixel size or GSD cannot be the only factor for scale determination.

As IKONOS images are degraded to lower resolution, however, the ability in object recognition starts to be surpassed by the unchanged TerraSAR-X image. For example, the repetitive structure in 4-m IKONOS (Figure 2.18 and Table 2.4) is not as visible as it is in TerraSAR-X, because the color contrast between the stripes is not strong enough and the dark stripes are relatively narrow for 4 m resolution. The freeway is barely identifiable in 4-m IKONOS because of the blurring of white

TABLE 2.3   Comparison of Detectable Terrain Features

Feature	Number of Pixels		Visibility		Identifiability	
	IKONOS	TSX	IKONOS	TSX	IKONOS	TSX
Power line	1-2	U	x	—	X	—
Dirt road	4	U	x	(-)	X	—
Pool	6	8	x	x	X	—
Pond	6	U	x	—	X	—
Hut	6	10	x	x	X	—
Main road	7	12	x	x	X	x
Single tree	8	U	x	x	X	—
Tree group	13	14	x	(—)	X	—
Country house	18	13	x	x	X	—

Note: x = possible, — = not possible, (—) = only possible under some conditions, U = undetermined number of pixels. The column pixel size shows the amount of pixel of the object in the respective image.

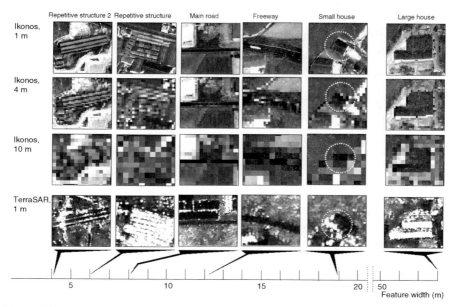

FIGURE 2.18 Comparison of features between TerraSAR-X and degraded IKONOS images.

features (lines and vehicles) on the dark background of the asphalt (Figure 2.18). In TerraSAR-X (1 m resolution), the freeway is uniform with a high contrast to the surrounding terrain. The small house is still visible in 4-m IKONOS data but no longer identifiable. In TerraSAR-X the small house is not clearly identifiable either, but the structural characteristics of the two-slope roof (bright spot inside the circle in Figure 2.18) make it possible to infer the presence of a sharp angle, which indicates the presence of a well-defined structure. The decreasing interpretability of artificial structures as IKONOS is degraded from 1 to 10 m can be more clearly seen in the large house (Figure 2.18). At the 10-m IKONOS data only the roads are clearly identified.

TABLE 2.4 Comparison of Identifiable Features

Feature	Number of Pixels		Visibility		Identifiability	
	IKONOS (4 m)	TSX	IKONOS (4 m)	TSX	IKONOS (4 m)	TSX
Repetitive structure 2	1-2	3-5	x	x	—	—
Repetitive structure	U	6	—	x	—	—
Main road	1–2	8	x	x	x	x
Freeway	7	12	x	x	x	x
Small house	6	18–20	x	—	—	—
Large house	15	53	x	x	—	—

Note: x = possible, — = not possible, U = undetermined number of pixels. The column pixel size shows the amount of pixel the object has in the specified image.

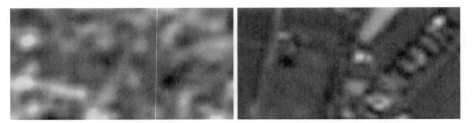

FIGURE 2.19   Two subsets of image of SPOT-4 (band 2, red) recorded on April 24, 2006. Both subsets are resampled to 1 m spatial resolution.

Based on these evaluations, we conclude that radar images should at least have a nominal scale ratio of 1 : 6 to provide a benefit for a SAR/multispectral EO fusion. For instance, this means that TerraSAR-X data can be used to sharpen SPOT or even Landsat TM image data but are of no or little use for improving the spatial resolution of modern multispectral very high resolution (VHR) sensors. This is demonstrated in Figures 2.19–2.21. A SPOT-4 scene recorded on April 24, 2006, is fused with the TerraSAR-X image from April 28, 2008, using the TerraSAR-X image as high-resolution input.

The fusion results show that the 20-m-resolution SPOT-4 image could be improved using the TerraSAR-X data. Many details such as streets, buildings, and also a few cars could be identified in the fused images. In addition, the original

FIGURE 2.20   Same two subsets as in Figure 2.19 but recorded with TerraSAR-X on April 28, 2008, in high-resolution spot light mode (1 m spatial resolution).

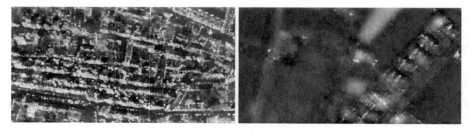

FIGURE 2.21   Results of Ehlers image fusion of two subsets. Spatial enhancement is clearly visible, and the spectral characteristics are well preserved.

spectral values are also well preserved. For complete details see Rosso et al. (2009). TerraSAR-X's capability of depicting features with certain structural characteristics may be of great value, for example, after a catastrophe where structural damage occurs. High-resolution radar images may not be able to show exactly what the damaged structures are but would clearly show if the structural characteristics are extensively modified. These advantages of radar over optical data are more evident as the resolution of the EO images decreases.

2.6 CONCLUSION

It is safe to conclude that iconic image fusion techniques can be used to overcome scale differences in multisensor remote sensing. It has to be noted, however, that for multisensor, multidate fusion spectral characteristic methods such as Ehlers fusion have to be applied so that the multispectral information can still be correctly analyzed. Our results indicate that for EO fusion scale ratios of $1:2$ to $1:10$ seem to be the reasonable range. This is very much in line with the findings reported by Ling et al. (2007b). For radar/EO image fusion, on the other hand, sensible scale ratios range from $1:6$ to $1:20$. For scale ratios outside these specified ranges, only noniconic fusion techniques such as feature-based (symbolic) or decision-based fusion techniques should be applied (Pohl and van Genderen, 1998).

REFERENCES

Carper, W. J., Lillesand, T. M., and Kiefer, R. W. 1990. The use of intensity-hue-saturation transformations for merging SPOT panchromatic and multispectral image data. *Photogrammetric Engineering and Remote Sensing* 56(4):459–467.

Chavez, W. J., Sides, S. C., and Anderson, J. A. 1991. Comparison of three different methods to merge multiresolution and multispectral data: TM & Spot Pan. *Photogrammetric Engineering and Remote Sensing* 57(3):295–303.

Cliche, G., Bonn, F., and Teillet, P. 1985. Integration of the SPOT pan channel into its multispectral mode for image sharpness enhancement. *Photogrammetric Engineering and Remote Sensing* 51(3):311–316.

Edwards, G., and Jeansoulin, R. 2004. Data fusion - from a logic perspective with a view to implementation, Guest Editorial. *International Journal of Geographical Information Science* 18(4):303–307.

Ehlers, M. 1991. Multisensor image fusion techniques in remote sensing. *ISPRS Journal of Photogrammetry and Remote Sensing* 46(1):19–30.

Ehlers, M. 2004. Spectral characteristics preserving image fusion based on Fourier domain filtering. In M. Ehlers, F. Posa, H. J. Kaufmann, U. Michel and G.De Carolis (Eds.), *Remote Sensing for Environmental Monitoring, GIS Applications, and Geology IV*, Proceedings of SPIE, 5574, Bellingham, WA: 1–13.

Ehlers, M., and Klonus, S. 2004. Erhalt der spektralen Charakteristika bei der Bildfusion durch FFT basierte Filterung. *Photogrammetrie-Fernerkundung-Geoinformation (PFG)* 6:495–506.

Ehlers, M., Klonus, S., Astrand, P., and Rosso, P. 2010. Multi-sensor image fusion for pansharpening in remote sensing. *International Journal for Image and Data Fusion (IJIDF)* 1(1):25–45.

Hall, D. L., and Llinas, J. 1997. An introduction to multisensor data fusion. *Proceedings of the IEEE* 85(1):6–23.

Harris, J. R., Murray, R., and Hirose, T. 1990. IHS transform for the integration of radar imagery with other remotely sensed data. *Photogrammetric Engineering and Remote Sensing* 56(12):1631–1641.

Joint Research Centre of the European Commission (JRC), 2008. *Common Technical Specifications for the 2009 Campaign of Remote-Sensing Control of Area-based Subsidies*, ITT no. 2008/S 228-3024 (http://mars.jrc.ec.europa.eu/Bulletins-Publications/Common-Technical-Specifications-for-the-2009-CwRS-Campaign – last accessed 21 March 2013).

Klein, L. A., 2004. *Sensor and Data Fusion: A Tool for Information Assessment and Decision Making*. SPIE, Bellingham, WA.

Klonus, S. 2011. Optimierung und Auswirkungen von ikonischen Bildfusionsverfahren zur Verbesserung von fernerkundlichen Auswerteverfahren. Ph.D. thesis, E-Dissertation, Universitaet Osnabrueck.

Klonus, S., and Ehlers, M. 2007. Image fusion using the Ehlers spectral characteristics preserving algorithm. *GIScience and Remote Sensing* 44(2):93–116.

Klonus, S., and Ehlers, M. 2008. Pansharpening with TerraSAR-X and Optical Data, Proceedings, 3rd TerraSAR-X Science Team Meeting, DLR, Oberhaffenhofen (http://sss.terrasar-x.dlr.de/papers_sci_meet_3/paper/MTH0299_klonus.pdf – last accessed 21 March 2013).

Ling, Y., Ehlers, M., Usery, E. L., and Madden, M. 2007a. FFT-enhanced IHS transform for fusing high-resolution satellite images. *ISPRS Journal of Photogrammetry and Remote Sensing* 61(6):381–392.

Ling, Y., Ehlers, M., Usery, E. L., and Madden, M. 2007b. Effects of spatial resolution ratio in image fusion. *International Journal of Remote Sensing* 29(7/8):2157–2168.

Mangolini, M. 1994. Apport de la fusion d'images satellitaires multicapteurs au niveau pixel en télédétection et photointerprétation. Thèse de Doctorat, Université Nice–Sophia Antipolis, France.

Pohl, C. 1999. Tools and methods for fusion of images of different spatial resolution. *International Archives of Photogrammetry and Remote Sensing* Vol. 32, Part 7-4-3 W6, Valladolid, Spain.

Pohl, C., and van Genderen, J. L. 1998. Multisensor image fusion in remote sensing: concepts, methods and applications. *International Journal of Remote Sensing* 19(5):823–854.

Rosso, P., Ehlers, M., and Klonus, S. 2009. Interpretability of TerraSAR-X fused data. In U. Michel and D. L. Civco (Eds.), *Remote Sensing for Environmental Monitoring, GIS Applications and Geology IX*, SPIE Proceedings, 7478: 74780H1-74780H12.

Wald, L. 1999. Some terms of reference in data fusion. *IEEE Transactions on Geoscience and Remote Sensing* 37:1190–1193.

Wald, L., Ranchin, T., and Magolini, M. 1997. Fusion of satellite images of different spatial resolutions—Assessing the quality of resulting images. *Photogrammetric Engineering and Remote Sensing* 63(6):691–699.

Welch, R., and Ehlers, M. 1987. Merging multiresolution SPOT HRV and Landsat TM data. *Photogrammetric Engineering and Remote Sensing* 53(3):301–303.

Welch, R., and Ehlers, M. 1988. Cartographic feature extraction from integrated SIR-B and Landsat TM images. *International Journal of Remote Sensing* 9(5):873–889.

Yuhendra, Alimuddin, I., Sumantyo, J. T. S., and Kuze, H. 2012. Assessment of pan-sharpening methods applied to image fusion of remotely sensed multi-band data. *International Journal of Applied Earth Observation and Geoinformation* 18:165–175.

Zhang, Y. 1999. A new merging method and its spectral and spatial effects. *International Journal of Remote Sensing* 20(10):2003–2014.

Zhang, Y. 2004. Understanding image fusion. *Photogrammetric Engineering and Remote Sensing* 70(6):657–661.

3

THERMAL INFRARED REMOTE SENSING FOR ANALYSIS OF LANDSCAPE ECOLOGICAL PROCESSES: CURRENT INSIGHTS AND TRENDS

DALE A. QUATTROCHI AND JEFFREY C. LUVALL

3.1 INTRODUCTION

Landscape ecology as a field of study requires data from broad spatial extents that cannot be collected using field-based methods alone. Remote sensing data and associated techniques have been used to address these needs, which include identifying and detailing the biophysical characteristics of species habitat, predicting the distribution of species and spatial variability of species richness, and detecting natural and human-caused change at scales ranging from individual landscapes to the entire world (Kerr and Ostrovsky, 2003). This has been exemplified in a growing number of special issues of journals and journal articles that have focused on remote sensing applications in landscape ecology (Cohen and Goward, 2004; Gillanders et al., 2008; Newton et al., 2009; Rocchini, 2010a,b). However, we believe that thermal remote sensing data have not been widely exploited to their full potential in landscape studies. Thermal data have important characteristics that can be used to derive quantitative measurements of surface energy balances and fluxes across the landscape, but widespread use of these data in landscape ecological research may still be somewhat enigmatic to some investigators.

Scale Issues in Remote Sensing, First Edition. Edited by Qihao Weng.
© 2014 John Wiley & Sons, Inc. Published 2014 by John Wiley & Sons, Inc.

In an article published in 1999, we examined the direct or indirect uses of thermal infrared (TIR) remote sensing data to analyze landscape biophysical characteristics to offer insight on how these data can be used more robustly for furthering the understanding and modeling of landscape ecological processes (Quattrochi and Luvall, 1999). As we noted at the time our article was published, we believed there was a perception that TIR data were difficult to use for applications in landscape characterization and analysis. Here we present a review and update of the literature related to TIR remote sensing in landscape ecological process studies both to further illustrate how the literature has grown and to expand upon research area themes that were not included in our original article. Additionally, as we noted in our 1999 article, accessing the literature related to TIR data and landscape ecological processes was difficult because of its fragmentation across a wide spectrum of journals or other scientific resources. Because of the interdisciplinary nature of research on TIR data and landscape processes, this is still true to some extent today; the literature on TIR remote sensing applications for land surface process analysis is being published in a wide range of publications, such as those focused strictly on remote sensing, or spread across numerous inter- or multidisciplinary publications such as hydrometeorology, climatology, meteorology, or agronomy.

As we related in 1999 and expounded upon in our edited volume *Thermal Remote Sensing in Land Surface Processes* (Quattrochi and Luvall, 2004), we foresaw that the application of TIR remote sensing data to landscape ecological studies has been limited for three primary reasons:

1. TIR data are little understood from both a theoretical and applications perspective within the landscape ecological community.
2. TIR data are perceived as being difficult to obtain and work with to those researchers who are uninitiated to the characteristics and attributes of these data in landscape ecological research.
3. The spatial resolution of TIR data, primarily from satellite data, is viewed as being too coarse for landscape ecological research, and calibration of these data for deriving measurements of landscape thermal energy fluxes is seen as being problematic.

Given the increase from 1999 (and even from 2004 when our book was published) in the TIR literature that has been published, these three issues have been considerably mitigated but not entirely mollified. It, therefore, is still useful to examine examples of the literature that have been published post-1999 to provide further evidence and review of how TIR data has been applied to landscape ecological and land processes research. As was described in our article and book, there are two fundamental ways that TIR data can contribute to an improved understanding of landscape processes: (1) through the measurement of surface temperatures as related to specific landscape and biophysical components and (2) through relating surface temperature with energy fluxes for specific phenomena or processes. This is not an exhaustive review; we wish only to provide further credence using selected studies taken from the literature that highlight and support the utility of TIR data to quantify and model land surface processes. We do so by providing citations that generally fall within several application areas that we believe are most critical for illustrating the virtues of TIR data and associated analysis methods.

3.2   SOME BACKGROUND ON NASA TIR SATELLITE INSTRUMENTS

Evaluation of Earth's radiation energy balance has been a primary design function of the meteorological and other Earth-sensing satellites since the launch of Explorer VII in 1959 (Diak et al., 2004). There has been considerable progress in estimating components of the land surface energy balance from orbit, particularly beginning with the National Aeronautics and Space Administration (NASA) Landsat series of satellites carrying the Thematic Mapper (TM) instrument first launched in 1984 and its successors. The TM sensor aboard Landsat 4 and 5 had spectral bands positioned between 0.45 and 12.5 μm in the electromagnetic spectrum. Six of these bands are in the visible and reflective infrared wave bands of the electromagnetic spectrum (0.45–2.35 μm), and there is one TIR spectral band in the 10.40–12.5-μm wave band range. All of the bands except for the TIR band have a spatial resolution of 30 m; the TIR has 120 m spatial resolution. The Enhanced Thematic Mapper+ (ETM+), which was launched onboard Landsat 7 in 1999, has the same spectral band configuration as the TM except the TIR band has a spatial resolution of 60 m. Landsat 8 launched in February 2013 has a sensor that is equivalent to the ETM+ both in spectral bandwidth and spatial resolution, except for the TIR band, which has a spatial resolution of 100 m.[1]

The collection of TIR data from space has been further augmented via the launch of the NASA Terra and Aqua missions in 1999 and 2002, respectively. Terra carries five sensor instruments, including the Moderate-Resolution Imaging Spectroradiometer (MODIS) and the Advanced Spaceborne Thermal Emission and Reflection Radiometer (ASTER), both of which have capabilities for imaging in TIR wavelengths. MODIS has multiple TIR bands, as does ASTER. The MODIS TIR bands are in the ranges 3.1–4.0 and 10.7–12.2 μm, and ASTER's are in the range 8.1–10.9 μm. The NASA Aqua mission also carries a MODIS instrument. Terra collects data twice daily at approximately 10:30 AM and 10:30 PM local time, while Aqua collects data twice daily, at approximately 1:30 AM and 1:30 PM local time. MODIS TIR data have a spatial resolution of 1km while ASTER thermal data are collected at 90 m spatial resolution. In-depth information on Terra and Aqua can be obtained at http://www.nasa.gov/mission_pages/terra/index.html and http://aqua .nasa.gov/index.php, respectively.

One recently launched (December 2011) joint NASA/NOAA (National Oceanic and Atmospheric Administration) mission that also offers TIR capabilities is the

---

[1] For complete information on the Landsat series of satellites, see http://landsat.usgs.gov/about_ldcm.php. Additional information on Landsat 8, known as the Landsat Data Continuity Mission (LDCM) prior to launch, can be accessed at http://ldcm.nasa.gov/.

Visible and Infrared Imaging Radiometer Suite (VIIRS) instrument onboard the National Polar-Orbiting Operational Environmental Satellite System (NPOESS), now called the Suomi National Polar-Orbiting Partnership (NPP) space platform. Although it has a coarse spatial resolution (approximately 0.39 km), VIIRS bears mention because it has four TIR spectral bands and it extends and improves upon a series of measurements initiated by the NOAA Advanced Very High Resolution Radiometer (AVHRR), which has been used in many past and present studies of land surface energy balance fluxes. NPP collects data at about 1:30 PM and 1:30 AM local time, similar to the temporal cycle of Aqua. More information on NPOESS/VIIRS is available at http://npp.gsfc.nasa.gov/index.html.

It must be noted that NASA or NOAA Earth-observing satellites are not the only space-based TIR platforms. The European Space Agency (ESA), the Chinese, and other countries have in orbit or plan to launch TIR remote sensing systems. However, a discussion of these systems will not be presented here for the sake of brevity.[2]

3.3   USE OF TIR DATA IN ANALYSIS LANDSCAPE ECOLOGICAL CHARACTERIZATION

Solar and thermal radiation within the Earth–atmosphere regime governs the energy available at Earth's surface for heating and cooling of the air (i.e., sensible heat), the evaporation of water from soil and vegetation (i.e., latent heat), and heating or cooling of natural (e.g., soil) and nonnatural (e.g., pavement) land surfaces. Earth's only significant source of energy is solar radiation, which is partitioned into various energy fluxes at the surface (Diak et al., 2004). The ultimate driving factor controlling surface characteristics such as soil moisture, land cover, and vegetation conditions is the energy transfer that occurs in land–atmosphere interactions. The simplest form of the surface energy balance (assuming no advection of energy across the land surface) is given by

$$R_{net} = G + H + LE \qquad (3.1)$$

where R_{net} is the net radiation balance, G refers to the soil heat flux (i.e., the energy used to warm the near-surface soil layers), H is the sensible heat flux, and LE is the latent heat flux. The ability to quantify the partitioning of available energy at the land surface into sensible and latent heat flux is key to understanding the impact of the land surface on atmospheric processes (Czajkowski et al., 2000). Understanding land–atmosphere energy exchange processes is important for improving short-term meteorological conditions (i.e., the weather) and in predicting the impacts of natural and anthropogenic changes in the land surface on long-term climate variability (Humes et al., 2000). Although land–atmosphere energy fluxes can be measured using in situ methods via surface thermal radiation measurements and soil moisture instruments, the synoptic view provided by remote sensing data from satellites can measure land surface

---

[2] Quattrochi et al. (2003) give a listing of the characteristics of U.S. and international imaging satellites either launched at that time or planned for future launch.

temperatures and energy fluxes over a wide area repetitively for multiple temporal periods (i.e., hours, days, weeks) for the same geographic area on Earth. This facilitates the modeling of surface energy fluxes for different land covers across the heterogeneous land surface to develop an understanding of how individual land covers with different thermal characteristics force energy exchanges between the land and atmosphere. There are numerous references that explain TIR theory and how it can be used to derive surface thermal energy balances using remote sensing data (see, e.g., Quattrochi and Luvall, 2009; Quattrochi et al., 2009) and this will not be explained here.

3.4 ESTIMATING LAND SURFACE ENERGY BUDGETS USING REMOTE SENSING DATA

Satellite remote sensing provides an excellent opportunity to study land–atmosphere energy exchanges at the regional scale. Many algorithms have been developed and tested using satellite TIR data to measure regional distributions of land surface temperature (LST), land surface reflectance, particularly that from vegetation, using the normalized vegetation index (NDVI), and fluxes of net radiation, soil heat, and sensible and latent heat flux. The NDVI has been used extensively to measure canopy density (or biomass content) to develop better regional estimates of energy fluxes for vegetation at the regional scale (see Quattrochi and Luvall, 1999, 2004). The NDVI has also been used to compare the energy fluxes of vegetation with other types of land covers (e.g., nonnatural surfaces) and to assess how energy dynamics of vegetation, especially evapotranspiration, affects surrounding land covers (NASA, 2013; Quattrochi et al., 2009).

Landsat TM and ETM data have been used extensively to derive land surface temperatures in conjunction with NDVIs. Fan et al. (2007) used ETM+ data to derive regional distribution of surface energy fluxes in conjunction with NDVI over a watershed in Inner Mongolia China. Distribution maps revealed strong contrasts in thermal energy responses of surface characteristics as a function of landscape features. Southworth (2004) investigated the utility of integrating Landsat data for differentiation between successive stages of forest growth in the Yucatan, Mexico. He found that the Landsat ETM+ thermal data contain considerable information for the discrimination of land cover classes in the dry tropical forest ecosystem. Li et al. (2004) used Landsat TM and ETM+ data to derive land surface temperatures as part of the Soil Moisture Experiments in 2002 (SMEX02) in central Iowa. Results from the study show that it is possible to extract accurate LSTs that vary from 0.98 to 1.47°C from Landsat 5 TM and Landsat 7 ETM+ data, respectively. Yves et al. (2006) used LST algorithms and NDVI values to estimate changes in vegetation in the European continent between 1982 and 1999 from the Pathfinder AVHRR (NOAA AVHRR) NDVI data set.[3] These data show a well-confirmed trend of increased NDVI values over Europe, with southern Europe seeing a decrease over the whole continent except

---

[3] The Pathfinder AVHRR NDVI data set is available from the NASA Goddard Earth Science Data and Information Services Center (GES DISC) at http://disc.sci.gsfc.nasa.gov/about-us.

for southern areas which show an increase in NDVI. LST averages stay stable or slightly decrease over the whole continent, except southern areas, which show an increase. These results provide evidence that arid and semiarid areas of southern Europe have become more arid, while the remainder of Europe has seen an in increase in vegetated lands. Wloczyk et al. (2011) used ETM+ data in conjunction with a temperature–vegetation index method (TVX) for areawide mapping of instantaneous air temperature. The TVX method was applied to a multitemporal data set of nine ETM+ scenes covering large parts of northeastern Germany. These satellite-derived measurements were compared with in situ measurements showing an average error of about 3 K whereas the mean error in LST estimation was about 2 K.[4] These results are comparable with previously reported results for the TVX method.

The MODIS and ASTER sensors have been a critical tool for providing regional estimates of LSTs. LSTs are warmer in the early afternoon than in the morning because this is a peak time for solar insolation. MODIS data from the Aqua mission, therefore, are more likely to be closer to the maximum daily LST than that acquired earlier in the day by Terra. Coops et al. (2007) investigated the differences in LST between Aqua and Terra to get an assessment of how large these differences are across Canada. Using MODIS Aqua and Terra data for 2000 through mid-2002, they found there are statistically significant differences between AM and PM LSTs ranging from 1.2 and 5°C, depending on the time of year. On the average, over 90% of the variation observed in the PM data can be explained by the AM LST land cover type and location.

Yang et al. (2011) employed several land cover indices, the soil-adjusted vegetation index (SAVI), the normalized multiband drought index (NMDI), the normalized difference built-up index (NSBI), and the normalized difference water index (NDWI), to investigate four land cover types (vegetation, bare soil, impervious, and water) in a suburban area of Beijing, China. They applied these indices to MODIS and ASTER data acquired in May 2001. The study was designed to evaluate differences in LST as a function of spatial scale differences between the 1-km MODIS TIR and the 90-m ASTER TIR data. They applied a disaggregation method for subpixel temperature analysis using a remote sensing end-member index-based technique to derive land surface temperature.[5] It was found that there was good agreement in LSTs between the two spatial resolutions. Another scaling study by Liu et al. (2006) used different scaling approaches to compare LSTs for MODIS and ASTER data over a part of the Loess Plateau in China. ASTER 90m TIR data were scaled up to match the 1km spatial resolution of the MODIS sensor to compare LST values between the two

---

[4] Kelvin is a measurement of heat energy or temperature which advances in the same increments as does Celsius. Its principal difference is that Kelvin measurements are written as K and have a much lower starting point; 0 K or 0 Kelvin is measured at −273.15°C, which is the point at which no heat energy exists in a substance (called absolute zero). Celsius converts to K by adding 273.15 to the Celsius number.

[5] Spectral mixture analysis provides an efficient mechanism for the interpretation and classification of remotely sensed multidimensional imagery. It aims to identify a set of reference signatures (also known as "end members") that can be used to model the reflectance spectrum at each pixel of the original of a finite number of ground components (Plaza et al., 2002).

instruments. They found that upscaled ASTER LSTs achieved an agreement of -0.2 ± 1.87 K in comparison to the MODIS LSTs.

As part of Soil Moisture-Atmosphere Coupling Experiment (SMACEX) (Kustas et al., 2005) conducted over Oklahoma, Kansas, and surrounding states, French et al. (2005) used ASTER data to detect and discern variations in surface temperature, emissivities, vegetation densities, and albedo for distinct land use types. They combined ASTER observations with two physically based surface energy flux models, the Two-Source Energy Balance (TSEB) and the Surface Energy Balance Algorithm for Land (SEBAL) models, to retrieve estimates of instantaneous surface energy fluxes. Intercomparison of results between all flux components indicated that the two models operate similarly when provided identical ASTER data inputs. Further assessment of a multiscale remote sensing model for disaggregating regional fluxes is given by Anderson et al. (2004). Here TIR data from 6 remote sensing satellites [including the NOAA Geostationary Operational Environmental Satellite (GOES)] are used in conjunction with the Atmosphere-Land Exchange Inverse (ALEXI) model and associated disaggregation technique (DisALEXI), in effecting regional to local downscaling of these data. An excellent reference that provides an overview of advances in thermal infrared-based land surface models is also provided by Kustas and Anderson (2009) and Anderson et al. (2004).

3.5   EVAPORATION/EVAPOTRANSPIRATION/SOIL MOISTURE

A predominant application of TIR data has been in inferring evaporation, evapotranspiration (ET), and soil moisture. This is verified by the numerous references in the literature relating to this application as we noted elsewhere (Quattrochi and Luvall, 1999, 2009). A good overview of remote sensing research in hydrometeorology and evapotranspiration, with particular emphasis on the major contributions that have been made by the U.S. Department of Agriculture's, Agricultural Research Service (ARS), is given by Kustas et al. (2003a). A review of surface temperature and vegetation indices remote sensing–based methods for retrieval of land surface energy fluxes and soil moisture is also proved by Petropoulos et al. (2009). An additional overview of remote sensing of evapotranspiration is given in Kustas et al. (2003a).

Landsat ETM+, MODIS, and ASTER data have been successfully used to derive parameters, such as surface temperature and emissivity, for input into soil moisture and ET models (Jacob et al., 2004). Liu et al. (2007) used ETM+ and meteorological data in a regional ET model for the Beijing, China, area. Comparisons of energy balance components (net radiation, soil heat flux, sensible and latent heat flux) with measured fluxes by the model were made, integrating the remotely sensed fluxes by the model. Results show that latent heat flux estimates with errors of mean bias error (MBE) $\pm$ root mean square error (RMSE) of -8.56 ± 23.79 W/m^2, sensible heat flux error of -8.56 ± 23.79 W/m^2, net radiation error of 25.16 ± 50.87 W/m^2, and soil heat flux error of 10.68 ± 22.81 W/m^2. The better agreement between the estimates and the measurements indicates that the remote sensing model is appropriate for estimating regional ET over heterogeneous surfaces.

Another study conducted as part of SMACEX by Su et al. (2005) used the Surface Energy Balance System (SEBS) to estimate land surface fluxes using remote sensing and meteorological data. SEBS consists of several separate modules to estimate the net radiation and soil heat flux and to partition the available energy into sensible and latent heat fluxes. Results from using SEBS show that the model can predict ET with accuracies approaching 10–15% of that of in situ measurements. To extend the field-based measurement of SEBS, information derived from Landsat ETM+ data and data from the North American Land Data Assimilation System (NLDAS)[6] were combined to determine regional surface energy fluxes for a clear day during the field experiment. Results from this analysis indicate that prediction accuracy was strongly related to crop type, with corn prediction showing improved estimates compared to those of soybean. This research found that differences between the mean values of observations and the SEBS Landsat-based predictions at in situ data collection sites were approximately 5%. Overall, results from their analysis indicate much potential toward routine prediction of surface heat fluxes using remote sensing data in conjunction with meteorological data.

In water-deficient areas, water resource management requires ET at high spatial and temporal resolutions. The use of remote spaceborne sensing data to do so, however, requires the assessment of trade-offs between spatial and temporal resolutions. The sharpening of remotely sensed data is one potential way to obviate the limitations posed by data from satellite platforms to derive surface temperature and NDVI at the spatiotemporal scales needed for water resource management applications. Yang et al. (2010) used the triangle algorithm to sharpen Landsat ETM+ data. Sharpened surface temperatures and reference temperatures were compared at 60 and 240 m spatial resolutions. The reflectance measurements are used to calculate the NDVI. The NDVI is then plotted as a function of surface temperature radiation (T_r) to evaluate the relationship between these two variables as well as providing and overlaying the index of moisture availability to establish a "warm edge" and a "cold edge" index (Figure 3.1). [A good overview of the triangle method is presented in Carlson (2007).]

It was found that RMSEs with the triangle algorithm are smaller than those with a functional relationship between surface temperature and NDVI. The triangle method combines measurements of T_r and reflectance in portions of the electromagnetic spectrum.

In another study focused on a water-deficient area, Landsat ETM+ data were used as input to a remote sensing–based ET algorithm called METRIC (Mapping Evapotranspiration at High Resolution Using Internalized Calibration) to provide accurate ET maps on actual crop water use over the Texas High Plains (THP) (Gowda et al., 2008). The performance of the ET model was evaluated by comparing the predicted daily ET with values derived from soil moisture budget at four commercial agricultural fields. Daily ET estimates resulted in a prediction error (RMSE) of $12.7 \pm 8.1\%$ when compared with ET derived from measured soil moisture through the soil water balance. Considering prevailing advection conditions in the THP, these results are

---

[6] Information on the NLDAS can be found at http://ldas.gsfc.nasa.gov/.

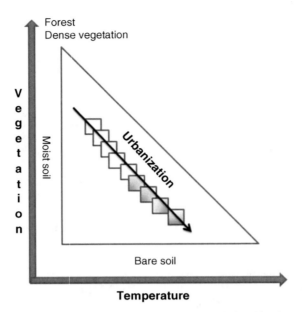

FIGURE 3.1   Schematic of triangle concept that illustrates relationships between temperature and vegetation within the overall perspective of NDVI, where the percentage of vegetated land cover and canopy density increases vertically and the that of bare ground increases horizontally. The example here shows that as the percentage of urbanized land cover and vegetation decreases, there is a corresponding increase in bare ground and higher surface temperatures (Quattrochi and Luvall, 2009). RMSE with the triangle algorithm is smaller than with a functional relationship between surface temperature and NDVI.

good. The investigators note that METRIC offers promise for use in mapping regional ET in the THP region.

In a study over the U.S. central Great Plains, Park et al. (2005) used surface temperatures (T_s) derived from MODIS data for correlation with concurrent water budget variables. Using a climate water budget program, four daily water budget factors (percentage of soil moisture, actual/potential ET ratio, moisture deficit, and moisture deficit potential ET ratio) were calculated at six weather station sites across western and central Kansas. Correlation analysis showed that T_s deviations from air temperature had a significant relationship with water budget factors. To do the analysis on a weekly basis, daily MODIS data were integrated into three different types of weekly composites, including maximum T_s driestday, and maximum T_s deviation from maximum air temperature, or max T_a. Results showed that the maximum T_s deviation (T_s–maxT_a) temperature composite had the largest correlation with the climatic water budget parameters. Correlation for different data acquisition times of MODIS TIR data improved the representativeness of signals for surface moisture conditions. The driest-day composite was most sensitive to time correction. After time correction, its relationship with soil moisture content improved by 11.1% on average, but the degree of correlation improvement varied spatially, but there was not a strong correlation with water budget factors in relation to the maximum T_s deviation composite method.

Three representative studies using MODIS data illustrate the potential of using these data for ET estimation. Modeling of actual daily ET in combination with MODIS data by Sánchez et al. (2007) allowed for the determination of surface fluxes over boreal forests on a daily basis from instantaneous information registered from a conventional meteorological tower as well as the canopy temperatures (T_c) retrieved from satellite. The comparison between T_c ground measured with a thermal infrared radiometer at the meteorological sites and T_c retrieved from MODIS showed an estimation error of $\pm 1.4°C$. Their modeling method was validated over the study site using 21 MODIS images from 2002 and 2003. The results were compared with eddy-correlation ground measurements; with an accuracy of ± 1.0 mm/day and an over-estimation of 0.3 mm/day shown in daily ET retrieval. Mallick et al. (2007) used MODIS optical and thermal band data and ground observations to estimate evaporative fraction and daily actual ET (AET) over agricultural areas in India. Five study regions, each covering a 10 km × 10 km area falling over agricultural land uses, were selected for ground observations at a time closest to MODIS overpasses. Eight MODIS scenes collected between August 2003 and January 2004 were resampled to 1 km and used to generate surface albedo, land surface temperature, and emissivity. Evaporative fraction and daily AET were generated using a fusion of MODIS-derived land surface variables coincident with ground observations. Land cover classes were assigned using a hierarchical decision rule applied to multidate NDVI and applied via a triangle method to estimate the relationship between NDVI and surface temperature. Energy balance daily AET from the fused MODIS data was found to deviate from water balance AET by between 4.3 and 24.5% across five study sites with a mean deviation of 11.6%. The RMSE from the energy balance AET was found to be 8% of the mean water balance AET. Thus, the satellite-based energy balance approach can be used to generate spatial AET, but as noted by the investigators, further refinement of this technique should produce more robust results.

Remote sensing with multispectral infrared can improve regional estimates of ET by providing new constraints on land surface energy balance. Current models use visible and near-infrared bands to obtain vegetated cover and in some cases utilize TIR data; these data together yield good ET estimates. However, it may be possible to enhance theseET models by using emissivity estimates derived from TIR emissivity, which is a property related to fractional vegetation cover but independent of plant greenness (French and Inamdar, 2010). This is demonstrated in a study using MODIS observations obtained over Oklahoma and Kansas which were compared with changes in NDVI for winter wheat and grazing land. It was found that emissivity changes were independent of NDVI and sensitive to standing canopies, regardless of growth stage or senescence. Therefore, emissivities were seasonally dynamic, able to detect wheat harvest timing, and helpful for modeling ET.

Data combinations from different satellite sensors potentially provide even more useful information on soil wetness than by using data from one satellite platform alone. Surface soil wetness determines moisture availability that controls the response and feedback mechanisms between land surface and atmospheric process. Mallick et al. (2009) performed a study to estimate volumetric surface soil moisture content in cropped areas in India at field ($<10^2$-m) to landscape ($\leq10^3$-m) scales. In situ data collected at the field scale were used to obtain a soil wetness index (SWI) from which soil moisture content (θ_V) was derived using ASTER data for the field scale and MODIS at the landscape scale.

Integration of satellite data with spatial data on vegetation and terrain features via GIS methods have also been used to map ET. Accurate estimation of ET is difficult to obtain over heterogeneous landscapes with diverse land covers and topographic terrains (Liu et al., 2007). Mariotto et al. (2011) performed a study to build advanced remote sensing and land surface energy balance algorithms to map ET in a heterogeneous semiarid area over the U.S. Department of Agriculture, Agricultural Research Service, Jornada Experimental Range that encompasses parts of southern Arizona, New Mexico, and Texas. ET of 12 different land covers was computed by applying the SEBAL model. A GIS raster/vector system was used to integrate multispectral TIR and reflectance imagery from ASTER with meteorological, terrain, and land cover data. The study showed that SEBAL run with all these input data provided the best agreement with ground measurements, in comparison with SEBAL run without any modification for terrain features and associated data, and it could significantly discriminate ET among 75.8% of vegetation types. SEBAL without ASTER integrated data set could not discriminate any vegetation types.

The influence of spatial scale on ET estimation using multiple satellite sensors collected over heterogeneous land surfaces is a critical research need. McCabe and Wood (2006) used Landsat ETM+ (60-m), ASTER (90-m), and MODIS (1020-m) data to independently estimate ET. The range of satellite sensor resolutions allows for analyses that span spatial scales from in situ measurements (i.e., pointscale) to the MODIS kilometer scale. ET estimates derived at these multiple resolutions were assessed against eddy covariance flux measurements during the SMACEX campaign over the Walnut Creek watershed in Iowa. Together, these data allow a comprehensive scale intercomparison of remotely sensed predictions that included intercomparison of the ET products from the various sensors as well as a statistical analysis for the retrievals at the watershed scale. A high degree of consistency was observed between the higher spatial resolution sensors (ETM+ and ASTER). The MODIS-based estimates were unable to discriminate the influence of land surface heterogeneity at the field scale but did effectively reproduce the average ET of the watershed response, which illustrated the utility of this sensor for regional scale ET estimates.

Further information on the assessment of ET and soil moisture content across different scales of observation that has implications in deriving ET from satellite-based data is given by Verstraeten et al. (2008). They provide a summary of the generally accepted theory of ET, a summary of ET assessment at different scales of observation, a summary of data assimilation schemes for estimating ET using reflectance and TIR remote sensing data, and a summary of soil moisture retrieval techniques at different spatial and temporal scales. Another useful reference on scaling of TIR data for evaporation estimation is given by Li et al. (2009). They provide an overview of the commonly applied ET models using remote sensing data at regional scales. They discuss the main inputs, assumptions, theories, advantages, and drawbacks of different ET models. They also provide insight into the limitations and promising aspects of the estimation of ET-based remotely sensed data and ground-based measurements.

3.6 DROUGHT MONITORING

In addition to using TIR data for ET and soil moisture analysis over vegetated surfaces, there is also a need for using these data for assessment of drought conditions. Anderson and Kustas (2008) illustrate that the ALEXI model can successfully be used with TIR data to model ET and drought at local to continental scales. They demonstrate this using GOES AVHRR data to produce ET soil moisture stress estimates at 10 km grid resolution over the coterminous United States. They also indicate that ALEXI run in a disaggregation mode (DisALEXI) can generate moderate- to high-resolution (10^0–10^3-m) ET flux maps using data satellite platforms such as Landsat and MODIS. This methodology is examined more extensively and reported on by Anderson et al. (2011. Min and Minghu (2010) show that combining a spectral vegetation index (NDVI) with TIR data (T_s) in a T_s/NDVI triangle model can provide a promising measure for drought monitoring. They use the T_s/NDVI triangle method using MODIS NDVI and LST data to explore dryness monitoring in Heilongjiang Province, China. The spatial pattern observed using this method demonstrates that the summer dryness is characterized with extensive and long-duration droughty conditions. They find that the T_s/NDVI method can provide near-real-time drought monitoring in the study area. Another study conducted in China by Wu et al. (2008) used MODIS TIR data within a GIS format to generate a soil moisture map based on the relationship between thermal inertia and soil moisture. Results indicate that thermal inertia derived from MODIS data is consistent with the actual dryness characteristics that occurred as verified with meteorological data.[7]

Karniell et al. (2010) provide an insightful analysis of the merits and limitations of the use of NDVI and LST for drought assessment. Their work investigates the generality of the LST–NDVI relationship over a wide range of moisture and climatic/radiation regimes encountered over the North American continent (up to 60° N) during the summer growing season. Information on LST and NDVI was obtained from long-term (21-year) data sets acquired with the AVHRR sensor. It was found that when water is the limiting factor for vegetation growth (which is the typical situation for low latitudes of the study area during the midseason), the LST–NDVI correlation is negative. However, when energy is the limiting factor for vegetation growth (in higher latitudes and elevations, particularly at the beginning of the growing season), a positive correlation exists between LST and NDVI. Multiple regression analysis revealed that during the beginning and end of the growing season solar

---

[7] Thermal inertia is the ability of a landscape to resist change in temperature. Because thermal inertia is related to surface composition or to near-surface moisture, remote sensing can be used to measure this property. We explain the utility of thermal inertia measurements in landscape analysis in our 1999 *Landscape Ecology* article on pages 583–584.

radiation is the predominant factor driving the correlation between LST and NDVI, whereas other biophysical variables play a lesser role. Air temperature is the primary factor in midsummer. They conclude that there is a need to use empirical LST–NDVI relationships with caution to restrict their application to drought monitoring to areas and periods where negative correlations are observed, primarily to conditions when water—and not energy—is the primary factor limiting vegetation growth.

3.7 DESERT OR ARID ENVIRONMENTS

Desert or arid environments occupy a significant portion of Earth's surface, and with the prospect of the spatial extent of arid lands possibly increasing due to global climate change, they are important areas for analysis in landscape ecology. In association with drought monitoring, TIR data have been used to study surface temperature characteristics over desert or arid environments. For example, Shamsipour et al. (2011) used AVHRR and meteorological data to assess drought events in the semiarid central plains of Iran. Drought recognition is based on the analysis of the standard precipitation index (SPI) derived from meteorological variables and NDVI obtained from AVHRR data. These variables include the vegetation condition index (VCI), LST, land surface moisture (LSM), temperature condition index (TCI), and vegetation health index (VHI). Analysis was restricted to the spring season from 1998 to 2004. Results show that indices derived from the AVHRR thermal band have a higher sensitivity to drought conditions than indices derived from the visible bands. Indices derived from reflective bands such as NDVI and VCI appear to be better correlated to meteorological parameters than thermal band indices such as TCI. Indices that are calculated from both the reflective and TIR bands like LSM and VHI do not seem to be a reliable measure of drought conditions in the study area.

Research by Qin et al. (2001, 2002, 2005) also used thermal data from the AVHRR to estimate LST and the variation of this parameter over the Israel–Sinai peninsula. As they note, the retrieval of LST from AVHRR data with two channels in the 10.5–11.3-μm bandwidth is usually derived using the split-window algorithm technique. In order to assess the spatial distribution of LST over the study area, they used a modified version of the split-window technique that only requires input on emissivity and transmittance, as opposed to other versions of this technique that require atmospheric parameters that are generally difficult to estimate due to absence of in situ atmospheric profile data.[8] An LST image in combination with a pseudocolor image generated from AVHRR reflective

---

[8] Extensive work has gone into the development of algorithms to estimate LST from AVHRR channels 4 and 5. The primary approach is the so-called split-window technique that uses the difference in brightness temperatures between AVHRR channels 4 and 5 to correct for atmospheric effects on sea surface and land surface temperatures. The split-window method corrects for atmospheric effects based on differential absorption in adjacent infrared bands in deriving LST from satellite data. The split-window technique works independently of other data sources and takes advantage of the differential effect of the atmosphere on the radiometric signal across the atmosphic window region (Czajkowski et al., 2004.). Two other informative sources on the split-window technique and retrieval of LST from satellite data are Wan and Dozier (1996) and Dash et al. (2001).

wavelength bands (1, 2, and 4). A sharp contrast in arid land characteristics can be identified on both sides of the Israel–Egypt border. This contrast is a direct result of different vegetation cover and biogenic crust percentage on both sides.

Miliaresis and Partsinevelos (2010) used monthly night-averaged LST derived from MODIS throughout a year period (2006) in an attempt to segment the terrain of Egypt into regions with different LST seasonal variability and represent them parametrically. Regions with distinct spatial and temporal LST patterns were identified using several clustering techniques that captured aspects of spatial, temporal, and temperature homogeneity or differentiation. Segmentation was augmented by taking elevation, morphological features, and land cover information into consideration. Analyses of these data showed that the lowland northern coast region of Egypt along the Mediterranean Sea corresponds to the coolest clusters, indicating a latitude/elevation dependency of seasonal LST variability. Conversely, for inland regions, elevation and terrain dissection plays a key role in LST seasonal variability, while an east-to-west variability of spatial distribution in clusters is evident. Lastly, elevation-biased clustering revealed annual LST differences among the regions with the same physiographic and terrain characteristics. Thermal terrain segmentation outlined temporal variation of LST during the year period as well as the spatial distribution of LST zones.

3.8   THERMAL ENERGY THEORY AS APPLIED TO ECOLOGICAL THERMODYNAMICS

The concept of ecological thermodynamics provides a quantification of surface energy fluxes for landscape characterization in relation to the overall amount of energy input and output from specific land cover types. Ecological thermodynamics was introduced in the Quattrochi and Luvall (1999) *Landscape Ecology* paper, but here we present a more thorough understanding of the techniques and methods embodied within this concept to offer an updated and clearer understanding of its utility and virtues

Terrestrial ecosystem surface temperatures have been measured using airborne and satellite sensors for several decades. Using NASA's Thermal Infrared Multispectral Scanner (TIMS), Luvall and his co-workers (Luvall and Holbo, 1989, 1991; Luvall et al. 1990) have documented ecosystem energy budgets for tropical forests, mid-latitude varied ecosystems, and semiarid ecosystems. These data show that within a given biome type and under similar environmental conditions (air temperature, relative humidity, winds, and solar irradiance), the more developed the ecosystem, the cooler its surface temperature and the more degraded the quality of its reradiated energy. These data suggest that ecosystems develop structure and function that degrade the quality of the incoming energy more effectively; that is, they degrade more exergy,[9] which agrees with the predictions of nonequilibrium thermodynamic

---

[9] In thermodynamics, the exergy of a system is the maximum work available through any process that brings the system into equilibrium with a heat reservoir (environment). Exergy is the energy available for use. See Fraser and Kay (2004) for a discussion of exergy in an ecological context.

theory (Schneider and Kay, 1994a; Kay and Schneider, 1994; Schneider and Sagan, 2005). This remote sensing work suggests that analysis of airborne remote sensing energy flux data is a valuable tool for measuring the energy budget and energy transformations in terrestrial ecosystems. Given the stated hypothesis, a more developed ecosystem degrades more exergy, and the ecosystem temperature, R_n/K^*, beta index, and thermal response number (TRN) are excellent candidates for indicators of ecological integrity. The potential for these methods to be used for remotely sensed ecosystem classification and ecosystem health/integrity evaluation is apparent.

Recent advances in applying principles of nonequilibrium thermodynamics to ecology provide fundamental insights into energy partitioning in ecosystems. Ecosystems are nonequilibrium systems, open to material and energy flows, which grow and develop structures and processes to increase energy degradation. More developed terrestrial ecosystems will be more effective at dissipating the solar gradient (degrading its exergy content).

Thermal energy theory results from work to understand ecological development. [See Kay (2000) for an overview.] The research in ecological thermodynamics has focused on linking physics and systems sciences with biology, especially linking the science of ecology with the laws of thermodynamics. This research follows on the observation that similar developmental processes are observed in ecosystems, from small laboratory microcosms, to prairie grass systems, to vast forest systems and ocean plankton systems. Such similar phenomenology has long suggested underlying processes and rules for the development of ecological patterns of structure and function (Odum, 1969). Furthermore, recent advances in nonequilibrium thermodynamics coupled with the investigation of self-organizing phenomena in different types of systems (from simple convection cell systems to forested ecosystems) have revealed that all self-organizing phenomena (including ecosystem development) involve similar processes, processes which are mandated by the second law of thermodynamics. This conclusion, as discussed below, provides a basis for a quantitative description of ecosystem development (Fraser and Kay, 2004; Kay, 1991; Kay and Schneider, 1992; Schneider and Kay, 1993, 1994a, 1994b; Regier and Kay, 1996).

The study of self-organization phenomena in thermodynamic systems is based on systems that are open to energy or material flows and which reside in quasi-stable states some distance from equilibrium, (Nicolis and Prigogine, 1977). Both nonliving self-organizing systems (like convection cells, tornadoes, and lasers) and living self-organizing systems (from cells to ecosystems) are dependent on exergy (high-quality energy) fluxes from outside sources to sustain their self-organizing processes. These processes are maintained by the destruction of the exergy, that is, the conversion of the high-quality energy flux into a flux of lower quality forms of energy. Consequently, these processes increase the entropy of the larger "global" system in which the self-organizing system is embedded. Crucial insights into the dynamics of self-organizing systems can be gained from examining the role of the second law of thermodynamics in determining these dynamics.

Using exergy, the second law of thermodynamics can be applied to nonequilibrium regions and processes. This system can be described in terms of the exergy fluxes setting up gradients (e.g., temperature and pressure differences in classical thermodynamic systems). With the establishment of these gradients, the system is no longer in equilibrium. The system responds to these imposed gradients by self-organizing in a way which resists the ability of the exergy fluxes to establish gradients and hence move the system further away from equilibrium. More formally, a restatement of the second law says that as systems are moved away from equilibrium, they will utilize all avenues available to counter the applied gradients. As the applied gradients increase, so does the system's ability to oppose further movement from equilibrium (Schneider and Kay, 1994a). The more a system self-organizes, the more effective it will become at exergy utilization. Kay and Schneider (1994) have focused on the application of this thermodynamic principle to the science of ecology. Ecosystems are viewed as open thermodynamic systems with a large gradient impressed on them by the exergy flux from the sun. Ecosystems, according to the restated second law, develop in ways that systematically increase their ability to degrade the incoming solar exergy, hence counteracting the sun's ability to set up even larger gradients. It is clear that for forested ecosystems by far the majority of the energy is processed through sensible and latent heat fluxes, as illustrated by the measurements taken from the Hubbard Brook Forest (Figure 3.2).

Thus it can be predicted that more mature ecosystems will degrade the exergy they capture more completely than a less developed ecosystem (Table 3.1). The degree to which incoming solar exergy is degraded is a function of the surface temperature of the ecosystem. [See Fraser and Kay (2004) for details.] If a group of ecosystems receives the same amount of incoming radiation, we would expect that the most

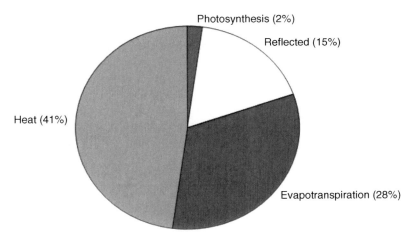

FIGURE 3.2 Partitioning of surface energy fluxes in Hubbard Brook (Bormann and Likens, 1979; Gosz et al., 1978; Kay, 1978). (See the color version of this figure in Color Plates section.)

TABLE 3.1   Radiative Transfer Estimates, Surface Temperatures, Beta Index, and TRN Measurements for Several Surface Types at Andrews Experimental Forest, Oregon

Radiative Flux Terms	Rock Quarry	Clearcut 2 yr Douglas Fir	Natural Regeneration 25 yr Douglas Fir	Plantation 25 yr Douglas Fir	Mature Forest 400 yr Douglas Fir
K^*, W/m^{-2}	718	799	895	854	1,005
L^*, W/m^{-2}	273	281	124	124	95
R_n, W/m^{-2}	445	517	771	730	830
R_n/K^*, %	62	65	86	85	90
T, °C	50.7	51.8	29.4	29.5	24.7
delta T	4.5	2.2	1.7	0.8	0.9
Beta index	-12.9	6.3	17.2	34.4	130.7
TRN, kJ/m^{-2} °C	168	406	788	1631	1549

Source: Modified from Luvall and Holbo (1989), Holbo and Luvall (1989), and Bishop et al. (2004).

Note: K^* = net incoming solar radiation, L^* = net long wave, R_n = net radiation, R_n/K^* = percent of net incoming solar radiation degraded into nonradiative processes.

mature ecosystem would reradiate its energy at the lowest quality level and thus would have the lowest surface temperature.

3.8.1   Beta Index as Measure of Surface Temperature Spatial Variation

Three measures characterize the thermodynamic performance of the ecosystem: the ratio R_n/K^*, the beta index, and the TRN. The average temperature for a forest canopy cannot express the spatial variability. However, as demonstrated by Holbo and Luvall (1989), the frequency distributions of temperatures can be used as a powerful model in differentiation and identification of land surface cover types and their properties. They found that a beta distribution closely resembles the observed temperature frequency distributions from forested landscapes. An advantage of using the beta distribution as a model is that it utilizes the pixel frequency distributions directly and no high-order, measurement-error-magnifying statistics are used (Figure 3.3). From this they developed the beta index, by which these forested landscapes could be classified and quantify the spatial temperature variability of the ecosystem. As ecosystems develop, nonequilibrium thermodynamic theory suggests that they would tend toward internal equilibrium. Therefore, we would expect the spatial variability of temperature to decrease as an ecosystem develops. Thus a large beta index should indicate a more developed ecosystem. These data are consistent with viewing ecosystems in terms of nonequilibrium structures and processes. Nonequilibrium thermodynamic theory suggests dissipative systems tend toward a steady state and develop homeostatic methods for maintaining the steady state and thus we expect temperature variability to decrease with ecosystem development.

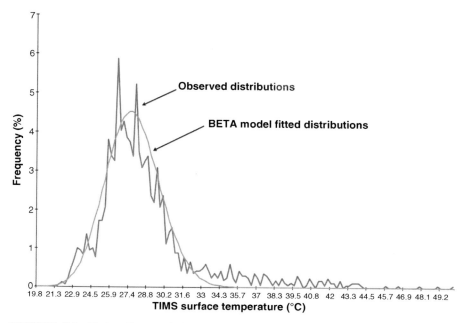

FIGURE 3.3 Fitting Beta probability distributions to observed frequency distributions (Holbo and Luvall, 1989).

3.8.2   Thermal Response Number

The second characterization measure, the TRN, can be applied whereever there are overlaps in adjacent flight lines. The TRN was developed in Luvall and Holbo (1989, 1991) as a remote sensing based technique for describing the surface energy budget within a forested landscape. This procedure treats changes in surface temperature as an aggregate response of the dissipated thermal energy fluxes (latent heat and sensible heat exchange and conduction heat exchange with biomass and soil). The TRN is therefore directly dependent on surface properties (canopy structure, amount and condition of biomass, heat capacity, and moisture). Surface net radiation integrates the effects of the nonradiative fluxes, and the rate of change in forest canopy temperature presents insight on how nonradiative fluxes are reacting to radiant energy inputs. The ratio of net radiation to change in temperature can be used to define a surface property referred to as the thermal response number:

$$\mathrm{TRN} = \sum_{t_1}^{t_2} R_n \Delta t / \Delta T \quad (\mathrm{kJ/m^{-2}\,K})$$

where $\sum_{t_1}^{t_2} R_n \Delta t$ represents the total amount of net radiation (R_n) for that surface over the time period between flights ($\Delta t = t_2 - t_1$) and ΔT is the change in mean temperature of that surface. Research by Luvall and Holbo (1989, 1991) and Luvall et al., (1990) using the TIMS showed that 15–30 min between overflights is sufficient time difference to obtain measurable and useful changes in forest canopy temperature due to the change in incoming solar radiation. The mean spatially averaged temperature for the surface elements at the times of imaging is estimated from $T = 1/(n \sum T_p)$, where each T_p is the temperature of a pixel in the thermal image and n is the number of pixels of the surface element. The TRN provides an analytical framework for studying the effects of surface thermal response for large spatial resolution map scales that can be aggregated for input to coarser scales as needed by climate models. The utility of TRN is that (1) it is a functional classifier of land cover types; (2) it provides an initial surface characterization for input to various climate models; (3) it is a physically based measurement; (4) it can be determined completely from a pixel by pixel measurement or for a polygon from a landscape feature which represents a group of pixels; and (5) surface topography and orientation of observation are not handicaps where adequate digital elevation data are available. The TRN can be used as an aggregate expression of both environmental energy fluxes and surface properties such as forest canopy structure and biomass, age, and physiological condition as well as urban structures and material types. A similar index, the thermal buffer capacity (TBC), was later proposed by Aerts et al. (2004) as a dissipation indicator:

$$\text{TBC} = \frac{t_2 - t_1}{T_s(t_2) - T_s(t_1)} = \frac{\Delta t}{\Delta T_s}$$

3.8.3   Ecological Complexity and Ecological Health

The use of ecosystem exergy theory with thermal remote sensing observations is beginning to be used to study other ecosystems throughout the world. Maes et al. (2011) used a series of DAIS (digital airborne imaging spectrometer) images collected over various forests, orchards, croplands, grazing lands, and urban areas in Northern and Central Belgium. They found that TRN and TBC have the highest discriminative power of all dissipative indices and were particularly suited for distinguishing differences in latent heat flux among the vegetation types. They also determined that TBC and TRN were the dissipation indicators that were least influenced by prevailing meteorological conditions.

Additional work by Lin et al. (2009) used TRN, TBC, and R_n/K^* to quantify plant community self-organization in a tropical seasonal rainforest, an artificial tropical rainforest, a rubber plantation, and two *Chromolaena odorata* (L.) R.M. King & H. Robinson communities aged 13 years and 1 year. These transects sampled the typical vegetation complexity and land use in Xishuangbanna, southwestern China. They concluded that these thermodynamic indices could discriminate differences in complexity among ecosystems studied both in the dry and wet seasons.

Norris et al. (2012) studied the application of ecological thermodynamics theory to ecosystem climate change adaptation and resilience. They concluded that using ecological thermodynamic indicators would significantly enhance the understanding of the characteristics of resilient and adaptable natural environments.

3.9 CONCLUDING REMARKS

In our final remarks in our 1999 paper, we noted that the incorporation of TIR data into landscape studies offers the prospect to measure the state and dynamics of energy fluxes across and between landscapes, from the patch to the regional levels. The utilization of TIR remote sensing in landscape ecological research has indeed, as illustrated by the numerous references cited herein, made meaningful and significant progress since the publication of our *Landscape Ecology* article. The utility of TIR data is now commonplace and a "known quantity" for application to landscape ecological research. This is a product of the increased amount of references in the literature to TIR data in landscape studies, but even more so because of the number of satellite platforms that have been launched in the last 14 years since our article was published. Of principal importance have been Landsat TM and ETM+ sensors and the TIR sensors onboard the NASA Terra Earth-observing platform (i.e., MODIS and ASTER), where the thermal data from these instruments have been widely used for analysis of a variety of landscape characteristics. Moreover, the launch of the VIIRS instrument in 2011 and Landsat 8 in 2013 has increased the overall availability of TIR data. New NASA missions that will be launched with TIR instruments are outlined in the National Research Council's report, "Earth Science and Applications from Space: National Imperative for the Next Decade and Beyond" (better known as the "Decadal Survey") (NAS, 2007), which established a roadmap for developing a suite of Earth-observing satellites in the future. The NRC provided further direction for NASA's Earth science missions in its recently published report, "Earth Science and Applications from Space: A Midterm Assessment of NASA's Implementation of the Decadal Survey" (NAS, 2012).

One future mission in particular is important for furthering the use and analysis of TIR data for landscape assessment—the Hyperspectral Infrared Imager (HyspIRI). This will be a combined hyperspectral/thermal instrument with 213 spectral channels between 380 and 2500 nm on 10-nm centers, and the TIR sensor will have eight spectral channels (seven between 7.5 and 12 μm and one at 4 μm). Both instruments will have a spatial resolution of 60 m and the revisit time for HyspIRI will be 19 and 5 days for the visible/shortwave infrared (VSWIR) and TIR instruments, respectively.[10] The HyspIRI mission, therefore, will offer an unprecedented opportunity to obtain high-spatial-resolution multispectral TIR data that can be used in landscape ecological and land surface processes research at revisit times wherein observations of the land surface can be made at repeat times that to date are unattainable by current NASA Earth science missions.

---

[10] More in-depth information on the HyspIRI mission can be accessed at the HyspIRI Mission Study website at http://hyspiri.jpl.nasa.gov/.

In the spirit of the Decadal Survey, HyspIRI is designed to address a number of thematic topics and underlying science questions related to the observation, measurement, and analysis of land surface characteristics and ecosystem functioning. Although the HyspIRI science questions, at least in part, all focus on land surface issues, three overarching science questions in particular are of significance to furthering and fostering data analysis and modeling with TIR data from a landscape ecological perspective:

- How does urbanization affect the local, regional, and global environment? Can we characterize this effect to help mitigate its impact on human health and welfare?
- What is the composition and temperature of the exposed surface of Earth? How do these factors change over time and affect land use and habitability?
- What is the composition of the exposed terrestrial surface of Earth, and how does it respond to anthropogenic and nonanthropogenic drivers?

Correspondingly, each of these overarching science questions has underlying science subquestions that elucidate issues associated with these larger questions. It is anticipated that HyspIRI with its hyperspectral/multispectral capabilities and improved revisit times will provide the VSWIR and TIR data needed for research to help address these questions

In summary, we have presented here a synopsis and description of relevant literature that has been published on thermal infrared remote sensing for analysis of landscape ecological and land surface processes research, since the publication of our article that appeared in *Landscape Ecology* in 1999, and which augments the information presented in our *Thermal Remote Sensing in Land Surfaced Processes* book, published in 2004. To this end, we believe that the review of the literature that is presented here annunciates the overall premise of the importance of TIR data for advancing the science of landscape ecological and land surface process research. TIR data are now readily available from a number of NASA Earth-observing satellite platforms at differing temporal, spectral, and spatial scales that lend themselves to the overall analyses and modeling of a host of landscape characteristics and processes. In actuality, the volume of TIR data that are now available is somewhat overwhelming, and the specific types of TIR data and the spatial/spectral scales of these data must be carefully matched with the landscape research in question. Dependent upon the objectives of specific research, it may in fact be prudent to utilize and compare TIR data collected at different spatial and spectral scales from different satellites to develop a more complete understanding of the thermal characteristics, energy balances, and fluxes that force or drive the landscape processes of interest. There is more TIR data to come in the future with the launch of NASA missions such as HyspIRI, as well as Earth-observing instruments launched by the European Space Agency (ESA) and other national entities. Through increased use and analyses, the science questions and research associated with developing a more comprehensive visual, qualitative, and quantitative insight into landscape characteristics, functioning, and processes that make the land surface, the" 'landscape," will be realized.

REFERENCES

Aerts, R., Wagendorp, T., November, E., Behailu, M., Deckers, J., and Muys, B. 2004. Ecosystem thermal buffer capacity as an indicator of the restoration status of protected areas in the Northern Ethiopian highlands. *Restoration Ecology* 12:586–596.

Anderson, M., and Kustas, W. P. 2008. Mapping evapotranspiration and drought at local to continental scales using thermal remote sensing. Proceedings, IEEE International Geoscience and Remote Sensing Symposium; July 6–11, Boston, MA, IV-121–123.

Anderson, M. C., Norman, J. M., Mecikalski, J. P., Torn, R. D., Kustas, W. P., and KusBsara, J. B. 2004. A multiscale remote sensing model for disaggregating regional fluxes to micrometeorological scales. *Journal of Hydrometeorology* 5:343–363.

Anderson, M.C., Hain, C., Wardlow, B., Pimstein, A., Mecikalski, J. R., and Kustas, W.P. 2011. Evaluation of drought indices based on thermal remote sensing of evapotranspiration over the Continental United States. *Journal of Climate* 24:2025–2044.

Bishop, M. P., Colby, J. D., Luvall, J. C., Quattrochi, D. A., and Rickman, D. 2004. Remote sensing science and technology for studying mountain environments. In M. Bishop and J. F. Shroder (Eds.), *Geographic Information Science and Mountain Geomorphology Remote Sensing*. New York: Praxis Scientific and Springer pp. 147–187.

Bormann, F. H., and Likens, G. E. 1979. *Pattern and Process in a Forested Ecosystem: Disturbance, Development and the Steady State Based on the Hubbard Brook Ecosystem Study.*, New York: Springer, 253 pp.

Carlson, T. 2007. An overview of the "Triangle Method" for estimating surface evapotranspiration and soil moisture from satellite imagery. *Sensors* 7:1612–1629.

Cohen, W. B., and Goward, S. N. 2004. Landsat's role in ecological applications of remote sensing. *Bioscience* 54:535–545.

Coops, N. C. C., Duro, D. C., Wulder, M. A., and Han, T. 2007. Estimating afternoon MODIS land surface temperature (LST) based on morning MODIS overpass, location and elevation information. *International Journal of Remote Sensing* 28: 2391–2396.

Czajkowski, K. P., Goward, S. N., Stadler, S. J., and Walz, A. 2000. Thermal remote sensing of near surface environmental variables: Applications over the Oklahoma Mesonet. *Professional Geographer* 52:345–357.

Czajkowski, K. P., Goward, S. N., Mulhern, T., Goetz, S. J., Watz, A., Shirey, D., Stadler, S., Prince, S. D., and Dubayah, R. O. 2004. Estimating environmental variables using thermal remote sensing. In D. A. Quattrochi and J. C. Luvall (Eds.) *Thermal Remote Sensing in Land Surface Processes*. Boca Raton, FL: CRC Press, pp. 11–32.

Dash, P., Göttsche, F. M., Olesen, F. S., and Fischer, H. 2001. Retrieval of land surface temperature and emissivity from satellite data: Physics, theoretical limitations and current methods. *Journal of the Indian Society of Remote Sensing* 29:23–30.

Diak, G. R., Mecikalski, J. R., Anderson, M. C., Norman, J. M., Kustas, W.P., Torn, R. D., and DeWolf, R.L. 2004. Estimating land surface energy budgets from space. *Bulletin of the American Meteorological Society* 85:65–78.

Fan, L., Liu, S., Bernhofer, C., Lliu, H., and Berger, F. H. 2007. Regional land surface energy fluxes by satellite remote sensing in the Upper Xilin River Watershed (Inner Mongolia, China). *Theroretical and Applied Climatology* 88:231–245.

Fraser, R., and Kay, J. J. 2004. Exergy analysis of ecosystems: Establishing a role for thermal remote sensing. In D. A. Quattrochi and L. C. Luvall (Eds.), *Thermal Remote Sensing in Land Surface Processes*. Boca Raton, FL: CRC Press, pp. 283–360.

French, A. N., and Inadar, A. 2010. Land cover characterization for hydrological modeling using thermal infrared emissivities. *International Journal of Remote Sensing* 31:3867–3883.

French, A. N., Jacob, F., Anderson, M. C., Kustas, W. P., Timmermans, W., Gieske, A., Su, Z., Su, H., McCabe, M. F., Li, F., Prueger, J., and Brunsell, N. 2005. Surface energy fluxes with the Advanced Spaceborne Thermal Emission and Reflection radiometer (ASTER) at the Iowa 2002 SMACEX site (USA). *Remote Sensing of Environment* 99:55–65.

Gillanders, S. N., Coops, N. C., Wulder, M. A., Gergel, S. E., and Nelson, T. 2008. Multitemporal remote sensing of landscape dynamics and pattern change: Describing natural and anthropogenic trends. *Progress in Physical Geography* 2:503–528.

Gosz, J. R., Holmes, R. T., Likens, G. E., and Bormann, F. H. 1978. The flow of energy in a forest ecosystem. *Scientific American* 3:92–102.

Gowda, P. H., Chávez, J. L., Howell, T. A., Marek, T. H., and New, L. L. 2008. Surface energy balance based evapotranspiration mapping in the Texas high plains. *Sensors* 8:5186–5201.

Holbo, H. R., and Luvall, J. C. 1989. Modeling surface temperature distributions in forest landscapes. *Remote Sensing of Environment* 27:11–24.

Humes, K. S., Hardy, R., and Kustas, W. P. 2000. Spatial patterns in surface energy balance components derived from remotely sensed data. *Professional Geographer* 52:272–288.

Jacob, F., Peiticolin, F., Schmgge, T., Vermote, É., French, A., and Ogawa, K. 2004. Comparison of land surface emissivity and radiometric temperature derived from MODIS and ASTER sensors. *Remote Sensing of Environment* 90:137–152.

Karniell, A., Agam, N., Pinker, R. T., Anderson, M., Imhoff, M. L., Gutman, G. G., and Panov, N. 2010. Use of NDVI and drought assessment: Merits and limitations. *Journal of Climate* 23:618–633.

Kay, J. J. 1991. A non-equilibrium thermodynamic framework for discussing ecosystem integrity. *Environmental Management* 15:483–495.

Kay, J. J. 1994. A non-equilibrium thermodynamic framework for discussing ecosystem integrity. *Environmental Management*. 15:483–495.

Kay, J. J. 2000. Ecosystems as self-organizing holarchic open systems: narratives and the second law of thermodynamics. In S. Jorgensen and F. Miller (Eds.), *Handbook of Ecosystem Theories and Management*. London: CRC Press, pp. 135–159.

Kay, J. J., and Schneider, E. D. 1994. Embracing complexity, the challenge of the ecosystem approach. *Alternatives* 20:32–38.

Kay, J., and Schneider, E. 1992. Thermodynamics and measures of ecosystem integrity. In: Ecological Indicators, Vol. 1, *Proceedings of the International Symposium on Ecological Indicators*, Fort Lauderdale, Florida, Elsevier, New York, 1992.

Kerr, J. T., and Ostrovsky, M. 2003. From space to species: Ecological applications for remote sensing. *Trends in Ecology and Evolution* 18:299–305.

Kolb, T. E., Wagner, M. R., and Covington, W. W. 1994. Utilitarian and ecosystem perspectives: Concepts of forest health. *Journal of Forestry* 92:10–15.

Kustas, W. P., and Anderson, M. 2009. Advances in thermal infrared remote sensing for land surface modeling. *Agricultural and Forest Meteorology* 149:2071–2081.

Kustas, W. P., Diak, G. R., and Moran, M. S. 2003a. Evaoptranspiration, remote sensing of. In B. A. Stewart and T. A. Howell (Eds.), *Encyclopedia of Water Science*. New York: Marcell Dekker, pp. 267–274.

Kustas, W. P., French, A. N., Hatfield, J. L., Jackson, T. J., Moran, M. S., Rango, A., Ritchie, J. C., and Schmugge, T. J. 2003b. Remote sensing research in hydrometeorology. *Photogrammetric Engineering and Remote Sensing* 69:631–646.

Kustas, W. P., Hatfield, J. L., and Prueger, J. H. 2005. The Soil Moisture—Atmosphere Coupling Experiment (SMACEX): Background, hydrometeorolgical conditions, and preliminary findsings. *Journal of Hydrometeorolgy* 6:791–804.

Li, F., Jackson, T. J., Kustas, W. P., Schmugge, T. J., French, A. N., Cosh, M. H., and Bindlish, R. 2004. Deriving land surface temperatures from Landsat 5 and 7 during the SMEX02/SMACEX. *Remote Sensing of Environment* 92:521–534.

Li, Z.-L., Tang, R., Wan, Z., Bi, Y., Zhou, C., Tang, B., Yan, G., and Zhang, X. 2009. A review of current methodologies for regional evapotranspiration estimation from remotely sensed data. *Sensors* 9:3801–3853.

Lin, H., Cao, M., Stoy, P. C., and Zhang, Y. 2009. Assessing self-organization of plant communities—A thermodynamic approach. *Ecological Modeling* 220:784–790.

Liu, S., Hu, G., Lu, L., and Mao, D. 2007. Estimation of regional evapotranspiration by TM/ETM+ data over heterogeneous surfaces. *Photogrammetric Engineering and Remote Sensing* 73:1169–1178.

Liu, Y., Hiyama, T., and Yamaguchi, Y. 2006. Scalling of land surface temperature using satellite data: A case examination on ASTER products over a heterogeneous terrain area. *Remote Sensing of Environment* 105:115–128.

Luvall, J. C., and Holbo, H. R. 1989. Measurements of short-term thermal responses of coniferous forest canopies using thermal scanner data. *Remote Sensing of Environment*. 27:1–10.

Luvall, J. C., and Holbo, H. R. 1991. Thermal remote sensing methods in landscape ecology. In M. Turner and H. Gardner (Eds.), *Quantitative Methods in Landscape Ecology*. New York: Springer, pp. 127–152.

Luvall, J. C., Lieberman, D., Lieberman, M., Hartshorn, G., and Peralta, R. 1990. Estimation of tropical forest canopy temperatures, thermal response numbers, and evapotranspiration using an aircraft-based thermal sensor. *Photogrammetric Engineering and Remote Sensing* 56:1393–1401.

Maes, W. H., Pashuysen, T., Trabucco, A., Veroustraete, F., and Muys, B. 2011. Does energy dissipation increase with ecosystem succession? Testing the ecosystem exergy theory combining theoretical simulations and thermal remote sensing observations. *Ecological Modeling* 222:3917–3941.

Mallick, K., Bhattacharya, B. K., Chaurasia, S., Dutta, S., Nigam, R., Mukherjee, J., Banerjee, S., Kar, G., Roa, V. U. M., Gadgil, A. S., and Parihar, J. S. 2007. Evapotranspiration using MODIS data and limited ground observations over selected agroecosystems in India. *International Journal of Remote Sensing* 28:2091–2110.

Mallick, K., Bhattacharya, B. K., and Patel, N. K. 2009. Estimating volumetric surface moisture content for cropped soils using a soil wetness index based on surface temperatures and NDVI. *Agricultural and Forest Meteorology* 149:1327–1342.

Mariotto, I., Gutschick, V. P., and Clason, D. L. 2011. Mapping evapotranspiration from ASTER data through GIS spatial integration of vegetation and terrain features. *Photogrammetric Engineering and Remote Sensing* 77:483–493.

McCabe, M. F., and Wood, E. F. 2006. Scale influences on the remote estimation of evapotranspiration using multiple satellite sensors. *Remote Sensing of Environment* 105:271–285.

Miliaresis, G., and Partsinevelos, P. 2010. Terrain segmentation of Egypt from multi-temporal night LST imagery and elevation data. *Sensors* 2:2083–2096.

Min, Y., and Minghu, C. 2010. Ts/NDVI space based drought monitoring study from satellite remote sensing data in Heilongjiang. Proceedings, 2010 World Automation Congress (WAC); September 19-23, 2010, Kobe, Japan, pp. 23–28.

National Academy of Sciences, (NAS). 2007. National imperative for the next decade and beyond. Washington, DC: National Academies Press. Available at http://www.nap.edu/catalog.php?record_id=11820. Accessed 2013 May 22.

National Academy of Sciences, (NAS). 2012. Earth science and applications from space: A midterm assessment of NASA's implementation of the decadal survey. Washington, DC: National Academies Press. Available at http://www.nap.edu/catalog.php?record_id=13405. Accessed 2013 May 22.

National Aeronautics and Space Administration (NASA). 2013. Measuring Vegetation (NDVI & EVI) Normalized Differnce Vegetation Index (NDVI). NASA Earth Observatory: Available at http://earthobservatory.nasa.gov/Features/MeasuringVegetation/measuring_vegetation_2.php. Accessed 2013 May 22.

Newton, A. C., Hil, R. A., Echeverria, C., Golicher, D., Rey Benayas, J. M., Cayuela, L., and Hinsley, S. A. 2009. Remote sensing and the future of landscape ecology. *Progress in Physical Geography* 33:529–546.

Nicolis, G., and Prigogine, I. 1977. *Self-Organization in Non-equilibrium Systems*. New York: Wiley.

Norris, C., Hobson, P., and Ibisch, P. L. 2012. Microclimate and vegetation function as indicators of forest thermodynamic efficiency. *Journal of Applied Ecology* 49: 562–570.

Odum, E. O. 1969. The strategy of ecosystem development. *Science* 164:262–270.

Park, S., Feddema, J. J., and Egbert, S. L. 2005. MODIS land surface temperature composite data and their relationships with climatic water budget factors in the Central Great Plains. *International Journal of Remote Sensing* 26:1127–1144.

Petropoulos, G., Carlson, T. N., Wooster, M. J., and Islam, S. 2009. A review of T_s/VI remote sensing based methods for the retrieval of land surface energy fluxes and soil moisture. *Progress in Physical Geography* 33:224–250.

Plaza, A., Martinez, P., Pérez, R., and Plaza, J. 2002. Spatial/spectral endmember extraction by multidimensional morphological operations. *IEEE Transactions on Geoscience and Remote Sensing* 40:2025–2041.

Qin, Z., Berlinger, P. R., and Karnieli, A. 2005. Ground temperature measurement and emissivity determiniation to understand the thermal anomaly and its significance on the development of an arid environmental ecosystem in the sand dunes across Israel-Egypt border. *Journal of Arid Environments* 60:27–52.

Qin, Z., Karnieli, A., and Berlinger, P. 2001. Thermal variation in Israel-Sinai (Egypt) peninsula region. *International Journal of Remote Sensing* 2:915–919.

Qin, Z., Karnieli, A., and Berliner, P. 2002. Remote sensing analysis of the land surface temperature anomaly in the sand-dune region across the Israel–Egypt border. *International Journal of Remote Sensing* 23:3991–4018.

Quattrochi, D. A., Anderson, M., Prakash, A., Wright, R., Pagano, T., Hall, D. K., Eneva, M., Coolbaugh, M. F., and Allen, R. G. 2009. Thermal remote sensing: Theory, sensors, and applications. In M. W. Jackson (Ed.), *Earth Observing Platforms & Sensors, Manual of Remote Sensing*, 3rd ed. Vol. 1.Falls Church, VA: American Society of Photogrammetry and Remote Sensing, pp. 107–187.

Quattrochi, D. A., and Luvall, J. C. 1999. Thermal infrared remote sensing data for analysis of landscape ecological processes: Methods and applications. *Landscape Ecology* 14:577–598.

Quattrochi, D. A., and Luvall, J. C. 2004. *Thermal Remote Sensing in Land Surface Processes*. Boca Raton, FL: CRC Press, 440 p.

Quattrochi, D. A., and Luvall, J. C. 2009. Thermal remote sensing in earth science research. In T. Warner, M. D., Nellis and G. Foody (Eds.), *Handbook of Remote Sensing*. London: Oxford University Press, pp. 64–78.

Quattrochi, D. A., Walsh, S. J., Jensen, J. R., and Ridd, M. K. 2003. Remote sensing. In G. L. Galle and C. J. Willmot (Eds.), *Geography in America at the Dawn of the 21st Century*. Oxford: Oxford University Press, pp. 376–416.

Regier, H., and Kay, J. J. 1996. An heuristic model of transformations of the aquatic ecosystems of the Great Lakes St. Lawrence River Basin. *Journal of Aquatic Ecosystem Health* 5:3–21.

Rocchini, D. (Ed.). 2010a. Special issue on ecological status and change by remote sensing. *Remote Sensing* 2:2072–4292.

Rocchini, D. 2010b. Ecological status and change by remote sensing. *Remote Sensing* 2:2424–2425.

Sánchez, J. M., Casselles, V., Niclós, R., Valor, E., Coll, C., and Laurila, T. 2007. Evaluation of the *B*-method for determining actual evapotranspiratikon in a boreal forest from MODIS data. *International Journal of Remote Sensing* 28:1231–1250.

Schneider, E. D., and Kay, J. J. 1993. Exergy degradation, thermodynamics and the development of ecosystems. Energy, systems, and ecology. In G. Tsatsaronis, J. Szargut, Z. Kolenda and A. Ziebik (Eds.), *Proceedings on an International Conference on Energy Systems and Ecology*. New York, NY: Advanced Energy Systems Division, American Society of Mechanical Engineers, Vol. 1, pp. 33–42.

Schneider, E. D., and Kay, J. J. 1994a. Life as a manifestation of the second law of thermodynamics. *Mathematical and Computer Modelling* 19:25–48.

Schneider, E. D., and Kay, J. J. 1994b. Complexity and thermodynamics: towards a new ecology. *Futures* 24:626–647.

Schneider, E. D., and Sagan, D. 2005. *In to the Cool: Energy Flow, Thermodynamics, and Life*. Chicago: University of Chicago Press.

Shamsipour, A., Zawar-Reza, P., Kasem, S., Panah, A., and Azizi, G. 2011. Analysis of drought events for the semi-arid central plains of Iran with satellite and meteorological based indicators. *International Journal of Remote Sensing* 32:9559–9569.

Southworth, J. 2004. An assessment of Landsat TM band 6 thermal data for analyzing land cover in tropical dry forest regions. *International Journal of Remote Sensing* 25:689–706.

Su, H., McCabe, M. F., and Wood, E. F. 2005. Modeling evapotranspiration during SMACEX: Comparing two approaches for local- and regional-scale prediction. *Journal of Hydrometeorology* 6:910–922.

Verstraeten, W. W., Veroustraete, F., and Feyen, J. 2008. Assessment of evapotranspiration and soil moisture content across different scales of observation. *Sensors* 34:70–117.

Wan, Z., and Dozier, J. 1996. A generalized split-window algorithm for retrieving land-surface temperature from space. *IEEE Transactions on Geoscience and Remote Sensing* 34:892–905.

Wang, K., and Liang, S. 2009. Evaluation of ASTER and MODIS land surface temperature and emissivity products using long-term surface longwave radiation observation at SURFRAD sites. *Remote Sensing of Environment* 113:1556–1565.

Wloczyk, C., Borg, E., Richter, R., and Miegel, B. K. 2011. Estimation of instantaneous air temperature above vegetation and soil surfaces from Landsat 7 ETM+ data in northern Germany. *International Journal of Remote Sensing* 32:9119–9136.

Wu, J., Cai, G., Xue, Y., and Du, M. 2008. Drought monitoring in northern China plains combining RS and GIS technology. Proceedings, IEEE International Geoscience and Remote Sensing Symposium;. July 7–11, 2008, Boston, MA, 1:331–334.

Yang, G., Pu, R., Zhano, C., Huang, W., and Wang, J. 2011. Estimation of subpixel land surface temperature using an endmembre index based technique: A case examination on ASTER and MODIS temperature products over a heterogenous area. *Remote Sensing of Environment* 115:1202–1219.

Yang, H., Cong, Z., Liu, Z., and Lei, Z. 2010. Estimating sub-pixel temperature using the triangle algorithm. *International Journal of Remote Sensing* 31:6047–6060.

Yves, J., Sobrino, J. A., and Verhoef, W. 2006. Changes in land surface temperature and NDVI values over Europe between 1982–1999. *Remote Sensing of Environment* 103:43–55.

4

ON THE ISSUE OF SCALE IN URBAN REMOTE SENSING

QIHAO WENG

4.1 INTRODUCTION

In a review article on the scale and resolution effects in remote sensing and geographic information systems (GISs), Cao and Lam (1997) raised two questions: (1) "how large an area should be covered to appropriately examine a geographic phenomenon, or at what scale and resolution should the study be conducted" and (2) "whether or not the results of the study at one scale can be extrapolated to other scales." These two questions are still valid today for most studies in urban remote sensing.

A key to scale- and resolution-related issues is to develop proper methods for "determining the most appropriate scale and resolution of study and assessing the effects of scale and resolution" (Cao and Lam, 1997). The majority of research efforts in urban remote sensing over the past decade have been made for mapping urban landscapes at various scales and on the spatial resolution requirements of such mapping (Weng, 2012). Previous models, methods, and image analysis algorithms in urban remote sensing have been largely developed for the imagery of medium resolution (10–100 m). In contrast, there is less interest in spectral and geometric properties of urban features. The advent of high-spatial-resolution satellite images and increased interests in spaceborne hyperspectral images, Lidar sensing, and their synergy with existing technologies are stimulating new research ideas in urban remote sensing and are driving the future research trends with new models and algorithms. The urban remote sensing community will need to address how these new frontiers in Earth observation technology since 1999 and advances in remote sensing imaging science—such as object-oriented image analysis, data fusion, and artificial neural

Scale Issues in Remote Sensing, First Edition. Edited by Qihao Weng.
© 2014 John Wiley & Sons, Inc. Published 2014 by John Wiley & Sons, Inc.

networks—have impacted our understanding of the scale and resolution issues. Because little has been done in the past, more researchs is also needed to better understand temporal resolution, change and evolution of urban features over time, and temporal requirements for urban mapping (Weng, 2012). There is not a simple answer to any of the questions discussed above. In this review, I start with the requirements for mapping three interrelated entities or substances in the urban space (i.e., material, land cover, and land use) and their relationships. Spectral resolution is a common consideration in imaging and mapping and is closely associated with the categorical scale. Then, the relationship between spatial resolution—which is termed the observational scale of remote sensing in this chapter—and the fabric of urban landscape is examined. Central to this relationship is the problem of mixed pixels in the urban areas. The pixel and subpixel approaches to urban analyses are thus discussed. Next, the author's two previous studies are discussed, both assessing the patterns of land surface temperature at different aggregation levels in order to find out the operational scale/optimal scale for the studies. Section 4.5 is developed to review the issue of scale dependency of urban phenomena and to discuss two case studies, one on LST variability across multiple census levels (block, block group, and tract) and the other on multiscale residential population estimation modeling. Section 4.6 provides a summary of the discussions and reflects on future developments.

4.2   URBAN LAND MAPPING AND CATEGORICAL SCALE

Urban remote sensing should consider the requirements for mapping three inter-related entities or substances on Earth's surface (i.e., material, land cover, and land use) and their relationships (Weng and Lu, 2009; Weng, 2012). Urban areas are composed of a variety of materials, including different types of artificial materials (i.e., impervious surfaces), soils, rocks and minerals, and green and nonphotosyn-thetic vegetation. These materials comprise land cover and are used in different manners for various purposes by human beings. Land cover can be defined as the biophysical state of Earth's surface and immediate subsurface, including biota, soil, topography, surface water and groundwater, and human structures (Turner et al., 1995). Land use can be defined as the human use of the land and involves both the manner in which the biophysical attributes of the land are manipulated and the purpose for which the land is used (Turner et al., 1995). Remote sensing technology has been applied to map urban land use, land cover, and materials. Their relation-ships are illustrated in Figure 4.1. Each type of land cover may possess unique surface properties (material). However, mapping land covers and materials have different requirements. Land cover mapping needs to consider characteristics in addition to those coming from the material (Herold et al., 2006). The surface structure (roughness) may influence the spectral response as much as the intraclass variability (Gong and Howarth, 1990; Myint, 2001; Shaban and Dikshit, 2001; Herold et al., 2006). Two different land covers, for example, asphalt roads and composite shingle/tar roofs, may have very similar materials (hydrocarbons) and thus are difficult to discern, although from a material perspective these surfaces can

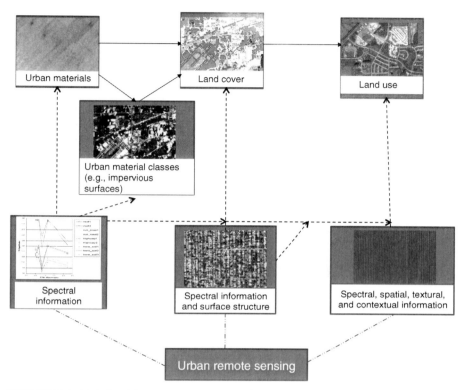

FIGURE 4.1   Relationship among remote sensing of urban materials, land cover, and land use (after Weng and Lu, 2009). (See the color version of this figure in Color Plates section.)

be mapped accurately with hyperspectral remote sensing techniques (Herold et al., 2006). Therefore, land cover mapping requires taking into account the intraclass variability and spectral separability. On the other hand, analysis of land use classes would nearly be impossible with spectral information alone. Additional information, such as spatial, textural, and contextual information, is usually required in order to have a successful land use classification in urban areas (Gong and Howarth, 1992; Stuckens et al., 2000; Herold et al., 2003).

The spectral characteristics of land surfaces are the fundamental principles for land imaging. Previous studies have examined the spectral properties of urban materials (Hepner et al., 1998; Ben-Dor et al., 2001; Herold et al., 2003; Heiden et al., 2007) and spectral resolution requirements for separating them (Jensen and Cowen, 1999). Jensen and Cowen (1999) focused their discussion mainly on multispectral imagery data and suggested that spatial resolution was more important than spectral resolution in urban mapping. The spectra from visible to near infrared (NIR), mid-infrared (MIR), and microwave are suitable for land use/land cover (LULC) classification at coarser categorical scales (e.g., levels I and II of the Anderson classification); however, at the finer categorical scales (e.g., levels III and IV of the Anderson classification) and for extraction of buildings and roads, panchromatic band is needed (Jensen and Cowen, 1999). Hyperspectral imagery data have been successfully used for urban land use/cover classification (Benediktsson et al., 1995; Hoffbeck and Landgrebe, 1996; Platt and Goetz, 2004; Thenkabail et al., 2004a,b), extraction of impervious surfaces (Weng et al., 2008), vegetation mapping (McGwire et al., 2000; Schmidt et al., 2004; Pu et al., 2008), and water mapping (Bagheri and Yu, 2008; Moses et al., 2009). A large number of spectral bands provide the potential to derive detailed information on the nature and properties of different surface materials on the ground, but it also means a difficulty in image processing and a large data redundancy due to high correlation among the adjacent bands. Increase of spectral bands may improve classification accuracy, only when those bands are useful in discriminating the classes (Thenkabail et al., 2004b).

Traditional classification methods of LULC based on detailed fieldwork suffered two major common drawbacks: confusion between land use and land cover and lack of uniformity or comparability in classification schemes, leaving behind a sheer difficulty for comparing land use patterns over time or between areas (Mather, 1986). The use of aerial photographs and satellite images after the late 1960s does not solve these problems, since these techniques are based on the formal expression of land use rather than on the actual activity itself (Mather, 1986). In fact, many land use types cannot be identified from the air. As a result, mapping of Earth's surface tends to present a mixture of land use and land cover data with an emphasis on the latter (Lo, 1986). This problem is reflected in the title of the classification developed in the United States for the mapping of the country at a scale of 1 : 100,000 or 1 : 250,000 commencing in 1974 (Anderson et al., 1976). Moreover, this U.S. Geological Survey (USGS) Land Use/Land Cover Classification System (so-called Anderson scheme) has been designed as a resource-oriented one. Therefore, eight out of nine in the first-level categories relate to nonurban areas. The success of most land use or land cover mapping efforts has typically been measured by the ability to match remote sensing spectral signatures to the Anderson scheme, which, in the urban areas, is mainly land use (Ridd, 1995). The confusion between land use and land cover contributes to the low classification accuracy (Foody, 2002). In addition, the spatial scale and categorical scale is not explicitly linked in the classification scheme. The former refers to the manner in which image information content is determined by spatial resolution and the way the spatial resolution is handled in the image processing, while the latter refers to the level of detail in classification categories (Ju et al., 2005). This disconnection leads to the problem of simply lumping classes into more general classes in multiscale LULC classifications, which may cause a great lost of categorical information. Since most classifications are conducted at a single spatial and categorical scale, there remains an important issue of matching an appropriate categorical scale of the Anderson scheme with the spatial resolution of the satellite image used (Welch, 1982; Jensen and Cowen, 1999). However, the nature of some applications requires LULC classification to be conducted at multiple spatial and/or categorical scales because a single scale cannot delineate all classes due to contrasting sizes, shapes, and internal variations of different landscape patches (Wu and David, 2002; Raptis et al., 2003). This is especially true for complex, heterogeneous landscapes, such as urban ecosystems. When statistical clusters are grouped into LULC classes, in which smaller areas (e.g., pixels) are combined into larger ones (e.g., patches), both spatial resolution and statistical information are lost (Clapham, 2003).

4.3 OBSERVATIONAL SCALE AND IMAGE SCENE MODELS

Spatial resolution is a function of sensor altitude, detector size, focal size, and system configuration (Jensen, 2005). Spatial resolution is closely related to the term of spatial scale (Ju et al., 2005). As a matter of fact, spatial resolution defines the "measurement scale" (Lam and Quattrochi, 1992) or the "observational scale" of a sensor.

Spatial resolution defines the level of spatial detail depicted in an image, and it is often related to the size of the smallest possible feature that can be detected from an image. This definition implies that only objects larger than the spatial resolution of a sensor can be picked out from an image. However, a smaller feature may sometimes be detectable if its reflectance dominates within a particular resolution cell or it has a unique shape (e.g., linear features). Another meaning of spatial resolution is that a ground feature should be distinguishable as a separate entity in the image. But the separation from neighbors or background is not always sufficient to identify the object. Therefore, the concept of spatial resolution includes both detectability and separability. For any feature to be resolvable in an image, it involves consideration of spatial resolution, spectral contrast, and feature shape. Jensen and Cowen (1999) suggested that the minimum spatial resolution requirement should be one-half the diameter of the smallest object of interest. For two major types of impervious surface, buildings (perimeter, area, height, and property line) and roads (width) are generally detectable with the minimum spatial resolution of 0.25–0.5 m, while road centerline can be detected at a lower resolution of 1–30 m (Jensen and Cowen, 1999). Before 1999, lack of high-spatial-resolution images (less than 10 m) is a main reason for scarce research on urban remote sensing before 2000 (Weng, 2012). The medium-spatial-resolution images (10–100 m), such as Landsat and SPOT, were not readily available and were expensive to most researchers from developing countries. For a remote sensing project, image spatial resolution should not be the only factor needed to be considered. The relationship between the geographical scale/extent of a study area and the spatial resolution of the remote sensing image has to be studied (Quattrochi and Goodchild, 1997). For mapping at the continental or global scale, coarse-spatial-resolution data are usually employed. Gamba and Herold (2009) assessed eight major research efforts in global urban extent mapping and found that most maps were produced at the spatial resolution of 1–2 km. When using coarse-resolution images, a threshold has to be defined with respect to what constitutes an urban/built-up pixel (Lu et al., 2008; Schneider et al., 2010).

With the advent of very high resolution satellite imagery, such as IKONOS (launched in 1999), QuickBird (2001), and OrbView (2003) images, great efforts have been made in the applications of these remote sensing images in urban studies. High-resolution satellite imagery has been applied in mapping impervious surfaces in urban areas (Cablk and Minor, 2003; Goetz et al., 2003; Lu and Weng, 2009; Wu,

2009; Hu and Weng, 2011). These fine-spatial-resolution images contain rich spatial information, providing a greater potential to extract much more detailed thematic information (e.g., LULC), cartographic features (buildings and roads), and metric information with stereo-images (e.g., height and area). These information and cartographic characteristics are highly beneficial to estimating and mapping impervious surfaces in the urban areas. However, some new problems come with these image data, notably shadows caused by topography, tall buildings, or trees (Dare, 2005) and the high spectral variation within the same land cover class (Hsieh et al., 2001). Shadows obscure impervious surfaces underneath and thus increase the difficulty to extract both thematic and cartographic information. These disadvantages may lower image classification accuracy if classifiers used cannot effectively handle them (Irons et al., 1985; Cushnie, 1987). In order to make full use of the rich spatial information inherent in fine-spatial-resolution data, it is necessary to minimize the negative impact of high intraspectral variation. Algorithms that use the combined spectral and spatial information may be especially effective for impervious surface extraction in the urban areas (Lu and Weng, 2007).

Per-pixel classifications prevail in the previous remote sensing literature, in which each pixel is assigned to one category and land cover (or other themes) classes are mutually exclusive. Per-pixel classification algorithms are sometimes referred to as "hard" classifiers. Due to the heterogeneity of landscapes (particularly in urban landscapes) and the limitation in spatial resolution of remote sensing imagery, mixed pixels are common in medium- and coarse-spatial-resolution data. However, the proportion of mixed pixels is significantly reduced in a high-resolution satellite image scene. The presence of mixed pixels has been recognized as a major problem affecting the effective use of per-pixel classifiers (Fisher, 1997; Cracknell, 1998). The mixed-pixel problem results from the fact that the observational scale (i.e., spatial resolution) fails to correspond to the spatial characteristics of the target (Mather, 1999). Strahler et al. (1986) defined H- and L-resolution scene models based on the relationship between the size of the scene elements and the resolution cell of the sensor. The scene elements in the L-resolution model are smaller than the resolution cells and are thus not detectable. When the objects in the scene become increasingly smaller than the resolution cell size, they may no longer be regarded as individual objects. Hence, the reflectance measured by the sensor may be treated as the sum of interactions among various types of scene elements as weighted by their relative proportions (Strahler et al., 1986). This is what happens with medium-resolution imagery, such as those of Landsat TM or ETM+, ASTER, SPOT, and Indian satellites, applied for urban mapping. As the spatial resolution interacts with the fabric of urban landscapes, the problem of mixed pixels is created. Such a mixture becomes especially prevalent in residential areas where buildings, roads, trees, lawns, and water can all lump together into a single pixel (Epstein et al., 2002). The low accuracy of image classification in urban areas reflects, to a certain degree, the inability of traditional per-pixel classifiers to handle composite signatures. Therefore, the "soft" approach of image classifications has been developed, in which each pixel is assigned a class membership of each land cover type rather than a single label (Wang, 1990). Different approaches have been used to derive a soft classifier, including fuzzy-set theory, Dempster–Shafer theory, certainty factors (Bloch, 1996), and neural networks (Foody, 1999; Mannan and Ray, 2003). Nevertheless, as Mather (1999) suggested, neither hard nor soft classification was an appropriate tool for the analysis of heterogeneous landscapes. To provide a better understanding of the compositions and processes of urban landscapes, Ridd (1995) proposed an interesting conceptual model for remote sensing analysis of urban landscapes, that is, the vegetation–impervious surface–soil (VIS) model. It assumes that land cover in urban environments is a linear combination of three components, namely, vegetation, impervious surface, and soil. Ridd suggested that this model can be applied to spatial–temporal analyses of urban morphology, biophysical, and human systems. While urban land use information may be more useful in socioeconomic and planning applications, biophysical information that can be directly derived from satellite data is more suitable for describing and quantifying urban structures and processes (Ridd, 1995). The VIS model was developed for Salt Lake City, Utah, but has been tested in other cities (Ward et al., 2000; Madhavan et al., 2001; Setiawan et al., 2006). All of these studies employed the VIS model as the conceptual framework to relate urban morphology to medium-resolution satellite imagery, but hard classification algorithms were applied. Therefore, the problem of mixed pixels cannot be addressed, and the analysis of urban landscapes was still based on "pixels" or "pixel groups." Weng and Lu (2009) suggested that linear spectral mixture analysis (LSMA) provided a suitable technique to detect and map urban materials and VIS component surfaces in repetitive and consistent ways and to solve the spectral mixing of medium-spatial-resolution imagery. The reconciliation between the VIS model and LSMA provided a continuum field model which offered an alternative, effective approach for characterizing and quantifying the spatial and temporal changes of the urban landscape compositions. However, Weng and Lu (2009) warned that the applicability of this continuum model must be further examined in terms of its spectral, spatial, and temporal variability.

4.4 OPERATIONAL SCALE

Urban landscape processes appear to be hierarchical in pattern and structure. A study of the relationship between the patterns at different levels in the hierarchy may help in obtaining a better understanding of the scale and resolution problem (Cao and Lam, 1997; Weng et al., 2004) and in finding the optimal scale for examining the relationship, that is, the operational scale (Frohn, 1998; Liu and Weng, 2009). Lo et al. (1997) suggested that the urban surface characteristics required a minimum thermal mapping resolution of 5–10 m based on a study in Huntsville, Alabama. Nichol (1996) confirmed that satellite-derived land surface temperature (LST) image data at the scale of 10^2 m was adequate for depicting most of the intraurban LST variations related to urban morphology based on a study in Singapore. The length of spatial scale that characterizes the overall distribution of scales of all objects in a collection may be compared among different cities. Small (2009) suggested that the modal scale length was 10–20 m based on a comparative analysis of 14 cities in the world.

Weng et al. (2004) suggested that the operational scale for the urban thermal landscape analysis yielded around 120 m in Indianapolis, Indiana. Remote sensing of urban heat islands has traditionally used the normalized difference vegetation index (NDVI) as the indicator of vegetation abundance to estimate the LST–vegetation relationship. Weng et al. (2004) investigated the applicability of vegetation fraction derived from LSMA as an alternative indicator of vegetation abundance. An experiment was conducted with a Landsat ETM+ image of Indianapolis City, Indiana, acquired on June 22, 2002. They found that LST possessed a slightly stronger negative correlation with the unmixed vegetation fraction than with NDVI for all land cover types across the spatial resolution from 30 to 960 m. Correlations reached their strongest at 120 m resolution. Fractal analysis of image texture further showed that the complexity of these images increased initially with pixel aggregation and peaked around 120 m, but decreased with further aggregation. It was, therefore, suggested that the operational scale for examining the relationship between LST and NDVI or vegetation fraction was around 120 m.

In another study, Liu and Weng (2009) examined the scaling effect between LST and LULC in Indianapolis and found that the optimal spatial resolution for assessing this unique relationship was 90 m. Four Terra ASTER images were used to derive LULC maps and LST patterns in four seasons. Each LULC and LST image was resampled to eight aggregation levels: 15, 30, 60, 90, 120, 250, 500, and 1000 m. The scaling-up effect on the spatial and ecological characteristics of landscape patterns and LSTs were examined by the use of landscape metrics. Optimal spatial scale was determined on the basis of the minimum distance in the landscape metric spaces. Their results showed that the patch percentages of LULC and LST patches were not strongly affected by the scaling-up process. The patch densities and landscape shape indices and LST patches kept decreasing across the scales without distinct seasonal differences. Ninety meters was found to be the optimal spatial resolution for assessing the landscape-level relationship between LULC and LST patterns.

4.5   SCALE DEPENDENCY OF URBAN PHENOMENA

The majority of urban phenomena are scale dependent, meaning that the urban patterns change with scale of observation. In reality, very few geographical phenomena are scale independent, in which the patterns do not change across scales (Cao and Lam, 1997). Geographic studies are frequently conducted on the basis of areal units such as states, counties, census tracts, block groups, and blocks. Across-scale analyses with such areal units may produce scale-related problems, most notably, the modifiable areal unit problem (MAUP). Fotheringham and Wong (1991) defined MAUP as "the sensitivity of analytical results to the definition of units for which data are collected." Specifically, the results may vary with the aggregation level (the "scale effect") and with the aggregation schemes (the "zoning effect"). A researcher who attempts to extrapolate the result of a study at one scale to other scales may meet three kinds of erroneous inferences: individualistic fallacy, cross-level fallacy, and ecological fallacy (Alker, 1969). There is no ideal solution to solve MAUP; it is, however, mainly examined by conducting statistical correlation and regression analysis

(Fotheringham and Wong, 1991). Empirical studies may be the only possible ways to explore the nature of MAUP (Fotheringham and Wong, 1991). Recently, there has been an increase in using spatial autocorrelation indices to indicate geographical patterns or geospatial data distributions. The underlying fundamental of spatial autocorrelation is best explained by Tobler's first law of geography—"everything is related to everything else, but near things are more related than distance things" (Lo and Yeung, 2002). The index is devised to measure the spatial ordering as well as the spatial covariance structure of geospatial data and provides insights into the degree of clustering, randomness, or fragmentation of a pattern (Read and Lam, 2002). However, spatial autocorrelation is subject to the influence of scale, meaning that at one scale its index may suggest a concentrated pattern while at another it may suggest a scattered pattern (Cao and Lam, 1997).

4.5.1 Spatial Variations of Land Surface Temperature at Multiple Census Scales

Liang and Weng (2008) conducted a multiscale analysis of census-based LST variations and determinants in Indianapolis, Indiana. Urban temperatures have a close relationship with many environmental, economic, and social issues in the urban areas. This study utilized LST data, derived from a Landsat ETM+ image of Indianapolis, Indiana, to examine census-based variations and to model their relationships with the parameters of urban morphology. The NDVI, buildings, roads, and water bodies were selected as the variables of the urban morphology. Correlation analysis and stepwise regression modeling at each census level, that is, block, block group, and tract, were performed. The sensitivity of the relationship to aggregation and thus the scale effect of MAUP were examined. Their results showed that LST had a strongest positive correlation with buildings but was negatively correlated with water at all scales. The correlation between LST and the four variables tended to become stronger as the scale increased. Table 4.1 indicates that the adjusted R^2 value increased with the scale, suggesting that more variations in LST could be explained by the regression models. The regression model for the tract level possessed the closest goodness of fit in the population of LST with the least estimation error. More independent variables are needed to predict LST at finer scales. Meanwhile, when the analytical scale altered, the contribution of each independent variable to the models changed. For example, an increase of 10 in P_{bldg} generated an increase in LST of 2.95 K at the block level. The same increase at the block group and tract levels generated increases of 5.21 and 5.95 K in the predicted LST values, respectively. However, the strength of the correlation between LST and four biophysical factors was not always consistent with their contributions to LST regression modeling. For example, roads contributed the least to the LST estimation, although it correlated stronger with LST than water. This is because in multivariate modeling regression estimates attached to any single independent variable could become inflated or deflated, since changes in the variable may be confused by variations in other independent variables.

In order to validate the relationship between input data and LST models, a residual map was produced for each model (Figure 4.2). The range of residuals decreased from

TABLE 4.1 Regression Coefficients for Three LST Models and Four Variables of Urban Morphology

Models	LST1	LST2	LST3
Adjusted R^2	0.752	0.921	0.948
Sig. of the F statistic	0.000	0.000	0.000
M_{NDVI}			
Standardized beta Coefficients	−0.654	−0.518	−0.429
t	−138.549	−35.450	−19.756
Sig.	0.000	0.000	0.000
P_{bildg}			
Standardized beta Coefficients	0.295	0.421	0.394
t	63.218	27.590	12.741
Sig.	0.000	0.000	0.000
P_{road}			
Standardized beta Coefficients	0.059	0.139	0.228
t	13.742	11.108	8.970
Sig.	0.000	0.000	0.000
P_{water}			
Standardized beta Coefficients	−0.276	−0.234	−0.241
t	−63.517	−19.624	−14.236
Sig.	0.000	0.000	0.000

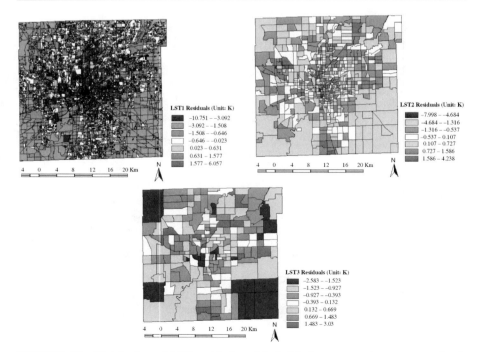

FIGURE 4.2 Distribution of LST residuals at black, black group, and census tract level. (See the color version of this figure in Color Plates section.)

16.81 to 12.24 to 5.61 with the increase in geographical scale. For any LST model, LST was consistently underestimated over the low-temperature areas, while in the high-temperature areas the distribution of residuals varied with scale. LST was predominantly overestimated at the block scale. As the scale increased, LST became generally underestimated. At the tract scale, LST was equally over- and under-estimated. In the downtown area where high temperatures existed, predicted LST values may be either over- or underestimated at any scale, implying the heterogeneity in the thermal landscape. The spatial patterns of residuals appear analogous at the block group and tract levels.

4.5.2 Population Estimation

Many remotely sensed images collected from different sensors have been utilized to estimate population. With various spatial resolutions, they are especially applicable at a certain scale for the study. For instance, high-spatial-resolution aerial photo-graphy is useful for population estimation at the microscale (Lo and Welch, 1977; Lo, 1986; Cowen et al. 1995), while low-spatial-resolution data, such as those from the Defense Meteorological Satellite Program Operational Linescan System (DMSP-OLS), are suitable for modeling at the global or regional scale. However, if a medium scale such as at a city level is concerned, images with medium spatial resolution, such as those obtained from Landsat TM/ETM+ and Terra ASTER sensors, should be considered. Research has proved that such data are efficient and effective in predicting population in city or county levels (Harvey, 2002a,b; Qiu et al., 2003, Li and Weng, 2005; Lu et al., 2006). Lo (1986) summarized several approaches commonly used in population estimation with remotely sensed data: counting the dwelling units, using per-pixel spectral reflectance, measuring urban areas, and using land use information. The application of these methods may be, in fact, considered to respond to different analytical scales, with the first two applicable at small areas (1 km^2 or less) and the last two for larger (or regional) and medium scales, respectively (Harvey, 2002a).

Because remotely sensed images are scale dependent, population models derived from such data are also subject to the impact of scale. Lo (2001) used DMSP-OLS nighttime light data to model the Chinese population and population densities at three different spatial scales: province, county, and city. Either allometric growth or linear regression models were found to be promising in estimating population at all three levels, but the best models were obtained at the city level. Qiu et al. (2003) carried out a biscale study of the decennium urban population growth from 1990 to 2000 in the north Dallas–Fort Worth metropolis using models developed with remote sensing and GIS techniques. Both models yielded comparable results with that obtained from a more complex commercial demographics model at the city as well as the census tract levels, yet the GIS model remained robust to the scale change because of its insensitiveness to the spatial scale. The remote sensing model was attenuated when moved to the census tract from the city level. Liang et al. (2007) estimated the residential population of Indianapolis at multiple scales of census units (block, block group, and tract) using remote sensing–derived impervious surfaces. The impervious surface has been emerging as a key indicator of urbanization in recent years. The map of impervious surfaces may be useful in revealing socioeconomic characteristics of a city. They found the impervious surface was an effective variable in modeling residential population at all three census scales. Table 4.2 indicates that by any measure (R^2, MRE, and MedRE), the three best population models were all found at the tract level. The difference between MRE and MedRE shows that the performance of population density models was strongly affected by extreme values (extreme low and high population density). A common problem found in all three models was that they significantly underestimated the high-density areas but overestimated the low-density areas. This is because population modeling developed in this study assumed a general uniform population density per census unit. Hence, none of the models predicted well for the extreme values.

TABLE 4.2   Population Estimation Models for Indianapolis at Each Census Level

Census Levels	Correlation	R^2	Regression Model	MRE	MedRE	ET (%)
Block	0.428	0.183	PD = −1219.335 + 6987.572 × MRImp	396.15	37.12	−12.69
	0.480	0.230	SPD = 2.096 + 85.737 × MRImp	51.48	19.45	−3.16
	0.466	0.218	LPD = 5.410 + 0.466 × MRImp	30.71	25.52	+1.42
BG	0.595	0.354	PD = −1995.631 + 7378.838 × MRImp	275.80	32.77	−3.91
	0.592	0.351	SPD = −12.403 + 104.295 × MRImp	44.75	18.71	+0.62
	0.555	0.308	LPD = 4.354 + 5.785 × MRImp	10.41	6.16	+1.14
Tract	0.710	0.504	PD = −3021.746 + 9550.366 × MRImp	55.96	27.15	+4.01
	0.706	0.499	SPD = −29.567 + 141.658 × MRImp	22.74	14.32	+3.53
	0.672	0.452	LPD = 3.134 + 8.492 × MRImp	6.36	4.28	+1.33

Source: Liang et al. (2007).

Notes: Corr., Pearson correlation coefficient between PD (SPD, LPD) and MRImp; MRE, mean value of relative error; MedRE, median value of relative error; ET (%), error of total, which is the total polulation estimation error based on the overall data set in the study area [an addition (+) symbol means overestimated while a minus (−) symbol means underestimated]; PD, populatin density; SPD, square root of population density; LPD, natural log of population density.

4.6 CONCLUSIONS

This review starts with a discussion of the requirements for mapping urban materials, land cover, and land use and their relationships. This examination allows us to understand what spectral property should and should not be considered in the imaging and mapping. The level of detail in classification categories is essentially modulated by spectral resolution. But, another important consideration in urban remote sensing is spatial resolution. Its interaction with the fabric of urban landscape and the problem of mixed pixels are the subject of the debate between the pixel and subpixel approaches. Research has yet to clearly reveal the linkage between the categorical scale and spatial resolution.

Scale influences the examination of landscape patterns (Liu and Weng, 2009). The change of scale is relevant to the issues of data aggregation, information transfer, and the identification of appropriate scales for analysis (Krönert et al., 2001; Wu and Hobbs, 2002). Sections 4.4 and 4.5 relate to this theme of study. The patterns of land surface temperature at different aggregation levels were assessed in order to find the operational scale/optimal scale through two case studies. The issues of data aggregation and information transfer were addressed by reviewing the concept of scale dependency and by discussing LST variability and residential population estimation modeling across multiple census levels.

Remote sensing technology has been evolving rapidly in the twenty-first century. New frontiers such as very high resolution sensing, hyperspectral sensing, lidar and their synergy with existing technologies and advances in image processing techniques (such as object-oriented image analysis, data fusion, artificial neural networks) are changing the image information content we obtain and the way we handle the image processing. Both aspects will change our understanding of this basic but pivotal issue—scale and therefore future research should be warranted in these aspects. For example, an increase in spatial resolution may not lead to a better observation since objects may be oversampled and their features may vary and be confusing (Hsieh et al., 2001; Aplin and Atkinson, 2004). Moreover, imagery with too fine resolution for specific purpose can be degraded in the process of image resampling (Ju et al., 2005). In the object-based image analysis, the extraction, representation, modeling, and analyses of image objects at multiple scales have become common concerns of many researchers (Hay et al. 2002a,b; Tzotsos et al., 2011).

REFERENCES

Alker, H. R. 1969. A typology of ecological fallacies. In M. Dogan and S. Rokkan (Eds.), *Quantitative Ecological Analysis in the Social Sciences.* Cambridge, MA: MIT Press, pp. 69–86.

Aplin, P., and Atkinson, P. M. 2004. Predicting missing field boundaries to increase per-field classification accuracy. *Photogrammetric Engineering & Remote Sensing* 70(1):141–149.

Anderson, J. R., Hardy, E. E., Roach, J. T., and Witmer, R. E. 1976. *A Land Use and Land Cover Classification Systems for Use with Remote Sensing Data*. USGS Professional Paper 964.

Bagheri, S., and Yu, T. 2008. Hyperspectral sensing for assessing nearshore water quality conditions of Hudson/Raritan estuary. *Journal of Environmental Informatics* 11(2):123–130.

Ben-Dor, E., Levin, N., and Saaroni, H. 2001. A spectral based recognition of the urban environment using the visible and near-infrared spectral region (0.4–1.1 m)—A case study over Tel-Aviv. *International Journal of Remote Sensing* 22(11):2193–2218.

Benediktsson, J. A., Sveinsson, J. R., and Arnason, K. 1995. Classification and feature extraction of AVIRIS data. *IEEE Transactions on Geoscience and Remote Sensing* 33:1194–1205.

Bloch, I. 1996. Information combination operators for data fusion: A comparative review with classification. *IEEE Transactions on Systems, Man, and Cybernetics* 26:52–67.

Cablk, M. E., and Minor, T. B. 2003. Detecting and discriminating impervious cover with high resolution IKONOS data using principal component analysis and morphological operators. *International Journal of Remote Sensing* 24:4627–4645.

Cao, C., and Lam, N. S.-N. 1997. Understanding the scale and resolution effects in remote sensing. In D. A. Quattrochi and M. F. Goodchild (Eds.), *Scale in Remote Sensing and GIS*. Boca Raton, FL: CRC Press, pp. 57–72.

Clapham, W. B. Jr. 2003. Continuum-based classification of remotely sensed imagery to describe urban sprawl on a watershed scale. *Remote Sensing of Environment* 86:322–340.

Cowen, D. J., Jensen, J. R., Bresnahan, G., Ehler, G. B., Graves, D., Huang, X., Wiesner, C., and Mackey, H. E. Jr. 1995. The design and implementation of an integrated GIS for environmental applications. *Photogrammetric Engineering & Remote Sensing* 61(11):1393–1404.

Cracknell, A. P. 1998. Synergy in remote sensing—what's in a pixel? *International Journal of Remote Sensing* 19:2025–2047.

Cushnie, J. L. 1987. The interactive effect of spatial resolution and degree of internal variability within land-cover types on classification accuracies. *International Journal of Remote Sensing* 8:15–29.

Dare, P. M. 2005. Shadow analysis in high-resolution satellite imagery of urban areas. *Photogrammetric Engineering & Remote Sensing* 71:169–177.

Epstein, J., Payne, K., and Kramer, E. 2002. Techniques for mapping suburban sprawl. *Photogrammetric Engineering and Remote Sensing* 63:913–918.

Fisher, P. 1997. The pixel: A snare and a delusion. *International Journal of Remote Sensing* 18:679–685.

Foody, G. M. 1999. Image classification with a neural network: From completely-crisp to fully-fuzzy situation. In P. M. Atkinson and N. J. Tate (Eds.), *Advances in Remote Sensing and GIS Analysis*. New York: Wiley, pp. 17–37.

Foody, G. M. 2002. Status of land cover classification accuracy assessment. *Remote Sensing of Environment* 80:185–201.

Fotheringham, A. S., and Wong, D. W. S. 1991. The modifiable areal unit problem in multivariate statistical analysis. *Environment and Planning A* 23:1025–1044.

Frohn, R. C. 1998. *Remote Sensing for Landscape Ecology: New Metric Indicators for Monitoring, Modelling, and Assessment of Ecosystems*. Boca Raton, FL: Lewis Publishers.

Gamba, P., and Herold, M. 2009. *Global Mapping of Human Settlements: Experiences, Datasets, and Prospects.* Boca Raton, FL: CRC Press.

Goetz, S. J., Wright, R. K., Smith, A. J., Zinecker, E., and Schaub, E. 2003. IKONOS imagery for resource management: tree cover, impervious surfaces, and riparian buffer analyses in the mid-Atlantic region. *Remote Sensing of Environment* 88:195–208.

Gong, P., and Howarth, P. J. 1990. The use of structure information for improving land-cover classification accuracies at the rural-urban fringe. *Photogrammetric Engineering & Remote Sensing* 56(1):67–73.

Gong, P., and Howarth, P. J. 1992. Frequency-based contextual classification and gray-level vector reduction for land-use identification. *Photogrammetric Engineering & Remote Sensing* 58(4):423–437.

Harvey, J. T. 2002a. Estimating census district populations from satellite imagery: Some approaches and limitations. *International Journal of Remote Sensing* 23(10):2071–2095.

Harvey, J. T. 2002b. Population estimation models based on individual TM pixels. *Photogrammetric Engineering & Remote Sensing* 68(11):1181–1192.

Hay, G. J., Dube, P., Bouchard, A., and Marceau, D. J. 2002a. A scale-space primer for exploring and quantifying complex landscapes. *Ecological Modelling* 153(1–2):2–49.

Hay, G. J., Marceau, D. J., and Bouchard, A. 2002b. Modelling multiscale landscape structure within a hierarchical scale-space framework. *International Archives of Photogrammetry and Remote Sensing* 34(4):532–536.

Heiden, U., Segl, K., Roessner, S., and Kaufmann, H. 2007. Determination of robust spectral features for identification of urban surface materials in hyperspectral remote sensing data. *Remote Sensing of Environment* 111:537–552.

Hepner, G. F., Houshmand, B., Kulikov, I., and Bryant, N. 1998. Investigation of the integration of AVIRIS and IFSAR for urban analysis. *Photogrammetric Engineering & Remote Sensing* 64(8):813–820.

Herold, M., Liu, X., and Clark, K. C. 2003. Spatial metrics and image texture for mapping urban land use. *Photogrammetric Engineering & Remote Sensing* 69(9):991–1001.

Herold, M., Schiefer, S., Hostert, P., and Roberts, D. A. 2006. Applying imaging spectrometry in urban areas. In Q. Weng and D. A. Quattrochi (Eds.), *Urban Remote Sensing*. Boca Raton, FL: CRC Press, pp. 137–161.

Hoffbeck, J. P., and Landgrebe, D. A. 1996. Classification of remote sensing having high spectral resolution images. *Remote Sensing of Environment* 57:119–126.

Hsieh, P.-F., Lee, L. C., and Chen, N.-Y. 2001. Effect of spatial resolution on classification errors of pure and mixed pixels in remote sensing. *IEEE Transactions on Geoscience and Remote Sensing* 39:2657–2663.

Hu, X., and Weng, Q. 2011. Impervious surface area extraction from IKONOS imagery using an object-based fuzzy method. *Geocarto International* 26(1):3–20.

Irons, J. R., Markham, B. L., Nelson, R. F., Toll, D. L., Williams, D. L., Latty, R. S., and Stauffer, M. L. 1985. The effects of spatial resolution on the classification of Thematic Mapper data. *International Journal of Remote Sensing* 6:1385–1403.

Jensen, J. R. 2005. *Introductory Digital Image Processing: A Remote Sensing Perspective*, 3rd ed. Upper Saddle River, NJ: Prentice Hall.

Jensen, J. R., and Cowen, D. C. 1999. Remote sensing of urban/suburban infrastructure and socioeconomic attributes. *Photogrammetric Engineering & Remote Sensing* 65:611–622.

Ju, J., Gopal, S., and Kolaczyk, E. D. 2005. On the choice of spatial and categorical scale in remote sensing land cover classification. *Remote Sensing of Environment* 96:62–77.

Krönert, R., Steinhardt, U., and Volk, M. 2001. *Landscape Balance and Landscape Assessment*. New York: Springer.

Lam, N. S.-N., and Quattrochi, D. A. 1992. On the issues of scale, resolution, and fractal analysis in the mapping sciences. *Professional Geographer* 44(1):88–98.

Li, G., and Weng, Q. 2005. Using Landsat ETM+ imagery to measure population density in Indianapolis, Indiana, USA. *Photogrammetric Engineering & Remote Sensing* 71(8):947–958.

Liang, B., and Weng, Q. 2008. A multi-scale analysis of census-based land surface temperature variations and determinants in Indianapolis, United States. *Journal of Urban Planning and Development* 134(3):129–139.

Liang, B., Weng, Q., and Lu, D. 2007. Census-based multiple scale residential population modeling with impervious surface data. In Q. Weng (Ed.), *Remote Sensing of Impervious Surfaces*. Boca Raton, FL: CRC/Taylor & Francis, pp. 409–430.

Liu, H., and Weng, Q. 2009. Scaling-up effect on the relationship between landscape pattern and land surface temperature. *Photogrammetric Engineering & Remote Sensing* 75(3):291–304.

Lo, C. P., 1986. *Applied Remote Sensing*. New York: Longman.

Lo, C. P., 2001. Modeling the population of China using DMSP operational linescan system nighttime data. *Photogrammetric Engineering & Remote Sensing* 67(9):1037–1047.

Lo, C. P., Quattrochi, D. A., and Luvall, J. C. 1997. Application of high-resolution thermal infrared remote sensing and GIS to assess the urban heat island effect. *International Journal of Remote Sensing* 18:287–304.

Lo, C. P., and Welch, R. 1977. Chinese urban population estimates. *Annals of the Association of American Geographers* 67(2):246–253.

Lo, C. P., and Yeung, A. K. W. 2002. *Concepts and Techniques of Geographic Information Systems*. Upper Saddle River, NJ: Prentice Hall, pp 351–353.

Lu, D., Tian, H., Zhou, G., and Ge, H. 2008. Regional mapping of human settlements in southeastern China with multisensor remotely sensed data. *Remote Sensing of Environment* 112(9):3668–3679.

Lu, D., and Weng, Q. 2007. A survey of image classification methods and techniques for improving classification performance. *International Journal of Remote Sensing* 28(5): 823–870.

Lu, D., and Weng, Q. 2009. Extraction of urban impervious surfaces from IKONOS imagery. *International Journal of Remote Sensing* 30(5):1297–1311.

Lu, D., Weng, Q., and Li, G. 2006. Residential population estimation using a remote sensing derived impervious surface approach. *International Journal of Remote Sensing* 27(16):3553–3570.

Madhavan, B. B., Kubo, S., Kurisaki, N., and Sivakumar, T. V. L. N. 2001. Appraising the anatomy and spatial growth of the Bangkok Metropolitan area using a vegetation-impervious-soil model through remote sensing. *International Journal of Remote Sensing* 22:789–806.

Mannan, B., and Ray, A. K. 2003. Crisp and fuzzy competitive learning networks for supervised classification of multispectral IRS scenes. *International Journal of Remote Sensing* 24:3491–3502.

Mather, A. S. 1986. *Land Use*. London: Longman.

Mather, P. M. 1999. Land cover classification revisited. In P. M. Atkinson and N. J. Tate (Eds.), *Advances in Remote Sensing and GIS*. New York: Wiley, pp. 7–16.

McGwire, K., Minor, T., and Fenstermaker, L. 2000. Hyperspectral mixture modeling for quantifying sparse vegetation cover in arid environments. *Remote Sensing of Environment* 72:360–374.

Moses, W. J., Gitelson, A. A., Berdnikov, S., and Povazhnyy, V. 2009. Satellite estimation of Chlorophyll-a concentration using the red and NIR bands of MERIS—The Azov Sea case study. *IEEE Geoscience and Remote Sensing Letters* 6:845–849.

Myint, S. W. 2001. A robust texture analysis and classification approach for urban land-use and land-cover feature discrimination. *Geocarto International* 16:27–38.

Nichol, J. E. 1996. High-resolution surface temperature patterns related to urban morphology in a tropical city: A satellite-based study. *Journal of Applied Meteorology* 35(1):135–146.

Platt, R. V., and Goetz, A. F. H. 2004. A comparison of AVIRIS and Landsat for land use classification at the urban fringe. *Photogrammetric Engineering & Remote Sensing* 70:813–819.

Pu, R., Kelly, M., Anderson, G. L., and Gong, P. 2008. Using CASI hyperspectral imagery to detect mortality and vegetation stress associated with a new hardwood forest disease. *Photogrammetric Engineering & Remote Sensing* 74(1):65–75.

Qiu, F., Woller, K. L., and Briggs, R. 2003. Modeling urban population growth from remotely sensed imagery and TIGER GIS road data. *Photogrammetric Engineering & Remote Sensing* 69:1031–1042.

Quattrochi, D. A., and Goodchild, M. F. 1997. *Scale in Remote Sensing and GIS*. New York: Lewis Publishers.

Raptis, V. S., Vaughan, R. A., and Wright, G. G. 2003. The effect of scaling on land cover classification from satellite data. *Computers & Geosciences* 29:705–714.

Read, J. M., and Lam, N. S-.N. 2002. Spatial methods for characterising land cover and detecting land-cover changes for the tropics. *International Journal of Remote Sensing* 12(2):2457–2474.

Ridd, M. K. 1995. Exploring a V-I-S (Vegetation-Impervious Surface-Soil) model for urban ecosystem analysis through remote sensing: comparative anatomy for cities. *International Journal of Remote Sensing* 16(12):2165–2185.

Schmidt, K. S., Skidmore, A. K., Kloosterman, E. H., van Oosten, H., Kumar, L., and Janssen, J. A. M. 2004. Mapping coastal vegetation using an expert system and hyperspectral imagery. *Photogrammetric Engineering & Remote Sensing* 70:703–715.

Schneider, A., Friedl, M. A., and Potere, D. 2010. Mapping global urban areas using MODIS 500-m data: New methods and datasets based on "urban ecoregions." *Remote Sensing of Environment* 114:1733–1746.

Setiawan, H., Mathieu, R., and Thompson-Fawcett, M. 2006. Assessing the applicability of the V-I-S model to map urban land use in the developing world: Case study of Yogyakarta, Indonesia. *Computers, Environment and Urban Systems* 30(4):503–522.

Shaban, M. A., and Dikshit, O. 2001. Improvement of classification in urban areas by the use of textural features: the case study of Lucknow city, Uttar Pradesh. *International Journal of Remote Sensing* 22:565–593.

Small, C. 2009. The color of cities: An overview of urban spectral diversity. In P. Gamba and M. Herold (Eds.), *Global Mapping of Human Settlement: Experiences, Datasets, and Prospects*. Boca Raton, FL: CRC Press, pp. 59–105.

Strahler, A. H., Woodcock, C. E., and Smith, J. A. 1986. On the nature of models in remote sensing. *Remote Sensing of Environment* 70:121–139.

Stuckens, J., Coppin, P. R., and Bauer, M. E. 2000. Integrating contextual information with per-pixel classification for improved land cover classification. *Remote Sensing of Environment* 71:282–296.

Thenkabail, P. S., Enclona, E. A., Ashton, M. S., and van der Meer, B. 2004a. Accuracy assessments of hyperspectral waveband performance for vegetation analysis applications. *Remote Sensing of Environment* 91:354–376.

Thenkabail, P. S., Enclona, E. A., Ashton, M. S., Legg, C., and de Dieu, M. J. 2004b. Hyperion, IKONOS, ALI, and ETM+ sensors in the study of African rainforests. *Remote Sensing of Environment* 90:23–43.

Turner, B. L. II., Skole, D., Sanderson, S., Fisher, G., Fresco, L., and Leemans, R. 1995. Land-use and land-cover change: Science and research plan. Stockholm and Geneva. International Geosphere-Biosphere Program and the Human Dimensions of Global Environmental Change Programme (IGBP Report No. 35 and HDP Report No. 7).

Tzotsos, A., Karantzalos, K., and Argialas, D. 2011. Object-based image analysis through nonlinear scale-space filtering. *ISPRS Journal of Photogrammetry and Remote Sensing* 66:2–16.

Wang, F. 1990. Fuzzy supervised classification of remote sensing images. *IEEE Transactions on Geoscience and Remote Sensing* 28(2):194–201.

Ward, D., Phinn, S. R., and Murray, A. T. 2000. Monitoring growth in rapidly urbanizing areas using remotely sensed data. *Professional Geographer* 53:371–386.

Welch, R. A. 1982. Spatial resolution requirements for urban studies. *International Journal of Remote Sensing* 3:139–146.

Weng, Q. 2012. Remote sensing of impervious surfaces in the urban areas: Requirements, methods, and trends. *Remote Sensing of Environment* 117(2):34–49.

Weng, Q., Hu, X., and Lu, D. 2008. Extracting impervious surface from medium spatial resolution multispectral and hyperspectral imagery: A comparison. *International Journal of Remote Sensing* 29(11):3209–3232.

Weng, Q., and Lu, D. 2009. Landscape as a continuum: An examination of the urban landscape structures and dynamics of Indianapolis city, 1991–2000. *International Journal of Remote Sensing* 30(10):2547–2577.

Weng, Q., Lu, D., and Schubring, J. 2004. Estimation of land surface temperature-vegetation abundance relationship for urban heat island studies. *Remote Sensing of Environment* 89(4):467–483.

Wu, C. 2009. Quantifying high-resolution impervious surfaces using spectral mixture analysis. *International Journal of Remote Sensing* 30(11):2915–2932.

Wu, J. G., and David, J. L. 2002. A spatially explicit hierarchical approach to modeling complex ecological systems: Theory and application. *Ecological Modelling* 153:7–26.

Wu, J., and Hobbs, R. 2002. Key issues and research priorities in landscape ecology: An idiosyncratic synthesis. *Landscape Ecology* 17(4):355–365.

PART II

SCALE IN REMOTE SENSING
OF PLANTS AND ECOSYSTEMS

5

CHANGE DETECTION USING VEGETATION INDICES AND MULTIPLATFORM SATELLITE IMAGERY AT MULTIPLE TEMPORAL AND SPATIAL SCALES

EDWARD P. GLENN, PAMELA L. NAGLER, AND ALFREDO R. HUETE

5.1 INTRODUCTION

Scale has different meanings in different disciplines. In ecology scale can refer either to the spatial or temporal *extent* of observations and models or to the *resolution* (grain size) of the observations or models (van Gardingen et al., 1997). In satellite remote sensing studies spatial scale refers either to the area covered by an image (i.e., extent of coverage) or the resolution of the individual image elements (pixels) (Woodcock and Strahler, 1987). On the other hand, temporal scale is largely determined by the return frequency of a particular satellite, which determines how closely spaced in time images of the same area can be acquired.

Spatial and temporal resolution are often inversely related. For example, the geostationary Geostationary Operational Environmental Satellites (GOES) have continuous coverage of most of the Western Hemisphere, but at low resolution (1–8 km) (De Pondeca et al., 2011). Sun-synchronous satellites such as Terra and Aqua (Tziziki et al., 2012) or the Advanced Very High Resolution Radiometer (AVHRR) series (Reed et al., 1994) have near-daily revisit intervals but also relatively coarse spatial resolution (250 m to 1 km), while the higher resolution Landsat (30-m) satellites have a 16-day return period, and for land applications the effective time

Scale Issues in Remote Sensing, First Edition. Edited by Qihao Weng.
© 2014 John Wiley & Sons, Inc. Published 2014 by John Wiley & Sons, Inc.

between usable images can be longer due to cloud cover (Song et al., 2001). Very high spatial resolution imagery such as GeoEye's IKONOS and Digital Globe's QuickBird or WorldView 2 images (~60 cm) have 1–7-day return times depending on latitude and tasking mode (Anderson and Marchisio, 2012) but in practice their temporal resolution can be limited by the high cost of images and scheduling constraints.

Also, as a generality, high spatial resolution at the pixel level often involves a trade-off in extent of coverage. As examples, the swath width of QuickBird is only 16.5 km, compared to 185 km for Landsat and 2330 km for Moderate Resolution Imaging Spectroradiometer (MODIS) images from the Terra satellite. In contrast to satellite imagery, tower-mounted, networked digital cameras (phenocams) provide near-surface images with nearly continuous temporal resolution and spatial resolution of 1 cm or less, but extent of coverage is generally limited to $1-100\,m^2$ (Richardson et al., 2007; Rundel et al., 2009; Sonnentag et al., 2012).

In an early review of issues of scale in remote sensing, Woodcock and Strahler (1987) pointed out that in most scientific endeavors the investigator selects the scale at which observations are collected, whereas investigators using spaceborne sensors are limited to the specific scales of observations inherent in the satellite imagery. At that time Landsat and AVHRR satellites were used for most Earth observation studies. However, with the increase in available sensor systems, as noted in the examples above, it is now possible to combine imagery to cover a wide range of temporal and spatial scales across image sets to conduct landscape change studies. More recent reviews reflect the much broader range of choices now available for obtaining data at multiple scales (e.g., Mulder et al., 2011; Skidmore et al., 2011; Turner 2011; Abraha and Savage, 2012; Pfeifer et al., 2012). These reviews emphasize the problems and opportunities in combining ecological observations and models across spatial and temporal scales, from the cellular level at response times of minutes to days (e.g., response of photosynthesis to temperature) to the global level at response times of decades (e.g., response of global net primary productivity to global warming).

Another scale issue is spectral resolution. Satellite images tend to have just a few band windows available and currently operational hyperspectral imagery is mainly acquired from aircraft and a few experimental satellites. Therefore, the detailed reflectance spectra that can be used to distinguish different types of soils and vegetation units are not available from application-oriented satellite imagery. Among other tools such as spectral mixture analysis, researchers often rely on vegetation indices (VIs), combinations of two or three bands that separate the landscape into soil, water, and different densities of green vegetation, based primarily on differences in their reflectance of light in the adjacent red and near-infrared (NIR) bands (Glenn et al., 2007). The task of monitoring vegetation by use of VIs is made feasible due to the convergent properties of plant spectral responses across functional types (Ustin and Gamon, 2010; Ollinger, 2011). VIs can be used to estimate biophysical parameters such as leaf area index (LAI) and fractional vegetation cover (f_c), but these relationships are not universal across different plant communities and need to be calibrated by ground measurements for each application.

The present review uses a case study approach to illustrate some of the opportunities now available for combining observations across scales, drawing on our own research over the past two decades. The review will explore the use of multiplatform sensor systems to characterize ecological change, as exemplified by efforts to scale the effects of a biocontrol insect (the leaf beetle *Diorhabda carinulata*) on the phenology and water use of *Tamarix* shrubs (*Tamarix ramosissima* and related species and hybrids) targeted for removal on western U.S. rivers, from the level of individual leaves to the regional level of measurement. A final section will summarize the lessons learned and will emphasize the need for ground data to calibrate and validate remote sensing data and the types of errors inherent in scaling point data over wide areas, illustrated with research on evapotranspiration (ET) of *Tamarix* using a wide range of ground measurement and remote sensing methods. The goal of the review is to describe emerging methods for using satellite imagery across temporal and spatial scales in enough detail that they can be applied in other landscape studies.

5.2 COMBINING PHENOCAMS, LANDSAT, AND MODIS IMAGERY TO MONITOR EFFECTS OF INSECT DEFOLIATION OF VEGETATION ACROSS SPATIAL AND TEMPORAL SCALES

5.2.1 Need for Multiplatform Methods in Detecting Insect Damage to Forests

Insect infestations cause widespread damage to forest ecosystems, and the damage is expected to increase in areas such as the southwestern United States where climates are expected to become warmer and drier (Dale et al., 2001; Seager et al., 2007). Insect effects appear at multiple spatial scales, from effects on individual plants to regional effects. Furthermore, insect infestation is often episodic, occurring over just a few weeks, so monitoring programs must have high temporal resolution as well cover a wide range of spatial scales. Wulder et al. (2006) pointed out that since forest management agencies require information at several scales of measurement, no single remote sensing approach is adequate. At the low-resolution end of the scale, Eklundh et al. (2009) tested a method for mapping Scots pine defoliation by the pine sawfly (*Neodiprion* spp.) in Norway using coarse-resolution 16-day composite imagery from the MODIS sensors on the Terra satellite. They point out that the normalized difference vegetation index (NDVI) values were useful in detecting areas of defoliation within the forest, but only weak relations were found between the degree of damage and MODIS change parameters. They concluded that MODIS could be used to detect damaged forest areas but that high-resolution imagery or fieldwork would be needed to estimate the intensity of the infestation.

Medium-resolution Landsat imagery has also been used to detect insect damage. Healey et al. (2005) combined six Landsat bands into a three-parameter forest disturbance index (FDI) using a tasseled cap transformation in which brightness, greenness, and wetness were evaluated by different band combinations. Areas of forest clear-cut were characterized by high brightness and low greenness and wetness values. Eshleman et al. (2009) tested the ability of the FDI to detect gypsy moth

(*Lymantria dispar*) damage to an Appalachian oak forest using stream nitrogen levels as a proxy for defoliation because nitrogen is released from damaged leaves into the watershed. The FDI was able to predict both the overall intensity and extent of forest damage; however, nearly complete defoliation was needed for the FDI to detect damage at the stand level (Healey et al., 2005).

High-resolution hyperspectral imagery has also been used to detect insect damage. Leckie et al. (2005) developed an automated procedure for detecting forest defoliation by the jack pine beetle (*Dendroctonus frontalis*) using airborne, 2.5-m-resolution hyperspectral imagery from the multispectral electro-optical imaging sensor. Information from six visible bands and one NIR band was combined to detect canopy discoloration due to beetle damage. Insect damage was detected with an overall accuracy of 84% in stands with damage levels ranging from light to heavy. Martin et al. (2008) used foliar nitrogen levels as a proxy for forest damage in 137 forested plots in North America, South America, and Australia. Most plant nitrogen is contained in pigments and enzymes in leaf chloroplasts, and defoliation results in a marked drop in both canopy nitrogen and chlorophyll, which can be detected spectrally. Imagery from the airborne visible/infrared imaging spectrometer accurately predicted canopy nitrogen levels measured by ground sampling. The results were valid regardless of the source of forest damage. However, high-resolution, hyperspectral imagery is not currently available for routine forest monitoring.

Monitoring the effects of insects on vegetation presents a challenge for remote sensing methods due to the variety of spatial and temporal scales at which damage is manifested. Rather than relying on a single remote sensing platform, as in the above studies, our approach was to collect data across several scales of measurements in an attempt to combine the strengths of each approach and minimize their individual weaknesses. For example, digital cameras provide very high spatial and temporal resolution over a very limited area, but they can be scaled over a wider area using Landsat imagery, while MODIS imagery can provide high temporal resolution once stand structure is mapped with higher resolution imagery.

5.2.2 Background on *Tamarix* and *Tamarix* Leaf Beetles on Western U.S. Rivers

Tamarix shrubs were introduced on western rivers as erosion control plants in the nineteenth century (Chew, 2009). They spread rapidly and now occupy several hundred thousand hectares of riparian habitat (Nagler et al., 2010a). Starting in the 1950s they were perceived as problem plants due to assumed high water use, low habitat value for birds, and competitive displacement of native trees (DiTomaso, 1998; Zavaleta, 2000). These defects have been challenged by more recent research (e.g., Stromberg et al., 2009; Hultine et al., 2009, 2010; Nagler et al., 2010a,b), but in the meantime, *Tamarix* leaf beetles have been released on western rivers starting in 2001 as biocontrol agents (Bean et al., 2007a,b; DeLoach and Carruthers, 2004; DeLoach et al., 2000, 2004; Lewis et al., 2003; Dudley et al., 2006; Moran et al., 2009), and they have also spread rapidly, raising new management issues in already stressed western riparian corridors (Hultine et al., 2010). Among these are the effects of *Tamarix* defoliation on riparian ET, the sum of evaporation and plant transpiration from the surface to the atmosphere. The prospect of water salvage was a prime motivation for introducing the beetles (e.g., Pattison et al., 2011).

As the beetles spread, three key questions amendable to remote sensing methods are (Nagler et al., 2012):

1. How do leaf beetles affect the phenology of *Tamarix* leaf development over multiple years of infestation (i.e., does it kill the shrubs or is damage only temporary)?
2. What magnitude of water salvage can be expected in infested river systems?
3. How far are beetles likely to spread in the southwestern United States?

We determined the effects of beetles on the phenology and ET of *Tamarix* stands at beetle release sites on six river systems in four western states (Hultine et al., 2009, 2010; Nagler et al., 2012) and have recently extended the findings to the Virgin River in southern Nevada as the beetles move south into new *Tamarix* habitat (Nagler et al., 2014). Defoliation is only partial and is usually temporary at any given site (Figure 5.1), complicating the use of remote sensing for monitoring beetle effects.

Three sensor systems differing in both spatial and temporal scales were employed and their strengths and weaknesses are described in the following sections.

FIGURE 5.1 Photographs showing effects of leaf beetles on *Tamarix* stands at Lower Dolores River in 2009 (a); Humbolt River in 2007 (b); Walker River in 2006 (c); and Upper Colorado River site in 2007 (d). (See the color version of this figure in Color Plates section.)

Networked multiband digital cameras (phenocams) (Richardson et al., 2007) deployed over individual shrubs documented the phenology of defoliation with high temporal and spatial resolution but over a very limited sample of shrubs. Then Landsat TM and MODIS satellite images were used to quantify the effects of defoliation on riparian green plant cover and ET from 2000 to 2010, and ET measured in years before beetle release and years after release were subjected to an analysis of variance (ANOVA) to test the hypothesis that beetles reduced riparian ET (Nagler et al., 2012).

5.2.3   Phenocams Combine High Spatial and Temporal Resolution with Limited Field of View

Phenocams are inexpensive digital cameras mounted over vegetation. They automatically acquire images on a frequent basis (every 15 min to daily or longer), which are transmitted by an onsite packet radio station to a computer at a data processing site. Phenocams and other tower-mounted, remote sensors have become practical continuous-monitoring tools due to advances in computers and sensor technologies and improvements in the Internet in the past 10 years (e.g., Richardson et al., 2007). Details of their use have been reviewed in Rundel et al. (2009), Zerger et al. (2010), and Sonnentag et al. (2012). In vegetation studies they are especially useful in determining the exact timing of phenological events such as leaf-out, peak greenness, flowering, and maturation of either natural stands of plants or agricultural crops. (Similar cameras are used to monitor the movement of animals.) Phenocam sites have been organized into state, regional, national, and international networks, for example, the USA National Phenology Network, to monitor effects of climate change on vegetation growth and flowering cycles.

The cameras provided multiband imagery with red, green, blue, and NIR bands which are combined to compute NDVI values:

$$NDVI = \frac{NIR - red}{NIR + red} \tag{5.1}$$

Unlike satellite sensors, band values from digital cameras are rarely calibrated to reflectance values, and digital number (DN) values vary widely across different cameras viewing the same scene. Many phenocams have only red, green, and blue bands available: greenness from these images is computed from these bands by algorithms such as the "excess greenness" index (Sonnentag et al., 2012):

$$ExG = \frac{2\,green}{red + green + blue} \tag{5.2}$$

Either NDVI or ExG or simple visual interpretation of images can be used for analyses of phonological phenomenon.

In our work with saltcedar leaf beetles, we used phenocams to track the defoliation process from its onset in early summer through the regreening of shrubs in late summer to determine if shrubs progressively weaken over years. We installed multiple high-resolution, wireless panchromatic and infrared cameras on 10-m towers at the University of Utah's Rio Mesa field station in northeastern Utah, an early site of *Tamarix* defoliation by leaf beetles, and on the Virgin River, near Mesquite, Nevada, one of the most recent rivers to experience defoliation. Cameras included nadir-mounted cameras trained on a 1-m^2 section of a shrub as well as wide-angle views encompassing several hundred square meters. Camera images were uploaded via a satellite Internet connection permitting us to monitor beetle and *Tamarix* impacts on riparian ecology and water usage in real time. All data were uploaded to servers where they could be viewed in near-real time using interactive software that allowed retrieval of individual images and summary graphs of trends in NDVI and fractional green cover over time.

Sensors were not intercalibrated and red and NIR DN values and consequently NDVI varied considerably between the cameras and NDVIs were negative in some cases. To produce a temporal data set responsive to changes in *Tamarix* canopy condition, NDVI was calculated, averaged, and scaled using the following steps:

1. A seven-day running average of all pixels within a subset area of the camera field of view ($\sim$1 m^2) was calculated.

2. Values for each camera were scaled between 0 and 1.0 using NDVI$_{max}$ and NDVI$_{min}$ values at each tower site over the three years of data collection:

$$NDVI_{PC}^* = 1 - \frac{NDVI_{max} - NDVI}{NDVI_{max} - NDVI_{min}} \qquad (5.3)$$

where NDVI$_{PC}^*$ is the scaled phenocam NDVI. This transformation allowed comparison of the relative amount of defoliation at each site but did not allow comparisons of the actual amount of foliage at each site due to differences between cameras (Nagler et al., 2012).

Visible-band digital cameras (NetCam SC-Multi-Megapixel Hybrid IP Camera, StarDot Technologies, Buena Park, CA) acquired 5 megapixels per image over a 1-m^2 field of view. Images acquired in early afternoon (13:45 local time, when shadow effects from nearby oject were minimal) were used for analyses. From the daily images, a subset of dates that captured the onset and extent of leaf greening in spring, defoliation in summer, regrowth of leaves in late summer, and senescence of leaves in fall were visually selected for analysis. *Tamarix* has minute, scalelike leaves attached to needlelike terminal stems. During defoliation, beetles eat the mesophyll cells of the leaves but leave the dead leaf remains and the terminal stems intact. Visually, leaves (needles) turn from green to brown during defoliation but remain on the plant. Individual green or brown leaves were easily visible on the images and were quantified by placing a 200-point grid over the image in Adobe Photoshop 8.0 (Adobe Systems, Inc., San Jose, CA) then scoring the fraction of grid intersections that covered green leaf material (Nagler et al., 2012).

Results were quantified as the percent of annual green leaf cover or $NDVI^*_{PC}$ lost due to defoliation for each year at each camera station. All cameras captured at least a portion of the initial greening of canopies which started on about April 1 each year and reached a maximum value by mid-June. This was followed by a rapid loss of nearly all green leaves during the defoliation period and in some cases a subsequent production of new green leaves in August and September. Plants then dropped their leaves and entered winter dormancy in early October. The extent of defoliation was estimated by projecting each curve over a complete growing season (April 1 to October 15), then inferring what the shape of the seasonal curve would have been in the absence of defoliation, by assuming that peak green cover had been reached before beetles were active, as documented by Hultine et al. (2009) at this site. The "missing" portion of the curves due to defoliation were then estimated on paper printouts of plots by weighing cut-out sections of the curves on an analytical balance, following a method used to determine the area of irregular shapes (e.g., Patil and Bodhe, 2001). For each curve, the weight of the lost-production cut-out were divided by the weight of the cut-out representing the total area under the curve times 100 to get the percentage of reduction.

The defoliation processes as viewed on the Virgin River from wide-angle cameras is shown in a time series of panchromatic images in Figure 5.2 taken in 2011, the first full season of defoliation at this site. Plants greened up in May 2011 but were completely defoliated by July 2011 and did not refoliate by September 1.

On the other hand, plants at the Dolores River (Rio Mesa) site responded to defoliation by producing new leaves in August of each year, and no net decrease in spring green-up was noted over three years of defoliation, 2008–2010 (Figure 5.3). Estimates of lost leaf production from visible-band and NDVI cameras were similar across years and sites: 32.0 and 30.8% per season, respectively ($F = 0.05$, $p = 0.838$, df $= 1,10$) (Nagler et al., 2012).

On balance, more information was obtained from the visual green leaf counts than from the NDVI data. The onset of defoliation was clearly visible on RGB images even in its early stages, when some of the green leaves abruptly turned brown, indicating the arrival of beetle. Visual analysis of images also allowed us to distinguish between leaves subjected to defoliation from those that were lost due to detachment of twigs by wind and to visually account for differences in greenness caused by shadows within the canopy or cloud cover. Furthermore, the results were expressed as fractional green cover, a meaningful biophysical variable, rather than as a derived vegetation index. Finally, the uncalibrated band reflectance values differed among cameras, and by scaling NDVI between 0 and 1.0 individually for each camera, it was not possible to compare values among cameras and sites, whereas the visual green leaf counts could be compared directly.

5.2.4 Landsat Imagery to Compare NDVI and ET Before and After Beetle Arrival along Six River Systems

The Landsat series of satellites has provided near-continuous coverage of Earth's surface since 1972, with a spatial resolution of 15–60 m in the visible and NIR bands

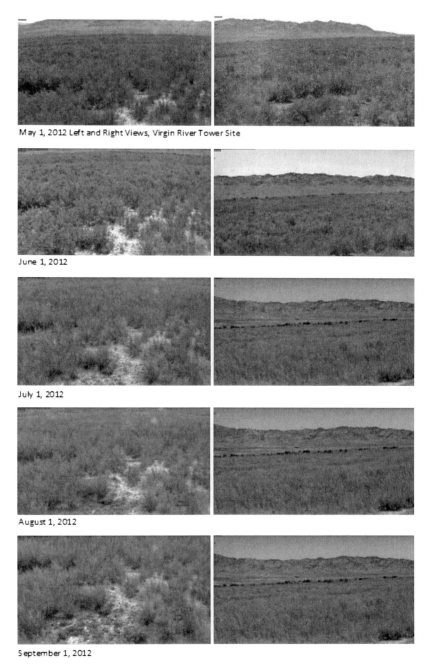

FIGURE 5.2 Series of wide-angle phenocam images showing progression of defoliation of *Tamarix* by leaf beetles on Virgin River. Two cameras mounted on the same tower capture different views of the floodplain. (See the color version of this figure in Color Plates section.)

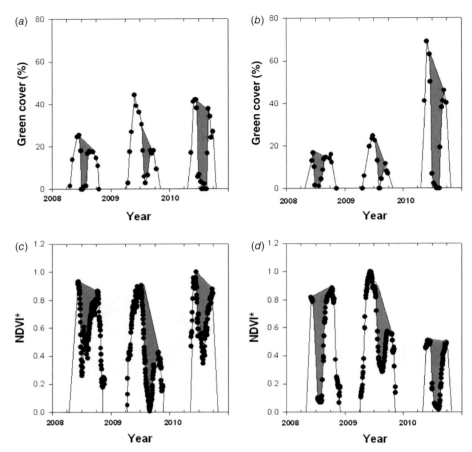

FIGURE 5.3 Data from phenocams showing defoliation of *Tamarix* by leaf beetles followed by new leaf production on Dolores River: (a, b) from visual analysis of panchromatic images at two sites on floodplain; (b, c) scaled NDVI values calculated from red and NIR bands from multiband cameras at same site.

(depending on satellite and band sensor) and a revisit time of 16 days for most of Earth's surface (Irish, 1999; Song et al., 2001). Landsat imagery has been especially useful in vegetation change detection studies. Typically, annual images obtained during the period of peak summer greenness are compared based on NDVI or other vegetation indices calculated from red and NIR reflectance values. Preprocessing of imagery by the U.S. Geological Survey (USGS) Earth Observation and Science Center includes georectification and terrain correction based on ground sample points (where available). Gain values for each sensor are internally calibrated and delivered to end users as calibrated DN values ranging from 1 to 256. To compare band values or vegetation index values across dates, the end users must first convert DNs to top-of-atmosphere reflectance values using data on sun angle and Earth–sun distance at the time of image acquisition (Irish, 1999). Most studies also correct images for atmospheric effects using one of several within-scene methods to account for effects of haze and moisture content on light transmission through the atmosphere (Song et al., 2001). Fortunately, a large body of literature is available on the use of Landsat images for change detection, and differences in NDVI of <10% attributable to vegetation changes between acquisition dates can be routinely detected (Gillanders et al., 2008).

Landsat images are also widely used to estimate ET. Two types of methods have been developed (Glenn et al., 2007): energy balance methods based on thermal band measurements of surface temperature (reviewed in Kalma et al., 2008; Kustas and Anderson, 2009) and VI estimates of ET by transpiring vegetation (reviewed in Glenn et al., 2007, 2008, 2010). VI methods lend themselves to change detection studies for three reasons:

1. They are based on the generally strong correlation between ET and vegetation indices in a variety of natural and agricultural ecosystems, with many methods producing coeffieints of determination (r^2) in the range of 0.75–0.95.
2. They characterize differences in ET that can be attributed directly to changes in foliage density between dates, which is the goal of this study.
3. They are easy to apply and do not require a large amount of ancillary ground data (Glenn et al., 2010).

Although many variations exist, most VI methods for ET are based on the crop coefficient method for estimating ET developed for agricultural crops:

$$ET = K_c ET_0 \qquad (5.4)$$

where K_c is a crop coefficient and ET_0 is potential or reference ET calculated from meteorological data. Generally ET_0 is defined as the maximum daily ET that could occur from a fully transpiring reference crop based on available energy and atmospheric water demand on the day of measurement (Allen et al., 1998). A wide variety of methods, from the simple to the complex, exist for estimating ET_0, but generally they give results within 10% of each other when calibrated for a given location (Allen et al., 1998). The simplest methods require only knowledge of mean monthly temperature and hours of daylight for calculation (e.g., the Blaney–Criddle method) (Brouwer and Heibloem, 1986), while the more complete Penman–Monteith method requires wind speed, temperature, humidity, and net radiation measurements, data which are usually collected at remotely operated micrometeorological stations (Allen et al., 1998, 2011).

Crop coefficients are generally developed through field experiments using lysimeters planted with the crop being studied and are typically at monthly intervals to conform to the main growth stages of the crop (Allen et al., 2011). These crop coefficients are usually developed for crops grown under optimum agronomic conditions and are therefore only useful approximations of the actual ET and water requirements for a given crop. However, the actual crop ET in the field can vary from estimated K_c-based ET for a number of reasons, including crop variety differences, planting density, climatic factors, nutrient status, irrigation management, salinity, and other conditions.

VI methods replace (or supplement) crop coefficients with a VI that reflects the actual growth stage of the crop at the time of measurement. VI-based crop coefficients $(K_{c,VI})$ have been developed for individual and mixed crops in agricultural regions starting nearly 30 years ago (e.g., Neale et al., 1989). More recently, the concept has been applied to natural ecosystems at local, regional, and global scales of measurement (Glenn et al., 2010):

$$ET = ET_o(VI^*) \tag{5.5}$$

where VI^* is a vegetation index scaled between bare soil ($VI^* = 0$) and full vegetation cover transpiring at the rate of ET_o ($VI^* = 1.0$) by the formula

$$VI^* = 1 - \frac{VI_{max} - VI}{VI_{max} - VI_{soil}} \tag{5.6}$$

where VI_{max} is the VI of fully transpiring vegetation and VI_{soil} is the VI of bare, dry soil with assumed ET of 0. The VI_{max} and VI_{soil} are usually determined by sampling areas of bare soil and dense vegetation within each image (Nagler et al., 2005a,e; Baugh and Groeneveld, 2006; Groeneveld and Baugh, 2007; Groeneveld et al., 2007).

It would be desirable to have paired sites, with and without beetle damage, to compare imagery taken at the same time. However, beetle damage was always widespread at a given site, so this sampling method was not possible. In our saltcedar beetle studies, we obtained one or two annual summer Landsat 5 images for each year from 2000 to 2010 for beetle release sites on six western U.S. rivers, encompassing years before and after beetle release at each site (Nagler et al., 2012). Images with no cloud cover over the sites of interest acquired from June 15 to August 15 were selected. Level 1T processed images referenced to fixed ground points were obtained from the U.S. Geological Survey Earth Explorer website (http://edcsns17.cr.usgs.gov/NewEarthExplorer/). Sites were chosen that encompassed one or more MODIS pixel footprints with the riparian area.

Methods for processing band data, converting NDVI values to scaled values ($NDVI^*_{TM}$), and calculating ET followed methods developed to estimate annual ET by western U.S. phreatophye communities, including *Tamarix* sites, from single summer Landsat images (Baugh and Groeneveld, 2006; Groeneveld and Baugh, 2007; Groeneveld et al., 2007). DN values were converted to apparent at-satellite reflectance values using data in the header files and equations and tabular information in the *Landsat 7 Science Data Users Handbook* (Irish, 1999). NDVI values were then scaled ($NDVI^*_{TM}$) between bare soil ($NDVI_{Soil}$) and maximum vegetation response ($NDVI_{max}$) on each image using the relationship in Equation (5.6) where $NDVI_{Max}$ was the highest NDVI value on the image determined from a display of the pixel statistics in ERDAS Imagine (ERDAS, Inc., Norcross, GA) and $NDVI_{Soil}$ was the NDVI of an area (1–2 ha) of dry lakebed or rock outcrop that was apparently unvegetated and stable year to year. This bare-soil (or rock) area was selected on the year 2000 image for each image series and the same area was resampled on subsequent images to evaluate the variability in the NDVI of the same scene across images in a time series. In a test of the value of scaling NDVI for each image, Baugh and Groeneveld (2006) compared $NDVI_{TM}^*$ with unscaled NDVI for its correlation with annual precipitation at 15 moisture flux tower sites in phreatophyte communities from 1986 to 2002. Foliage density as determined by NDVI in desert plant communities is expected to be correlated with annual rainfall. The coefficient of determination (r^2) between NDVI and precipitation was 0.37, compared to 0.77 for precipitation and $NDVI_{TM}^*$, and $NDVI_{TM}^*$ was better correlated with precipitation than any of 13 other VIs tested in the same study. They concluded that $NDVI_{TM}^*$ corrected for both atmospheric and soil-induced effects in an image series and eliminated the need to correct for these two effects separately in this application.

Following Groeneveld et al. (2007), ET was calculated as in Equation (5.5). Groeneveld et al. (2007) reported an r^2 of 0.94 for ET determined from single summer Landsat images and annual ET measured at 15 moisture flux tower sites set in western U.S. pheatophyte communities.

In this study, ET_o was calculated by the Blaney–Criddle formula (Brouwer and Heibloem, 1986):

$$ET_o = p(0.46T_{mean} + 8) \tag{5.7}$$

where p is daylight hours determined from a table by month and latitude and T_{mean} is mean monthly air temperature, obtained from NOAA cooperative reporting stations near each site (http://lwf.ncdc.noaa.gov/oa/climate/climatedata.html). To determine the effects of defoliation on ET at each site, an area-of-interest (AOI) file was prepared in ERDAS that encompassed the area of maximum defoliation reported from ground surveys at each site, and the same AOI file was used to extract $NDVI_{TM}^*$ values for each Landsat image from 2000 to 2010 and used to calculate annual ET. Annual ET values were divided into two groups representing years before and after widespread defoliation was noted at each site. Data were subjected to two-way ANOVA in which ET was the dependent variable and site and before/after defoliation were the categorical variables (Stevens, 1996). ET values for individual years before or after defoliation were treated as replicates in the ANOVA.

The coefficient of variation (CV) for $NDVI_{max}$ was low among image series, ranging from 3.0 to 4.2%, while the CV for $NDVI_{Soil}$ was higher (7.9–26.0%). The use of NDVI* rather than NDVI is designed to minimize soil effects on ET estimates (Baugh and Groeneveld, 2006). ET rates tended to be variable year to year and differed markedly among sites (Table 5.1). The lowest ET rates were at the Lower Dolores River site (100–350 mm yr^{-1}), while rates were as high as 600–800 mm yr^{-1} at the Humbolt River and Big Horn River sites. All sites except for the Middle-Upper Dolores River site showed a marked reduction in ET during the first year of active defoliation. The Middle-Upper Dolores River site had a mixed riparian community whereas the other sites were dominated by *Tamarix*. All sites showed at least a partial recovery in subsequent years, and at the Humbolt River site the highest ET rates over the decade were in postrelease years.

TABLE 5.1 Means, Standard Errors[a], and ANOVA Statistics for ET Measured Each Year by Landsat Imagery in *Tamarix* Stands Before and After Beetle Infestation (2000–2010) on Western U.S. River Systems

Site	ET Before (mm yr^{-1})	ET After (mm yr^{-1})	F	p	df
Lower Dolores	232 (17)	149 (102)	1.69	0.23	1,10
Middle-Upper Dolores	328 (21)	370 (39)	1.10	0.32	1,10
Upper Colorado	472 (23)	325 (25)	17.6	0.002	1,10
Bighorn	551 (24)	484 (117)	0.81	0.39	1,10
Humbolt	538 (17)	503 (62)	0.15	0.70	1,10
Walker	242 (37)	176 (37)	1.60	0.24	1,10
Mean	394 (59)	335 (61)			
Two-Way ANOVA					
Before/after			5.37	0.024	1,59
Site			20.0	<0.001	5,59
Interaction			1.18	0.331	5,59

[a]In parentheses.

When sites were analyzed individually, only the Upper Colorado River site had a significant ($p = 0.002$) decrease in ET after beetle release (Table 5.1).

However, a two-way ANOVA showed that ET was significantly different among sites ($p < 0.001$) and was higher before beetle release than after ($p = 0.024$) across sites. The interaction term (site × before/after) was not significant ($p = 0.331$). Annual ET was 394 mm yr^{-1} before beetle release and 335 mm yr^{-1} after release, 15% lower. The results are consistent with concurrent ground observations at the Lower Dolores River site and the Middle-Upper Dolores site, which showed that defoliation was relatively brief each year and did not affect all plants.

Landsat NDVI and ET estimates extended the phenocam observations over whole river reaches and added confidence to the conclusion that beetle damage tended to be temporary and that *Tamarix* stands could potentially recover each year by producing new leaves. However, Landsat had poor temporal coverage. Despite the theoretical revisit time of 16 days, in practice it was difficult to find images that captured the three stages of defoliation within a single year: pre–beetle arrival; the 6–8 weeks of defoliation; and postdefoliation, when beetles entered dormancy and *Tamarix* produced new leaves. Therefore, Landsat imagery by itself was not sufficient for monitoring defoliation events.

5.2.5 MODIS Imagery to Compare EVI and ET Before and After Beetle Arrival

The Terra and Aqua satellites were specifically designed as frequent-return Earth observation satellites with VIs intercalibrated with other satellite VI series (Huete et al., 2002). Data are preprocessed to deliver reflectance-based, atmospherically corrected VI values to end users. Imagery is composited on a pixel-by-pixel basis to combine the best 3–5-day composite values for each 8- or 16-day measurement interval. "Best" values are defined as the most cloud free (highest VI values) pixels with the closest to nadir view angle. Both the NDVI and enhanced vegetation index (EVI) are available to end users, as well as several other derived products, including LAI, net primary productivity, and other biophysical variables. Previous studies (Nagler et al., 2005a,e) have shown that the EVI is significantly ($p < 0.05$) better correlated with ground-measured ET than NDVI. Hence we used EVI to predict ET in the *Tamarix* beetle studies. EVI is calculated from red, blue, and NIR bands as described in Huete et al. (2002):

$$\text{EVI} = \frac{G(\text{NIR} - \text{red})}{\text{NIR} + C_1 \times \text{red} + C_2 \times \text{blue} + L} \tag{5.8}$$

where C_1 and C_2 are coefficients designed to correct for aerosol resistance, which uses the blue band to correct for aerosol influences in the red band, and have been set at -6 and 7.5, while G is a gain factor (set at 2.5) and L is a canopy background adjustment (set at 1.0) (Huete et al., 2002). Pixel size is 250 m and each image is a composite of three to five cloud-free images during each 16-day collection period. Pixel footprints were projected on high-resolution images using Google Earth to ensure it encompassed the beetle release site. In a few cases the only available MODIS pixels were wider than the riparian corridor and contained areas of adjacent uplands, which were sparsely vegetated. For those pixels, the approximate percentage of riparian habitat was estimated visually. These estimates are only approximations, as the center point of the pixels is somewhat indeterminate (Tan et al., 2006) and the area covered by each pixel is variable, depending on view angle [Oak Ridge National Laboratory (ORNL), 2010]. Similar to phenocam $\text{NDVI}_{\text{PC}}^*$ and Landsat $\text{NDVI}_{\text{TM}}^*$ values, EVI values were stretched between 0 (bare) and 1.0 (full riparian vegetation cover) by the equation

$$\text{EVI}^* = 1 - \frac{0.542 - \text{EVI}}{0.542 - 0.091} \tag{5.9}$$

where EVI* is scaled EVI and 0.542 and 0.091 are maximum and minimum EVI values from a large data set of riparian values in the western United States (Nagler et al., 2005b).

EVI* values were transformed to estimates of ET (mm d^{-1}) by the equation

$$\text{ET} = 1.22 \, \text{EVI}^*(\text{ET}_\text{o}) \tag{5.10}$$

Equation (5.10) was developed by regressing measurements of riparian and crop ET on the Lower Colorado River (measured by sap flux, moisture flux tower, and soil water depletion methods) with meteorological and remote sensing data, and it has a root mean-square error of about 20% of the mean value (Nagler et al., 2009b).

As with Landsat data, mean annual ET values for the years 2000–2009 at each site were divided into two groups: years before wide-area defoliation was noted and years after defoliation was noted. Data for mean values across pixels at each site were then

TABLE 5.2   Means, Standard Errors[a], and ANOVA Statistics for ET Measured Each Year by MODIS Imagery in *Tamarix* Stands Before and After Beetle Infestation (2000–2009) on Western U.S. River Systems.

Site	ET Before (mm yr^{-1})	ET After (mm yr^{-1})	F	p	df
Lower Dolores	270 (12)	251 (11)	0.702	0.406	1,48
Middle-Upper Dolores	224 (15)	232 (23)	0.203	0.653	1,88
Upper Colorado	349 (12)	275 (9)	10.8	0.001	1,8
Walker	169 (20)	109 (13)	7.04	0.029	1,8
Humbolt	587 (30)	466 (19)	11.4	0.002	1,28
Bighorn	282 (23)	275 (35)	0.041	0.846	1,8
Mean	314 (60)	269 (47)			
Two-Way ANOVA					
Before/after			4.73	0.031	1,188
Site			66.5	<0.001	5,188
Interaction			2.54	0.030	5,188

[a]In parentheses.

subjected to two-way ANOVA (Stevens, 1996) with site and before/after defoliation as categorical variables. Sites were defined as the six river systems for which pixels were collected. Individual pixels represented nonoverlapping and nonadjacent sample points within each site. Variances in pixel values within each of the six sites showed no significant departure from normality by the Shapiro–Wilks test ($p > 0.05$), so they were treated as replicated samples of ground conditions within each site.

Results (Table 5.2) were similar in magnitude and direction to those from Landsat estimates. Baseline ET values varied widely over sites, from peak summer rates of just 2 mm d^{-1} on the Middle-Upper Dolores River site, to 6–8 mm d^{-1} at the Humbolt River site, where peak *Tamarix* ET was equal to ET$_0$ in some years. ANOVA results were significant by both site ($p < 0.001$) and before/after defoliation ($p = 0.031$), but the interaction term was also significant ($p = 0.03$). Hence, a one-way ANOVA with before/after defoliation as the categorical variable was also conducted for each site. ET was significantly lower at the Upper Colorado River, Walker River, and Humbolt River sites in years after defoliation, but differences were not significant ($p > 0.05$) at the Dolores River or Big Horn River sites. Over all sites, mean annual ET was 314 mm yr^{-1} before beetle infestation (about 23% of ET$_o$) and 269 mm yr^{-1} after infestation, a 14% reduction. MODIS ET estimates were 15% lower than Landsat ET estimates across sites and years. This was likely due to the fact that MODIS pixels captured some adjacent, lower density vegetation outside the riparian zone.

Beetle effects varied considerably among river systems, requiring wide-area monitoring to gain a true picture of the extent of beetle damage. Figure 5.4 shows the initial stages of beetle damage on the Virgin River, Nevada, after first arrival of the beetles in 2010. ET was estimated by MODIS, and beetles reduced annual ET by over 50% during the second year of defoliation.

Much better temporal coverage was obtained with MODIS imagery compared to Landsat imagery. Figure 5.5 illustrates how a single summer image from Landsat can fail to capture the details of the defoliation processes.

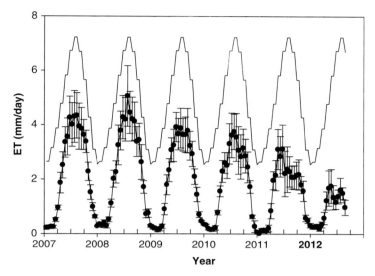

FIGURE 5.4   Impact of leaf beetles on *Tamarix* ET estimated by MODIS EVI on Virgin River, NV. Data points are the mean and standard error of six sites on the river; solid line is potential ET. Beetles first arrived in 2010 and defoliation and reduction in ET were widespread by 2012.

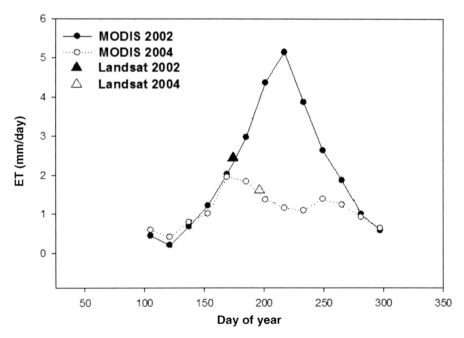

FIGURE 5.5   Comparison of Landsat and MODIS Tamarix ET estimates on Humbolt River, NV, before (2002) and after (2004) arrival of leaf beetles.

However, many of the riparian corridors were too narrow to be captured by MODIS imagery, and a single MODIS pixel could not capture the variability of defoliation within a site. Hence, imagery with different spatial and temporal scales were needed to get a clear picture of defoliation effects.

5.3   LESSONS LEARNED FROM THESE
AND OTHER CHANGE STUDIES

5.3.1   Importance of Ground Measurement as Basis
for Remote Sensing Scaling Procedures

Although not the primary subject of this chapter, in our research we have found it essential to build remote sensing methods on a foundation of ground measurements. Often the goal of remote sensing studies is stated as developing "stand-alone" methods that do not depend on ground data. However, in our research this has proven to be impractical, leading to error and uncertainty in the remote sensing products. Furthermore, the ground methods used for calibration or validation of remote sensing algorithms frequently contain rather large amounts of error or uncertainty, as discussed below.

In our studies, algorithms relating VIs to fractional f_c cover and LAI were developed through extensive ground campaigns in which f_c and LAI were measured in habitats of interest (e.g., riparian zones and arid and semiarid uplands), then scaled to progressively broader scales using multiband cameras in aircraft, high-resolution QuickBird imagery, Landsat imagery, and finally MODIS imagery, with each progressive scale linked to the ones below and above through regression equations (e.g., Nagler et al., 2001, 2005b–d, 2008; Glenn et al., 2008). In conducting the calibration studies, we came to realize that ground measurements of f_c and LAI are subject to errors and uncertainties on the order of 20–30%, constraining the potential accuracy of results scaled by remote sensing. There is also a propagation error in moving from one scale to another across methods, each of which has an individual error term. Furthermore, regressions between f_c and LAI and VIs can be species specific, due to differences in canopy properties such as leaf angle and shape and pigment content (e.g., Nagler et al., 2005c), requiring separate calibration equations for each plant community of interest.

Similar conclusions were reached concerning ET. Optimal algorithms relating VIs and meteorological data to ET were different for riparian and upland plant communities (Nagler et al., 2005a,c,d, 2009a,b; Glenn et al., 2008), and ground methods for measuring ET, for example, with moisture flux towers and sap flow sensors, have potential errors of 10–30% (Glenn et al., 2007, 2010; Allen et al., 2011).

5.3.2   Precision versus Accuracy: Importance of Multiple Independent
Methods for Measuring Biophysical Variables

In evaluating the validity of ground-calibrated remote sensing data, it is important to distinguish between precision and accuracy of measurements (Grubbs, 1973). Precision refers to the variance of measurements around a mean value; it can be increased by increasing the sample size, with the standard error of the mean decreasing in proportion to the square root of the sample size. Accuracy is the deviation of the measured results from the true value, but in the case of some biophysical variables, the true value is unknown. Therefore, increasing the precision of a set of measurements by increasing sample size will not improve accuracy.

As an example, LAI is usually measured by optical devices such as the LICOR 2000 LAI meter, but this instrument has a built-in assumption that the canopy is uniform and continuous over the measuring device. This assumption is violated in many natural stands of plants, particularly when plants are widely spaced as in arid zones. Errors of 50% or greater are common in riparian zones and upland desert habitats (Nagler et al., 2005c). Sap flux measurements of ET are subject to errors of up to 50% if not individually calibrated for the plant and ecosystem of interest (Allen et al., 2011). Moisture flux towers are subject to errors of up to 30% due to advection of energy from adjacent ecosystems and the so-called closure error in the energy balance equations by which ET is calculated from tower data (Allen et al., 2011). These are all examples of "accuracy" errors since true LAI or ET values are not available. They cannot be reduced by increasing sample size, but their magnitude can be assessed by using two or more independent methods for estimating the same parameter.

5.3.3 Example of Multiple Sources of Measurements to Constrain Accuracy of ET Estimates

Figure 5.6 compares ET estimated by four different ground methods at a *Tamarix* site named Diablo on the Colorado River.

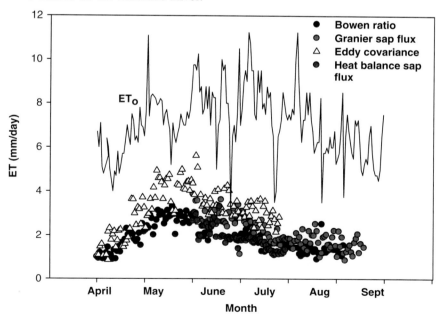

FIGURE 5.6 Comparison of *Tamarix* ET estimates in 2009 at stressed site (Diablo) on Lower Colorado River, CA. (See the color version of this figure in Color Plates section.)

The site was 1 km from the active river channel, depth to groundwater was 4 m, and the aquifer was saline, producing stress on the plant stand (Nagler et al. 2008, 2009a,b). Two methods used moisture flux towers to measure moisture losses from the canopy, but they differed in their principles of operation and underlying assumptions. Bowen ratio towers recorded air temperature and humidity at two heights over the canopy, and ET was calculated indirectly as a residual in the energy balance equation. Eddy covariance towers measured moisture fluxes directly with a precision hygrometer placed over the canopy. The other two methods used sap flux sensors, also with different sources of error and uncertainty. Granier-type sap flow sensors were placed in the trunk or main branch of saltcedar shrubs. ET was estimated by introducing a source of heat into the xylem tissue and measuring the rate at which it was dissipated away from the heat source in the transpiration stream. This required an estimate of the conducting area of sapwood in the stem and its specific conductance, which are problematic to measure. Heat balance sap flux sensors were placed around small side branches of shrubs, and ET was calculated by solving a heat balance equation based on temperature measurements upstream and downstream of the heat coil. This is a more direct method compared to Granier sensors but requires more scaling steps to estimate ET at the stand level. The four methods cover the range of commonly used field methods for ET and illustrate that all methods have sources of error and uncertainty in their application.

Eddy covariance ET results were about 50% higher than Bowen ratio results (Figure 5.6), probably because the two methods measure ET over a different footprint area and the *Tamarix* stands are not homogeneous over the site. Sap flow results fell within the range of tower results, but Granier ET results were 30% higher than heat balance results, attributed to different sets of assumptions in the calibration procedures, and possibly also because different sets of plants were sampled. Despite these differences, all four methods showed that ET was only about a third of ET_0 despite the fact that LAI was 2.66, indicating full plant cover. Hence, stress effects were confirmed by all methods. Furthermore, ET measured by all methods decreased over the growing season, attributed to an increase in depth to groundwater as water was withdrawn from the aquifer faster than it could be replenished from the river. All methods also showed a marked midday depression of ET which increased as depth to groundwater increased over the growing season (data not shown). Hence, despite variances in results, the combined methods produced a robust picture of *Tamarix* water relations under stress conditions, with more confidence than any single method could have produced.

Ground results were then compared to remote sensing results for the Diablo site and for an unstressed *Tamarix* site (Slytherin) nearer the river and with lower salinity in the aquifer (Figures 5.7a,b).

Remote sensing methods included a VI method (MODIS) and two thermal-band methods, the surface energy balance algorithm (SEBAL) and the two-source energy balance method (TSEB). The remote sensing methods were unable to detect stress effects revealed by ground methods at the Diablo site, whereas both ground and remote sensing methods produced similar results at the unstressed Slytherin site. Reasons for the failure of the remote sensing methods at the Diablo site were negative

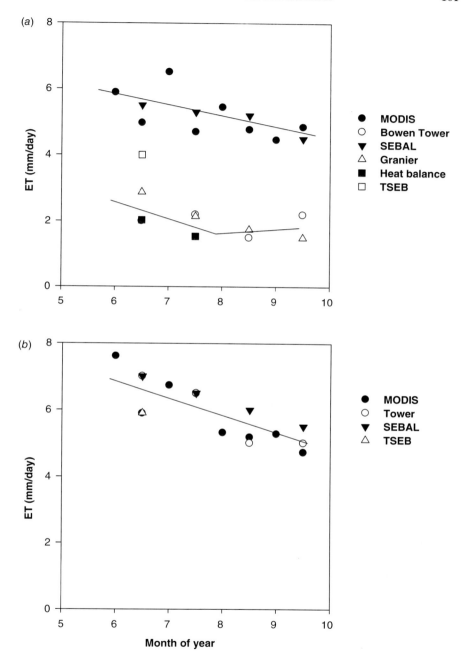

FIGURE 5.7 Comparison of *Tamarix* ET at stressed (a, Diablo) and unstressed (b, Slytherin) by ground and remote sensing methods. Ground methods were Granier and heat balance sap flow sensors and Bowen ratio flux towers; remote sensing methods were the MODIS VI method and two thermal-band methods, SEBAL and TSEB.

effects of stress on stomatal conductance, which impacted the VI method, and midday depression of ET, which impacted the thermal-band methods because they assume a constant relationship between ET and net radiation over the daylight hours. These results illustrate the need to combine and compare both ground and remote sensing method in scaling point measurements to larger landscape units and to exercise caution in accepting remote sensing results if not validated by ground data.

The case studies in this chapter show that multiscale remote sensing methods are useful in change detection and landscape monitoring programs, but they should be combined with ground data for validation, and multiple independent methods for measuring biophysical variables by both ground and remote sensing methods should be employed to discover sources of error and uncertainty that will affect interpretation of results. No single scale or method of measurement was suitable for monitoring riparian vegetation in these studies.

REFERENCES

Abraha, M. G., and Savage, M. J. 2012. Energy and mass exchange over incomplete vegetation cover. *Critical Reviews in Plant Sciences* 31:321–341.

Allen, R. G., Pereira, L., Rais, D., and Smith, M. 1998. *Crop Evapotranspiration—Guidelines for Computing Crop Water Requirements*. FAO Irrigation and Drainage Paper No. 56, Food and Agricultural Organization of the United Nations, Rome.

Allen, R. G., Pereira, L. S., Howell, T. A., and Jensen, M. E. 2011. Evapotranspiration information reporting: I. Factors governing measurement accuracy. *Agricultural Water Management* 98:899–920.

Anderson, N. T., and Marchisio, G. B. 2012. WorldView-2 and the evolution of the Digital-Globe remote sensing satellite constellation: Introductory paper for the special session on WorldView-2. In S. S. Shen and P. E. Lewis (Eds.), *Algorithms and Technologies for Multispectral, Hyperspectral, and Ultraspectral Imagery XVIII*. SPIE Digital Library, http://proceedings.spiedigitallibrary.org/proceeding.aspx?articleid=1354542.

Baugh, W. M., and Groeneveld, D. P. 2006. Broadband vegetation index performance evaluated for a low-cover environment. *International Journal of Remote Sensing* 27:4715–4730.

Bean, D. W., Dudley, T. D., and Keller, J. C. 2007a. Seasonal timing of diapause induction limits the effective range of *Diorhabda elongata deserticola* (Coleoptera: Chrysomelidae) as a biological control agent for tamarisk (*Tamarix* spp.). *Environmental Entomology*, 36:15–25.

Bean, D. W., Wang, T., Bartelt, R. J., and Zilkowski, B. W. 2007b. Diapause in the leaf beetle *Diorhabda elongata* (Coleoptera: Chrysomelidae), a biological control agent for tamarisk (*Tamarix* spp.). *Environmental Entomolology* 36:531–540.

Brouwer, C., and Heibloem, M. 1986. *Irrigation Water Management Training Manual No. 3*. FAO: Rome.

Chew, M. K. 2009. The monstering of Tamarisk: How scientists made a plant into a problem. *Journal of the History of Biology* 42:231–266.

Dale, V. H., Joyce, L. A., McNulty, S., Neilson, R. P., Ayres, M. P., Flannigan, M. D., Hanson, P. J., Irland, L. C., Lugo, A. E., Peterson, C. J., Simberloff, D., Swanson, F. J., Stocks, B. J., and Wotton, B. M. 2001. Climate change and forest disturbances. *BioScience* 51:734–734.

DeLoach, C. J., and Carruthers, R. 2004. Biological control programs for integrated invasive plant management. In: *Proceedings of Weed Society of America Meeting*. Weed Science Society of America: Kansas City, MO (CD-ROM).

DeLoach, C. J., Carruthers, R., Dudley, T., Eberts, D., Kazmer, D., Knutson, A., Bean, D., Knight, J., Lewis, P., Tracy, J., Herr, J. Abbot, G., Prestwich, S., Adams, G., Mityaev, I., Jashenko, R., Li, B., Sobhian, R., Kirk, A., Robbins, T., and Delfosse, E. 2004. First results for control of saltcedar (Tamarix spp.) in the open field in the western United States. In J.Cullen (Ed.), *Eleventh International Symposium on Biological Control of Weeds* Canberra, Australia: CSIRO.

DeLoach, C. J., Curruthers, R. I., Lovich, J. E., Dudley, T. L., and Smith, S. D. 2000. Ecological interactions in the biological control of saltcedar (Tamarix spp.) in the United States: Toward a new understanding. In N. R. Spencer (Ed.), *Proceedings of the X International Symposium on Biological Control of Weeds*. Bozeman: Montana State University, pp. 819–873.

De Pondeca, M., Manikin, G. S., Dimego, G., Benjamin, S. G., Parrish, D. F., Purser, J., Wu, W., Horel, J. D., Myrick, D. T., Ying, L., Aune, R. M., Keyser, D., Colman, B., Mann, G., and Vavra, J. 2011. The real-time mesoscale analysis at NOAA's National Centers for Environmental Prediction: Current status and development. *Weather and Forecasting* 26:593–611.

Di Tomaso, J. 1998. Impact, biology, and ecology of saltcedar (*Tamarix* spp.) in the southwestern United States. *Weed Technology* 12:326–336.

Dudley, T. L., Dalin, P., and Bean, D. W. 2006. Status of biological control of Tamarix spp. in California. In M. S. Hoddle and M. W. Johnson (Eds.), *Proceedings of the Fifth California Conference on Biological Control*. Riverside, CA: University of California, pp. 137–140.

Eklundh, L., Johansson, T., and Soldberg, S. 2009. Mapping insect defoliation in Scots pine with MODIS time-series data. *Remote Sensing of Environment* 113:1566–1573.

Eshleman, K. N., McNeil, B. E., and Townsend, P. A. 2009. Validation of a remote sensing based index of forest disturbance using streamwater nitrogen data. *Ecological Indicators* 9:476–484.

Gillanders, S. N., Coops, N. C., Wulder, M. A., Gergel, S. E., and Nelson, T. 2008. Multitemporal remote sensing of landscape dynamics and pattern change: describing natural and anthropogenic trends. *Progress in Physical Geography* 32:503–528.

Glenn, E. P., Huete, A., Nagler, P., Hirschboek, K., and Brown, P. 2007. Integrating remote sensing and ground methods to estimate evapotranspiration. *Critical Reviews in Plant Sciences* 26:139–168.

Glenn, E. P., Morino, K., Didan, K., Jordan, F., Carroll, K., Nagler, P., Hultine, K., Sheader, L., and Waugh, J. 2008. Scaling sap flux measurements of grazed and ungrazed shrub communities with fine and coarse-resolution remote sensing. *Ecohydrology* 1:316–329.

Glenn, E. P., Huete, A. R., Nagler, P. L. 2010. Vegetation index methods for estimating evapotranspiration by remote sensing. *Surveys in Geophysics* 31:531–555.

Glenn, E. P., Huete, A. R., Nagler, P. L., Nelson, S. G. 2008. Relationship between remotely sensed vegetation indices, canopy attributes and plant physiological processes: What vegetation indices can and cannot tell us about the landscape. *Sensors* 8:2136–2160.

Groeneveld, D. P., and Baugh, W. M. 2007. Correcting satellite data to detect vegetation signal for eco-hydrologic analyses. *Journal of Hydrology* 344:135–145.

Groeneveld, D. P., Baugh, W. M., Sanderson, J. S., and Cooper, D. J., 2007. Annual groundwater evapotranspiration mapped from single satellite scenes. *Journal of Hydrology* 344:146–156.

Grubbs, F. 1973. Errors of measurement, precision, accuracy and the statistical comparison of measuring instruments. *Technometrics* 15:53–66.

Healy, S. P., Cohen, W. B., Yang, Z. Q., and Krankina, O. N. 2005. Comparison of tasseled cap-based Landsat data structures for use in forest disturbance detection. *Remote Sensing of Environment* 97:301–310.

Huete, A., Didan, K., Miura, T., Rodriquez, E., Gao, X., and Ferreira, L. 2002. Overview of the radiometric and biophysical performance of the MODIS vegetation indices. *Remote Sensing of Environment* 83:195–213.

Hultine, K. R., Belnap, J., van Riper, C., Ehleringer, J. R., Dennison, P. E., Lee, M. E., Nagler, P. L., Snyder, K. A., Uselman, S. M., and West, J. B. 2009. Tamarisk biocontrol in the western United States: Ecological and societal implications. *Frontiers in Ecology and the Environment* 9:467–474.

Hultine, K. R., Nagler, P. L., Morino, K., Bush, S. E., Burtch, K. G., Dennison, P. E., Glenn, E. P., and Ehlringer, J. R. 2010. Sap flux-scaled transpiration by tamarisk (*Tamarix* spp.) before, during and after episodic defoliation by the saltcedar leaf beetle (*Diorhabda carinulata*). *Agricultural and Forest Meteorology* 150:1467–1475.

Irish, R. 1999. *Landsat 7 Science Data Users Handbook*. Landsat Project Science Office, Goddard Space Flight Center, http://lipwww.gsfc.nasa.gov/IAS/handbook.html.

Kalma, J. D., McVicar, T. R., and McCabe, M. F. 2008. Estimating land surface evaporation: A review of methods using remotely sensed surface temperature data. *Surveys in Geophysics* 29:421–469.

Kustas, W., and Anderson, M. 2009. Advances in thermal infrared remote sensing for land surface modeling. *Agricultural and Forest Meteorology* 149:2071–2081.

Leckie, D. G., Cloney, E., and Joyce, S. P. 2005. Automated detection and mapping of crown discolouration caused by jack pine budworm with 2.5 m resolution multispectral imagery. *International Journal of Applied Earth Observation and Geoinformation* 7:61–77.

Lewis, P. A., DeLoach, C. J., Knutson, A. E., and Tracy, J. L. 2003. Biology of *Diorhabda elongata deserticola* (Coleoptera: Chrysomelidae), an Asian leaf beetle for biological control of saltcedars (*Tamarix* spp.) in the United States. *Biological Control* 27:101–116.

Martin, M. E., Plourde, L. C., Ollinger, S. V., Smith, M. L., and McNeil, B. E. 2008. A generalizable method for remote sensing of canopy nitrogen across a wide range of forest ecosystems. *Remote Sensing of Environment* 112:3511–3519.

Moran, P. J., DeLoach, C. J., Dudley, T. L., and Sanabria, J. 2009. Open field host selection and behavior by tamarisk beetles (*Diorhabda* spp.) (Coleoptera: Chrysomelidae) in biological control of exotic saltcedars (*Tamarix* spp.) and risks to non-target athel (*T aphylla*) and native *Frankenia* spp. *Biological Control* 50:243–261.

Mulder, V. L., de Bruin, S., Schaepman, M. E., and Mayr, T. R. 2011. The use of remotes ensing in soil and terrain mapping—A review. *Geoderma* 62:1–19.

Nagler, P. L., Brown, T., Hultine, K. R., van Riper, C., Bean, D. W., Dennison, P. E., Murray, R. S., and Glenn, E. P. 2012. Regional scale impacts of *Tamarix* leaf beetles (*Diorhabda carinulata*) on the water availability of western U.S. rivers as determined by multi-scale remote sensing methods. *Remote Sensing of Environment* 118:227–240.

Nagler, P. L., Cleverly, J., Glenn, E., Lampkin, D., Huete, A., and Wan, Z. M. 2005a. Predicting riparian evapotranspiration from MODIS vegetation indices and meteorological data. *Remote Sensing of Environment* 94:17–30.

Nagler, P. L., Glenn, E. P., Didan, K., Osterberg, J., Jordan, F., and Cunningham, J. 2008. Wide-area estimates of stand structure and water use of *Tamarix* spp. on the Lower Colorado River: Implications for restoration and water management projects. *Restoration Ecology* 16:136–145.

Nagler, P. L., Glenn, E. P., and Huete, A. R. 2001. Assessment of spectral vegetation indices for riparian vegetation in the Colorado River delta, *Mexico. Journal of Arid Environments* 49:91–110.

Nagler, P. L., Glenn, E. P., Hursh, K., Curtis, C., and Huete, A. R. 2005b. Vegetation mapping for change detection on an arid-zone river. *Environmental Monitoring and Assessment* 109:255–274.

Nagler, P. L., Glenn, E. P., Jamevich, C. S., and Shafroth, P. B. 2010a. Distribution and abundance of saltcedar and Russian olive in the western United States. In P. B. Shafroth, C. A. Brown, and D. M. Merritt (Eds.), *Saltcedar and Russian Olive Control Demonstration Act Science Assessment.* U.S. Geological Survey Scientific Investigations Report 2009-5247. Washington, DC: USGS, pp. 7–32.

Nagler, P. L., Glenn, E. P., Thompson, T. L., and Huete, A. R. 2005d. Leaf area index and normalized difference vegetation index as predictors of canopy characteristics and light interception by riparian species on the Lower Colorado River. *Agricultural and Forest Meteorology* 125:1–17.

Nagler, P. L., Hinojosa-Huerta, O., Glenn, E. P., Garcia-Hernandez, J., Romo, R., Curtis, R., Huete, A. R. and Nelson, S. G. 2005c. Regeneration of native trees in the presence of invasive saltcedar in the Colorado River delta, *Mexico. Conservation Biology* 19:1842–1852.

Nagler, P. L., Morrino, K., Didan, K., Osterberg, J., Hultine, K., and Glenn, E. P. 2009a. Wide area estimates of saltcedar (*Tamarix* spp.) evapotranspiration on the lower Colorado River measured by heat balance and remote sensing methods. *Ecohydrology* 2:18–33.

Nagler P. L., Morino, K., Murray, R., Osterberg, J., and Glenn, E. P. 2009b. An empirical algorithm for estimating agricultural and riparian evapotranspiration using MODIS Enhanced Vegetation Index and ground ground measurements of ET. I. Descpription of method. *Remote Sensing* 1:1273–1297.

Nagler, P. L., Pearlstein, S., Glenn, E. P., Brown, T. B., Batement, H. L., Bean, D. W., and Hultine, K. R. 2014. Rapid dispersalt of saltcedar (*Tamarix* spp.) biocontrol beetles (*Diorhabda carinulata*) on a desert river detected by phenocams, MODIS imagery and ground obeservations. *Remote Sensing of Environment* 140:206–219.

Nagler, P., Scott, R., Westenberg, C., Cleverly. J., Glenn, E., and Huete, A. 2005e. Evapotranspiration on western U.S. rivers estimated using the Enhanced Vegetation Index from MODIS and data from eddy covariance and Bowen ratio flux towers. *Remote Sensing of Environment* 97:337–351.

Nagler, P. L., Shafroth, P. B., LaBaugh, J. W., Snyder, K. A., Scott, R. L., Merritt, D. M., and Osterberg, J. 2010b. The potential for water savings through the control of saltcedar and Russian olive. In P. B. Shafroth, C. A. Brown, and D. M. Merritt (Eds.), *Saltcedar and Russian Olive Control Demonstration Act Science Assessment.* U.S. Geological Survey Scientific Investigations Report 2009-5247. Washington, DC: USGS, pp. 33–49.

Neale, C. M. U., Bausch, W. C., and Heermann, D. F. 1989. Development of reflectance-based crop coefficients for corn. *Transactions of the ASAE* 32:1891–1899.

Ollinger, S.V. 2011. Sources of variability in canopy reflectance and the convergent properties of plants. *New Phytologist* 189:375–394.

O'Meara, S., Larsen, D., and Owens, C. 2010. Methods to control saltcedar and Russian olive. In P. B. Shafroth, C. A. Brown, and D. M. Merritt (Eds.), *Saltcedar and Russian Olive Control Demonstration Act Science Assessment*. U.S. Geological Survey Scientific Investigations Report 2009-5247. Washington, DC: USGS, pp. 65–102.

Patil, S., and Bohde, S. 2001. Image processing method to measure sugarcane leaf area. *International Journal of Engineering and Science Technology* 3:6393–6400.

Pattison, R. R., D'Antonio, C. M., Duley, T. L., Kip, K., and Allander, B. R. 2011. Early impacts of biological control on canopy cover and water use of the invasive saltcedar tree (*Tamarix* spp) in western Nevada, USA. *Oecologia* 165:605–616.

Pfeifer, M., Disney, M., Quaife, T., and Marchant, R. 2012. Terrestrial ecosystems from space: A review of earth observation products for macroecology applications. *Global Ecology and Biogeography* 21:603–624.

Reed, B. C., Brown, J. F., VanderZee, D., Loveland, T. R., Merchant, J. W., and Ohlen, D. O. 1994. Measuring phonological variability from satellite imagery. *Journal of Vegetation Science* 5:703–714.

Richardson, A. D., Jenkins, J. P., Braswell, B. H., Hollinger, D. Y., Ollinger, S. V., and Smith, L.-M. 2007. Use of digital webcam images to track spring green-up in a deciduous broadleaf forest. *Oecologia* 152:323–334.

Rundel, P. W., Graham, E. A., Allen, M. E., Fisher, J. C., and Harmon, T. C., 2009. Environmental sensor networks in ecological research. *New Phytologist* 182:589–607.

Seager R., Ting, M. F., Held, I. M., Kushnir, Y., Lu, J., Vecchi, G., Huang, H.-P., Harnik, N., Leetmaa, A., Lau, N.-C., Li, C., Velez, J., and Naik, N. 2007. Model projections of an imminent transition to a more arid climate in Southwestern North America. *Science* 316:1181–1184.

Skidmore, A. K., Franklin, J., and Dawson, T. P. 2011. Geospatial tools address emerging issues in spatial ecology: A review and commentary on the Special Issue. *International Journal of Geographical Information* 25:337–365.

Song, C., Woodcock, C. E., Seto, K. C., Lenney, M. P., and Macombe, S. A. 2001. Classification and change detection using Landsat TM data: When and how to correct atmospheric effects? *Remote Sensing of Environment* 75:230–244.

Sonnentag, O., Hulfkens, K., Teshera-Sterne, C., Young, A. M., Friedl, M., Braswell, B. H., Milliman, T., O'Keefe, J., and Richardson, A.D. 2012. Digital repeat photography for phonological research in forest ecosystems. *Agricultural and Forest Meteorology* 152:159–177.

Stevens, J. P. 1996. *Applied Multivariate Statistics for the Social Sciences, 3rd ed.* Mahway, NJ: Lawrence Erlbaum.

Stromberg, J. C., Chew, M. K., Nagler, P. L., and Glenn, E. P. 2009. Changing perceptions of change: The role of scientists in Tamarix and river management. *Restoration Ecology* 17:177–186.

Tan, B., Woodcock, C. E., Hu, J., Zhang, P., Ozdogan, M., Huang, D., Yang, W., Knyazikhin, Y., and Myneni, R. B. 2006. The impact of gridding artifacts on the local spatial properties of MODIS data: Implications for validation, compositing, and band-to-band registration across resolutions. *Remote Sensing of Environment* 105:98–114.

Tziziki, J., Mas, J., and Hinkley, E. A. 2012. Land cover mapping applications with MODIS: A literature review. *International Journal of Digital Earth*, 5:63–87.

Ustin, S. L., and Gamon, J. A. 2010. Remote sensing of plant functional types. *New Phytologist* 186:795–816.

van Gardingen, P. R., Foody, G. M., and Curran, P. J. (Eds.). 1997. *Scaling Up: From Cell to Landscape.* Cambridge, UK: Cambridge University Press.

Woodcock, C. E., and Strahler, A. H. 1987. The factor of scale in remote sensing. *Remote Sensing of Environment* 21:311–332.

Wulder, M. A., Dymond, C. C., White, J. C., Leckie, D. G., and Carroll, A. L. 2006. Surveying mountain pine beetle damage of forests: A review of remote sensing opportunities. *Forest Ecology and Management* 221:27–41.

Zavaleta, E. 2000. The economic value of controlling an invasive shrub. *Ambio* 29:462–467.

Zerger, A., Viscarra Rossel, R. A., Swain, D. L., Wark, T., Handcock, R. N., Doerr, V. A. J., Bishop-Hurley, G. J., Doerr, E. D., Givvons, P. G., and Lobsey, C. 2010. Enivronmental sensor networks for vegetation, animal and soil sciences. *International Journal of Applied Earth Observation and Geoinformation* 12:303–316.

6

UPSCALING WITH CONDITIONAL COSIMULATION FOR MAPPING ABOVE-GROUND FOREST CARBON

GUANGXING WANG AND MAOZHEN ZHANG

6.1 INTRODUCTION

Forest inventory sample plot data are often combined with remotely sensed images by regression modeling, neural networks, and K-nearest neighbors to map forest carbon, that is, generate spatially explicit estimates at a desirable spatial resolution (Lu et al., 2012; Wang et al., 2009). In these methods, forest carbon observations are available only at the sample plot locations, while remotely sensed data are available everywhere. Forest carbon at unobserved locations is interpolated by combining the sample plot data and remotely sensed images. The image data provide the linkage of forest carbon from the sample plot locations to the unobserved locations. Generally, the spatial resolutions of used sample plot data and images are consistent with the sizes of units of output maps. In practice, however, the sizes of forest inventory sample plots vary from $10\,\text{m} \times 10\,\text{m}$ to $50\,\text{m} \times 50\,\text{m}$ because of limitation of high cost to collect field data, while forest carbon maps at regional, national, and global scales are often required to have spatial resolutions that range from $90\,\text{m} \times 90\,\text{m}$ to $1\,\text{km} \times 1\,\text{km}$. Scaling up or aggregating the sample plot data from a finer spatial resolution to a coarser one has to be thus conducted and the existing methods lack the ability (Wang et al., 2009, 2011).

Upscaling of spatial data has been widely studied (Marceau, 1999; Marceau and Hay, 1999; Wang et al., 2009; Wu and Qi, 2000) and was first conducted in the field of socioeconomy. A typical example is that Gehlke and Biehl (1934) found out that

Scale Issues in Remote Sensing, First Edition. Edited by Qihao Weng.
© 2014 John Wiley & Sons, Inc. Published 2014 by John Wiley & Sons, Inc.

when spatial data were scaled up, coefficients of correlation between variables varied greatly depending on the number and size of area units. This implies that any results and conclusions from studies are dependent on the spatial resolutions that are used and can be described by the widely known modifiable area unit problem (MAUP) (Openshaw and Taylor, 1979). The MAUP means that different results can be obtained when data are aggregated from a finer spatial resolution to a coarser one. Given a spatial resolution to create the output maps, different upscaling methods can also lead to different results. Spatial patterns of forest carbon are related to the used spatial resolutions and upscaling methods to map it. The spatial patterns may show up in an optimal spatial resolution but disappear in others (Wang et al., 2008). Given the optimal spatial resolution, on the other hand, the spatial patterns may be captured by one upscaling method but missed by others. Thus, developing an accurate upscaling method for multiple-resolution analysis and spatial data aggregation is becoming very important to detect the spatial patterns of forest carbon, its dynamics over time, and its relationships with spatial resolutions (Marceau, 1999; Marceau and Hay, 1999; Pontius and Cheuk, 2006; Wang et al., 2009; Wu and Qi, 2000).

There are several widely used methods to scale up spatial data of a variable across scales (Jarvis, 1995; Marceau and Hay, 1999; Moellering and Tobler, 1972; Wu, 1999). The simplest method is the nearest neighbor. If a pixel at a coarser spatial resolution is regarded as a block that consists of smaller pixels, the nearest-neighbor method assigns the block with the value of the smaller pixel close to the block center. Obviously, this method misses the dominant feature determined by a distribution of values of smaller pixels within a block and may lead to the disappearance of the dominant pattern and process.

Window or block averaging is also a simple and widely used upscaling method. In this method, a mean value from the smaller pixels within the window or block is calculated and assigned to this block. This method is accurate and straightforward to aggregate spatial data from a finer spatial resolution to a coarser one. But it is assumed that the values of pixels have a normal distribution and linear relationships of spatial features exist when the spatial data are scaled up from a finer to a coarser resolution. In fact, the window averaging is a specific case of window filtering, that is, when the weight used for all the pixels within the window is the same. On the other hand, a different weight for each pixel can be used, which leads to various window-weighted averaging methods.

Wang et al. (2004a) pointed out that the key to accurately infer spatial information from a finer spatial resolution to a coarser one is to capture dominant spatial features, patterns, and processes of an interest variable. If equal weights are given to each pixel within a window, the window averaging method neglects the differences in the spatial autocorrelation between the pixel values. If normal distribution of data does not hold, a window average will lead to a misinterpretation of dominant spatial features. A typical example is that few extremely large values within a window will impede the window averaging method to capture the dominant feature. In addition, in this method it is assumed that the values of smaller pixels within each block are available. Combining forest inventory sample plot data and remotely sensed images to map forest carbon is not the case.

King (1991) presented an upscaling method in which it is assumed interest variables will be random variables, their outputs will vary spatially across a landscape, and the spatial heterogeneity of the landscape will be defined by a joint probability distribution of the variables. The expected values of the variables at any locations are then estimated by Monte Carlo simulation and are weighted with corresponding areas. This method simulates window means as realizations of random variables from a joint probability distribution and provides the potential to capture dominant features within windows.

Hay et al. (1997) developed an object-specific resampling method. In this method, it is assumed that the closer the locations, the more similar the data. Furthermore, each ground object can be represented with a corresponding image object and consists of pixels that have similar spectral features. When the values of pixels are aggregated using different window sizes such as 3×3 pixels, 5×5 pixels, 7×7 pixels, and so on, window variances can be calculated. As the window size increases, the window variance varies and a distinct threshold of area in the variance of pixel values will thus be observed. If the spatial resolution is fine enough, the area within the neighborhood of the threshold can then be determined for each pixel and used to determine the pixel's weight to calculate the window value. This method is similar to the window averaging mentioned previously and cannot be used to aggregate the sampling plot data that are not available everywhere.

Wang et al. (2004a) developed and assessed five upscaling methods in a study for aggregating and using Landsat Thematic Mapper (TM) images for mapping vegetation covers and inferring a topographic factor related to soil erosion from finer to coarser spatial resolutions. These methods included three spatial variability-weighted methods and two simulation methods. The idea behind the methods is that the spatial autocorrelation of variables measured as the square difference of two data values that are separated by a distance h can be used to derive weights of pixels within a window. The larger the square difference, the less the weight for this pixel is. The weight is thus defined as the reciprocal of the square difference. Based on the different definitions of the square difference, Wang et al. (2004a) obtained (i) the center-pixel variability-weighted method in which the square difference is quantified based on the differences of spectral values between the central pixel and each pixel within the window; (ii) the dominant-class variability-weighted method in which the square difference between the window dominant-class value and each pixel value is used; and (iii) the arithmetic-average variability-weighted method in which the square difference between the window arithmetic average and each pixel value is defined.

Moreover, Wang et al. (2004a) assumed a window dominant feature be the expectation of a random variable and be determined by a set of realizations by randomly drawing from its distribution at the location. The local distribution can be obtained by calculating a conditional mean and a conditional variance of the pixel values within the window or neighborhood. In addition, the spatial autocorrelation between the pixel values can also be incorporated into the conditional variance. From the obtained distribution, a value can be randomly drawn and used as a realization of the window average. This process is repeated many times, which results in more than one value. These values are used to calculate a window mean. The obtained distribution varies and can be modeled using a beta or normal probability density function (Hastings and Peacock, 1975), which leads to two upscaling methods called the beta distribution simulation and the normal distribution simulation.

Using the upscaling methods, Wang et al. (2004a) scaled up Landsat TM images from a pixel size of 30 m × 30 m to a coarser spatial resolution of 90 m × 90 m in which field observations of the interest variables were available. The aggregated images were then employed to map vegetation cover percentage that had a normal distribution and infer a soil erosion relevant topographic factor that was characterized by a reversed J-shape distribution. The results showed that the beta distribution simulation method was the best regardless of the distributions of the spatial data. The reason is mainly because the beta distribution simulation method has the strong ability to accommodate the variable distribution.

All the upscaling methods mentioned previously can be applied to aggregate spatial data of a map or image from a finer spatial resolution to a coarser one when spatial data are available everywhere. However, when forest carbon is mapped by combining forest inventory sample plot data and remotely sensed image, the plot data only exist at the sample locations and cannot be directly aggregated by simply calculating a window mean. First, field observations of interest variables are available only at the sample plots that have finer spatial resolutions. Second, few or no plot data may exist within a window that corresponds to the desirable coarser spatial resolution. Third, the plot data have to be combined with remotely sensed images in order to generate spatially explicit estimates of forest carbon at the unobserved locations, but the relationships of the plot data with the used images may differ at the finer and coarser spatial resolution, respectively (Simic et al., 2004; Wang et al., 2004b). Moreover, when spatial data are scaled from a finer spatial resolution to a coarser one, the uncertainties of the input data are propagated to the aggregated data. However, the existing methods neglect modeling the propagation of the uncertainty to the coarser spatial resolution. In addition, these methods also discard spatial dependency of neighboring sample locations and image pixels.

Block cokriging in geostatistics can be used to generate spatially explicit estimates of an interest variable at a coarser spatial resolution by combining sample plot data and remotely sensed images at a finer spatial resolution. Using this method, Atkinson and Kelly (1997) aggregated point snow depth data to larger blocks and at the same time obtained the variances of the block estimates as a measure of uncertainty. This implies that cokriging, to some extent, can meet some of the aforementioned challenges. However, the cokriging variances vary depending on only the spatial configuration of spatial data, not on the data values themselves. That is, the variances do not really reflect the uncertainties of block estimates.

Wang et al. (2004b) developed and compared two spatial variability based algorithms for scaling up vegetation cover percentage from a spatial resolution of 30 m × 30 m to a pixel size of 90 m × 90 m in the aid of TM images as secondary variables. Both methods are based on a collocated simple cokriging estimator and a sequential Gaussian cosimulation algorithm and take spatial autocorrelation of variables into account in the upscaling process of spatial data. This makes it possible to simultaneously and accurately obtain estimates and estimation variances of larger blocks from the sample and image data of finer spatial resolution. In both methods, it is assumed that an interest variable is a random process, its values at any locations are realizations of the variable, and they can be obtained by calculating conditional distributions and generating random numbers from the distributions. The first method developed by Wang et al. (2004b) creates the conditional distributions and estimates of the variable at a finer spatial resolution that is consistent with those of the used sample plot and image data, and the estimates at a coarser spatial resolution are then calculated using the window averaging method. The second method by Wang et al. (2004b) directly generates conditional distributions and estimates of this variable at blocks, a coarser spatial resolution, using the sample plot and image data at a finer spatial resolution. The former is called point simple cokriging point cosimulation (PSCPS), while the latter is called point simple cokriging block cosimulation (PSCBS).

Both simulation-based upscaling methods overcome the gaps mentioned previously. Wang et al. (2004b) pointed out that the methods worked very well for multiple continuous variables that had any distribution and how to choose them depends on the users' emphasis on accuracy of estimates and variances and computational time. The objective of this study is to further assess and compare these two methods to create maps of above-ground forest carbon at a spatial resolution of 990 m × 990 m using forest inventory sample plot data and TM images at a spatial resolution of 30 m × 30 m. This selected coarser spatial resolution is desirable partly because it is close to 1 km × 1 km, a widely used map unit size for mapping forest carbon at regional, national, and global scales, and partly because it is a window of 33 × 33 pixels of the original spatial resolution 30 m × 30 m and can be easily obtained. The above-ground forest carbon is the carbon equivalent of the above-ground tree biomass, including standing trees and deadwoods, but not stumps and ground vegetation. It is expected that this study can provide some suggestions and guidelines on how to use national forest inventory plot data collected at smaller permanent sample plots to generate forest carbon maps at any coarser spatial resolutions for large regions, nations, and even the whole world.

6.2   METHODS

Wang et al. (2004b) introduced both PSCPS and PSCBS upscaling methods in detail. In this study, they were briefly described. The first part of PSCPS deals with the generation of forest carbon pixel values using a sequential Gaussian cosimulation algorithm in which the sample plot data and remotely sensed images have consistent spatial resolution with the output forest carbon maps. Because of this, the pixels to be predicted are regarded as points. First of all, the cosimulation divides a study area into square cells or pixels and follows a random path to visit the pixels. The random path is determined using a random-number generator. At each pixel, a neighborhood is set up based on the maximum distance of the spatial autocorrelation of above-ground forest carbon that is modeled by a variogram model. A collocated simple cokriging unbiased estimator is employed to create a conditional mean and a conditional variance of the forest carbon by weighting the sample plot data, collocated image data, and previous estimates if any within the given neighborhood. The weights vary depending on the spatial variability determined by the spatial configuration of the data (Goovaerts, 1997). The point collocated simple cokriging estimator and cokriging variance are

$$z^{\text{sck}}(u) = \sum_{\alpha=1}^{n(u)} \lambda_\alpha^{\text{sck}}(u)[z(u_\alpha) - m_z] + \lambda_y^{\text{sck}}(u)[y(u) - m_y] + m_z \qquad (6.1)$$

$$\sigma^{2(\text{sck})}(u) = C_{zz}(0) - \sum_{\alpha=1}^{n(u)} \lambda_\alpha^{\text{sck}}(u)C_{zz}(u_\alpha - u) - \lambda_y^{\text{sck}}(u)C_{zy}(0) \qquad (6.2)$$

where $z^{\text{sck}}(u)$ is the estimate of a pixel, $z(u_\alpha)$ is a plot above-ground forest carbon observation, and $y(u)$ is the data of a spectral variable at a pixel u whose value is estimated; m_z and m_y are the means of sample plot and image data; and $\lambda_\alpha^{\text{sck}}(u)$ and $\lambda_y^{\text{sck}}(u)$ are the weights for the sample plot data and the collocated image data, respectively. The number of the used sample plot data within the neighborhood, $n(u)$, varies from location to location; $C_{zz}(0)$ is the variance of the sample plot data for the forest carbon and $C_{zy}(0)$ is the covariance between the forest carbon and spectral variable; and $C_{zz}(u_\alpha - u)$ is the spatial covariance function of the forest carbon and $C_{zy}(0)$ is the cross covariance between the forest carbon and the spectral variable.

The conditional mean and conditional variance obtained are then used to determine a condition probability distribution and, from it, a value is randomly drawn and considered as a realization of the forest carbon at this location. This value is also regarded as conditional data for the next simulations. Follow the random path and predict each pixel. Once all the pixels are predicted, a map of above-ground forest carbon is obtained. The above process can be repeated many times by setting up different random paths to visit the pixels and many predicted values can thus be created for each pixel. From the predicted values, a sample mean and a sample variance can be calculated and used as the predicted value and its uncertainty measure for each pixel. How many simulations should be run depends on the global variation of above-ground forest carbon and can be determined through use of different numbers of runs and by plotting the global variances of the predicted values against the number of simulations. As the number of runs increases, generally, the global variance decreases rapidly at the beginning and then slowly and eventually stabilizes. The number of runs when the global variance starts to get stable is considered reasonable.

The above procedure results in predicted values of above-ground forest carbon at a spatial resolution that is the same as those of the used sample plot and remotely sensed data. The second part of the PSCPS is to create estimates of the forest carbon and their variances for blocks, the desirable and coarser spatial resolution, from the above predicted values at the finer spatial resolution using the window averaging mentioned previously. Obviously, this upscaling method produces not only the aggregated estimates but also their variances as uncertainty measures at the coarser spatial resolution. The block variances vary spatially depending on not only the spatial configuration of the used data but also the data values themselves and reveal the spatial uncertainties of the obtained estimates. This algorithm requires normal score transformations of all the sample plot and image data in order to meet the assumptions of the multivariate Gaussian distribution.

The procedure of PSCBS is similar to that of PSCPS described above. The difference lies in that PSCPS first creates estimates at the same spatial resolution as the input data and then scales up the estimates to any coarser spatial resolution, meaning indirect upscaling of spatial data, while PSCBS directly scales up the spatial data from a finer spatial resolution to a coarser one. In PSCBS, the map units at the desirable and coarser spatial resolution are considered as a block and each block consists of smaller pixels that have the same size as the sample plots and image pixels. Following a simulation procedure similar to the above, the predicted value and its variance of each smaller pixel within a block are obtained using the collocated simple cokriging estimator. Using the block mean and block variance by averaging the estimates and cokriging variances of the pixels within the block, a conditional distribution of above-ground forest carbon for the block is determined. The block variance is calculated by modeling the propagation of both cokriging variances of smaller pixels and cova-riances among them. From the distribution for the block, a value is then randomly drawn and considered to be a realization of above-ground forest carbon at this block. With this method, cokriging is conducted on the basis of smaller pixels and the simulation is made on the basis of larger blocks. The spatial data and their uncertainties are directly scaled up. The block cosimulation can also be run many times, resulting in many predicted values for each block. From the predicted values, a sample average and a sample variance for each block are then obtained. This method provides the potential to improve the block conditional distribution by modeling the propagation of uncertainty from the sample plot and image data at the finer spatial resolution and the estimates of smaller pixels to the block and thus improve the quality of the block estimates and their variances.

6.3   STUDY AREA AND DATA SETS

This study was conducted in Lin-An County, ZheJiang Province of East China (Wang et al., 2011) (Figure 6.1*b*) . Its latitude and longitude range from 29°56′ N to 30°23′ N and from 118°51′ E to 119°52′ E, respectively. It is characterized by a typical subtropical climate with average annual temperature and precipitation of 16.4 °C and 1628 mm, respectively. This study area consists of 312,680 ha and has features of a mountainous area with an elevation range of 1770 m. The main forest types include Chinese fir and *Pinus massoniana* plantations and evergreen broad-leaf forests, deciduous and evergreen broad-leaf mixed forests, bamboo, and shrubs (Wang et al., 2011; Zhang et al., 2009).

Since the 1950s continuous forest inventory has been conducted in Lin-An County [Chinese Ministry of Forestry (CMF), 1996; Wang et al., 2011]. A total of 50 national permanent sample plots were established and repeatedly measured every five years.

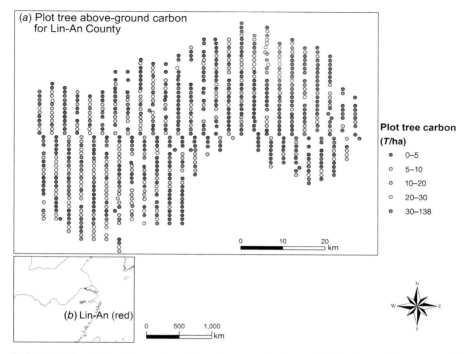

FIGURE 6.1 (*a*) Locations of forest inventory sample plots and their values of above-ground forest carbon and (*b*) location of study area. (See the color version of this figure in Color Plates section.)

In 2004, an additional more than 918 temporary forest inventory sample plots were added by the Lin-An County government. Thus, a total of 968 sample plots were available. Lin-An County was covered by two scenes of Landsat TM images. In order to simplify this study and reduce the errors due to mosaicing two images, only 776 sample plots that fell in the right one of the TM image scenes and measured in 2004 were used in this study (Figure 6.1*a*).

The permanent and temporary plots had the same size of 28.3 m × 28.3 m and the same methods were used to measure tree diameters at breast height (1.3 m) and tree heights. Based on the measurements of diameters and heights, the volumes of trees within each plot were calculated using the empirical regression models developed by the national forest inventory program for various species in this area (CMF, 1996). The tree volumes were then converted to the values of biomass by species or species groups using biomass conversion coefficients from the literature (Fang, 1999; Zhang, 1999; Fang and Wu, 2006; Hu, 2007; Ma et al., 2007; Zhang and Wang 2008). The values of tree biomass were summed to obtain plot biomass that was further converted to above-ground forest carbon pools (trunk, branch, and leaf) using a biomass-to-carbon coefficient of 0.5. The biomass conversion coefficients of the main tree species in this region varied from 0.3067 to 0.6391 g/cm^3. They were 0.444 g/cm^3 for Chinese fir, 0.447 g/cm^3 for *P. massoniana*, 0.387 g/cm^3 for softwood broad-leaf trees, and 0.447 g/cm^3 for hardwood broad-leaf trees. The values of plot above-ground forest carbon (tons per hectare, T/ha) are shown in Figure 6.1a.

A Landsat Enhanced TM Plus (ETM+) image dated August 3, 2004, was acquired and it covered 80% of Lin-An County (Wang et al., 2011). The image consisted of bands 1–5 and 7 that had a spatial resolution of 30 m × 30 m and was georeferenced to the Universal Transverse Mercator (UTM) coordinate system based on a digital topographic map and with a root mean-square error less than one pixel. The radiometric correction of the image was neglected because of good atmospheric condition when the image was acquired. In order to increase the Pearson product moment correlation coefficients of the images with above-ground forest carbon, various image transformations were conducted, including 6 inversions of original bands—1/TM1, 1/TM2, 1/TM3, 1/TM4, 1/TM5, 1/TM7—and 10 band or band group ratios—TM1/TM4, TM2/TM4, TM3/TM4, TM5/TM4, TM7/TM4, TM8/TM4, (TM4-TM3)/(TM4+TM3), (TM3+TM5)/TM7, (TM3+TM5+TM7)/TM4, and (TM2+TM3+TM5)/TM7. The significance of the coefficients was tested using the equation

$$r_\alpha = \frac{t_{n-2,1-\alpha/2}}{\sqrt{(n-2)+t^2_{n-2,1-\alpha/2}}}$$

(6.3)

where n is the number of sample plots and α (=5%) is the significance level.

6.4 RESULTS

The Pearson product moment correlation coefficients of plot above-ground forest carbon with 6 TM bands and 16 transformations ranged from −0.3359 to 0.3521 (Table 6.1). At a significance level of 0.05 and with $n = 776$ plots, all the coefficients except for that from TM1/TM4 were significantly different from zero. The

TABLE 6.1 Pearson Product Moment Correlation Coefficients of Plot Above-Ground Forest Carbon with Landsat TM Images and Their Transformations[a]

Variable	TM1	TM2	TM3	TM4	TM5	TM7
Correlation	−0.29134	−0.32703	−0.33586	−0.1575	−0.27852	−0.29767
Variable	1/TM1	1/TM2	1/TM3	1/TM4	1/TM5	1/TM7
Correlation	0.291732	0.331884	0.352074	0.115122	0.224013	0.29166
Variable	TM1/TM4	TM2/TM4	TM3/TM4	TM5/TM4	TM7/TM4	NDVI
Correlation	−0.00439	−0.10549	−0.21596	−0.26576	−0.26931	0.223298
Variable	BGR1	BGR2	BGR3	BGR4		
Correlation	0.208139	−0.25516	0.171384	−0.16705		

[a]Including inversions of original bands, band ratios, and band group ratios (BGRs): (NDVI = (TM4-TM3)/(TM4+TM3), BGR1 = (TM2+TM3+TM5)/TM7, BGR2 = (TM2+TM3+TM5)/TM4, BGR3 = (TM2+TM)/TM7, and BGR4 = (TM2+TM3)/TM4

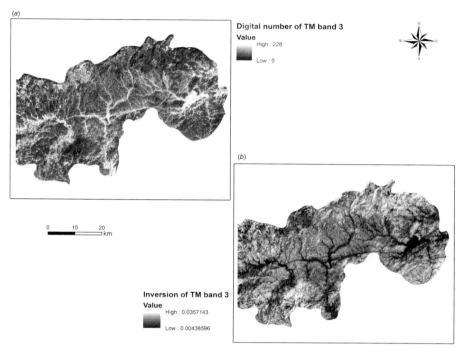

FIGURE 6.2 (*a*) Landsat TM band 3 and (*b*) its inversion, which had highest correlation with above-ground forest carbon.

complicated transformations obtained by band group ratios did not improve the correlation with above-ground forest carbon compared to the original TM bands and their simple band ratios. The TM band 3 and its inversion were most highly correlated with above-ground forest carbon with correlation coefficients of -0.3359 and 0.3521, respectively. That is, in the areas with higher values of above-ground forest carbon, there was more light absorbed and less light reflected in the red channel (Figure 6.2). The above-ground forest carbon thus had negative correlation with the red channel and positive correlation with its inversion. The inversion of TM band 3 was used in the cosimulation to generate above-ground forest carbon maps.

Various spatial autocorrelation models were used and compared to fit the sample variogram based on goodness of fit, a measure for quantifying how well the model fits the sample data based on the sum of squares due to error. The closer to zero the goodness of fit, the better the fit was. It was found that the spherical model fit the trend of spatial autocorrelation best (Figures 6.3a, *b*). When the original data set was used, the obtained model (Figure 6.3*a*) was

$$\gamma(h) = 229.4 + 65.1 \left[1.5\frac{h}{4} - 0.5\left(\frac{h}{4}\right)^3 \right] \tag{6.4}$$

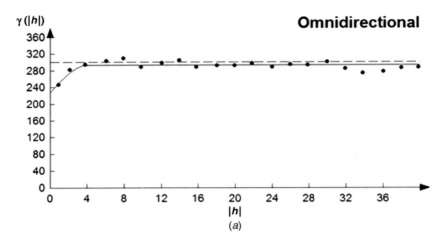

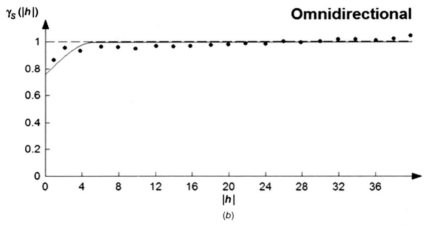

FIGURE 6.3 Spatial autocorrelation $\gamma(h)$ of above-ground forest carbon modeled using spherical model for (a) original data and (b) standardized data, where h is distance in km.

Its goodness of fit was 0.00164. In this study, furthermore, the two cosimulation procedures used required that the parameters of the model be standardized so that the sill value, that is, nugget plus structure parameter, equaled one unit (Figure 6.3b):

$$\gamma(h) = 0.76 + 0.24 \left[1.5 \frac{h}{5.2} - 0.5 \left(\frac{h}{5.2} \right)^3 \right] \qquad (6.5)$$

Its goodness of fit was 0.0032. The nugget and structure parameters were 0.76 and 0.24, respectively. These relatively large nugget parameters meant that in this study the spatial autocorrelation of above-ground forest carbon was relatively weak.

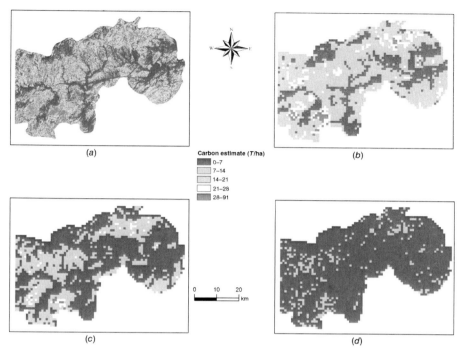

FIGURE 6.4 (*a*) Estimation map of above-ground forest carbon at spatial resolution (SR) of 30 m × 30 m using sequential Gaussian cosimulation; (*b*) above-ground forest carbon map at spatial resolution of 990 × 990 m by scaling up sample plot and image data from spatial resolution of 30 m × 30 m using PSCPS upscaling method; (*c*) above-ground forest carbon map at spatial resolution of 990 × 990 m by scaling up sample plot and image data from spatial resolution of 30 m × 30 m using PSCBS; and (*d*) difference map of forest carbon estimates between two methods. (See the color version of this figure in Color Plates section.)

The reason might be due to large sampling distances (1–3 km) of the used permanent and temporary plots.

Figure 6.4*a* shows the estimation map of above-ground forest carbon using sequential Gaussian cosimulation, that is, the first part of PSCPS. The pixel size of this map was 30 m × 30 m, similar to the sample plots and TM image used. The above-ground forest carbon had larger predicted values in the northeast, northwest, and southwest parts of the study area. The spatial distributions of the predicted values were similar to those of the plot above-ground forest carbon values (Figure 6.1*a*). The predicted values at the pixel size of 30 m × 30 m were then scaled up to a spatial resolution of 990 m × 990 m using window averaging (Figure 6.4*b*). The spatial distribution of the aggregated forest carbon map was similar to that before the up-scaling. That is, the upscaling captured the spatial patterns of above-ground forest carbon.

In Figure 6.4*c*, the map of the aggregated above-ground forest carbon values at the pixel size of 990 m × 990 m was directly obtained using the PSCBS upscaling method. The spatial distributions of the block predicted values were similar to those obtained using the PSCPS upscaling method (Figure 6.4b). But, all the differences of the aggregated block predicted values by PSCBS from those by PSCPS were positive, implying the PSCBS smoothed the block values more than the PSCPS, although differences that ranged from 0 to 12.7 tons/ha with a mean of 4.9 tons/ha were relatively small (Figure 6.4d). Because of the lack of field observations at the coarser spatial resolution, this study was not able to verify which upscaling method led to smaller root mean-square error for the aggregated block values.

Moreover, both upscaling methods PSCPS and PSCBS directly output the variances of predicted values at the coarser spatial resolution of 990 m × 990 m (Figures 6.5a,b). The spatial distributions of the block variances were similar to that of the block predicted values (Figure 6.4c). That is, in the areas where the block predicted values were greater, the variances of the block predicted values were also greater. However, PSCBS (Figure 6.5b) resulted in smaller variances of the block values compared to PSCPS (Figure 6.5a). Using the block predicted values and variances (Figure 6.4b vs. Figure 6.5a, Figure 6.4c vs. Figure 6.5b), the coefficients of variation for the block predicted values were calculated in Figure 6.5c for PSCPS and Figure 6.5d for PSCBS. Both upscaling methods led to similar spatial distributions of variation coefficients of the block values. Compared to the spatial distributions of the block predicted values and variances, the spatial patterns of the variation coefficients showed up in an opposite way. In the areas where the block estimates and variances were larger, the coefficients of variation were smaller and vice versa. In addition, these upscaling methods also produced the probability values for the block values larger than a given threshold value. In Figures 6.5e for PSCPS and Figure 6.5f for PSCBS, as examples, the maps of probability values for the block predicted values larger than the sample mean were presented. In the areas where there were larger block predicted values, the probability values for the predicted values larger than the sample mean were also higher.

6.5   CONCLUSIONS AND DISCUSSION

Combining sample plot data and remotely sensed images has been widely used to map forest carbon stocks through spatial interpolation methods, including regression modeling, neural network, and K-nearest neighbors. However, these methods lack the ability to directly aggregate the spatial data and their uncertainties from a finer spatial resolution to a coarser one when the sizes of forest inventory sample plots (i.e., 30 m × 30 m) are inconsistent with the sizes of map units (i.e., 1 km × 1 km) that are often required to map forest carbon stocks at regional, national, and global scales (Wang et al., 2004b, 2009, 2011). More important is that these methods neglect the spatial autocorrelation of variables and the propagation of input spatial uncertainties across scales. Although various data aggregation methods have been developed, none of them can be directly used to scale up the sample plot data and remotely sensed images from a finer spatial resolution to a coarser one because the plot data are available only at the sampled locations.

In this study, two upscaling methods developed by Wang et al. (2004b), PSCPS and PSCBS, were demonstrated to map above-ground forest carbon stocks in Lin-An

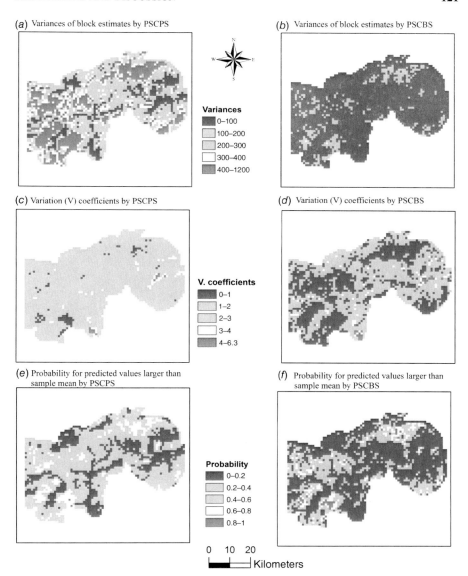

FIGURE 6.5 Variances of block predicted values for above-ground forest carbon at spatial resolution of 990 m × 990 m by (*a*) PSCPS and (*b*) PSCBS, respectively; their coefficients of variation by (*c*) PSCPS and (*d*) PSCBS, respectively; and probability maps for block predicted values larger than a given threshold—sample mean by (*e*) PSCPS and (*f*) PSCBS, respectively. (See the color version of this figure in Color Plates section.)

County, Zhejiang of East China. Both PSCPS and PSCBS are based on sequential Gaussian cosimulation and take spatial variability of variables into account in the upscaling process, which makes it possible to simultaneously and accurately obtain estimates and estimation variances at any coarser spatial resolutions from the sample plot and image data at finer spatial resolutions. The results showed both methods accurately captured the spatial distributions and patterns of above-ground forest carbon at a spatial resolution of 990 m × 990 m by combining and scaling up the forest inventory sample plot data at a spatial resolution of 28.3 m × 28.3 m and TM images at a pixel size of 30 m × 30 m. Moreover, both upscaling methods output the spatial uncertainties of the block predicted values measured as estimation variances and probabilities for the block predicted values larger than a given threshold. This implied both PSCPS and PSCBS not only scaled up the spatial data but also modeled the propagation of input spatial uncertainties from spatial resolutions of 28.3 m × 28.3 m and 30 m × 30 m to a spatial resolution of 990 m × 990 m. These uncertainties varied depending not only on the configuration of the sample locations but also on the sample plot and image data values themselves. Thus, the block variances and probabilities reflect well the spatial uncertainties of the input spatial data and their spatial configuration.

In this study, a root mean-square error (RMSE) of 12.9 tons/ha for the estimates of above-ground forest carbon at a spatial resolution of 30 m × 30 m was obtained for the PSCPS method. But, because of a lack of field observation of above-ground forest carbon at a spatial resolution of 990 m × 990 m, the RMSE of the aggregated above-ground forest carbon estimates were not obtained. That is, this study did not provide users with guidelines on how to choose these two methods for upscaling of spatial data based on RMSE. But the results did show that the PSCBS produced more smoothed block estimates and their variances than the PSCPS. This finding differed from that obtained in the Wang et al. (2004b) study in which the PSCPS led to more smoothing of the block estimates. The reason might be that in Wang et al. (2004) the spatial data were only scaled up from a spatial resolution of 30 m × 30 m to 90 m × 90 m, while in this study the spatial data were aggregated from a spatial resolution of 30 m × 30 m to 990 m × 990 m. The second reason might be that in PSCPS the conditional distribution of above-ground forest carbon was determined and the simulation was conducted at the pixel size of 30 m × 30 m, while in PSCBS these were done at the pixel size of 990 m × 990 m. Another reason might be that Wang et al. (2004b) used sample plots that had much smaller sampling distances compared to those in this study.

Instead of using field observations, an alternative to assess the accuracy of above-ground forest carbon estimates at a coarse spatial resolution such as 1 km × 1 km could be the use and upscaling of the estimates as references that are more accurate and obtained by high-spatial-resolution remotely sensed data. In fact, comparison of the PSCPS and PSCBS methods used in this study implied this idea. That is, PSCPS first predicted the values of above-ground forest carbon at a spatial resolution of 30 m × 30 m and then scaled up the predicted values to a spatial resolution of 990 m × 990 m blocks. The block estimates were finally used to assess the accuracy of the predicted values at the pixel size of 990 m × 990 m by the PSCBS. The finer spatial resolution images used were from Landsat TM. A better choice for such higher spatial resolution remotely sensed data could be lidar. However, the cost to acquire lidar data for a large study area is very high and the use of lidar thus is limited in many cases, especially in developing countries.

Moreover, these two methods are limited to mapping and scaling up of spatial data for continuous variables and cannot be used to mapping and aggregation of spatial data for categorical variables. The potential improvement thus lies at developing image-based indicator cosimulation. In addition, these methods require the normal score transformation of spatial data to meet the assumption of the multivariate Gaussian distribution. In practice, however, there is no effective way to ensure that spatial data have multivariate Gaussian distribution. Therefore, developing a new method that does not require multivariate Gaussian distribution is necessary.

6.6 SUMMARY

By combining plot data and TM images at 30 m spatial resolution, this study compared two upscaling methods—point simple cokriging point cosimulation and point simple cokriging block cosimulation—to map above-ground forest carbon at 990 m pixel size in Lin-An County in China. The results showed both methods not only scaled up the spatial data but also modeled the propagation of input uncertainties from a finer spatial resolution to a coarser one. The output uncertainties reflected the spatial variability of the estimation accuracy due to the spatial configuration of the input data locations and their values.

ACKNOWLEDGMENT

This research was partly funded by the National Natural Science Foundation of China (No. 30972360).

REFERENCES

Atkinson, P. M., and Kelly R. E. 1997. Scaling up point snow depth data in the U. K. for comparison with SSM/I imagery. *International Journal of Remote Sensing* 18:437–443.

Chinese Ministry of Forestry (CMF), Department of Forest Resource and Management. 1996. *Forest Resources of China 1949 to 1993*. Liu-hai-hu-tong 7, Beijing: Chinese Forestry Press.

Fang, W., and Wu, Y. 2006. A study on the rate of hardwood tissue. *Journal of Fujian College of Forestry* 26(3):224–228.

Fang, Y. 1999. Variation of wood gravity for *Pinus Massoniana* plantations. *Forestry Science and Technology* 6:37.

Gehlke, C. E., and Biehl, K. 1934. Certain effects of grouping upon the size of the correlation coefficient in census tract material. *Journal of the American Statistical Association Supplement* 29:169–170.

Goovaerts, P. 1997. *Geostatistics for Natural Resources Evaluation*. New York: Oxford University Press.

Hastings, N. A. and Peacock, J. B. 1975, *Statistical distributions*. London. UK: Butterworth & Co (Publishers) Ltd.

Hay, G. J., Niemann, K. O., and Goodenough, D. G. 1997. Spatial thresholds, image-objects, and up-scaling: A multi-scale evaluation. *Remote Sensing of Environment* 62:1–19.

Hu, J. 2007. The study on the clone selection and assessment for the seed orchard of *Pinus massoniana. Journal of Fujian Forestry Science and Technology* 34(2):32–35.

Jarvis, P. G. 1995. Scaling process and problems. *Plant, Cell and Environment* 18:1079–1089.

King, A. 1991. Translating models cross scales in the landscape. In M. G. Turner and R. H. Gardner (Eds.) *Quantitative Methods in Landscape Ecology*. New York: Springer.

Lu, D., Chen, Q., Wang, G., Moran, E., Batistella, M., Zhang, M., Laurin, G. V., and Saah, D. 2012. Estimation and uncertainty analysis of above-ground forest biomass with Landsat and LiDAR data: Brief overview and case studies. *International Journal of Forestry Research* 1:1–16.

Ma, Z., Liu, Q., Xu, W., Li, X., and Liu, Y. 2007. Carbon storage of artificial forest in Qianyanzhou, Jiangxi Province. *Scientia Silvae Sinicae* 43(11):1–7.

Marceau, D. J. 1999. The scale issue in social and natural sciences. *Canadian Journal of Remote Sensing* 25:347–356.

Marceau, D. J., and Hay, G. J. 1999. Remote sensing contributions to the scale issue. *Canadian Journal of Remote Sensing* 25:357–366.

Moellering, H., and Tobler, W. 1972. Geographical variances. *Geographical Analysis* 4:34–64.

Openshaw, S., and Taylor, P. 1979. A million or so correlation coefficients: Three experiments on the modifiable area unit problems. In N. Wrigley (Ed.) *Statistical Applications in the Spatial Science*. London: Pion.

Pontius, Jr. R. G., and Cheuk. M. L. 2006. A generalized cross-tabulation matrix to compare soft-classified maps at multiple resolutions. *International Journal of Geographical Information Science* 20(1):1–30.

Simic, A., Chen, J. M., Liu, J., and Acillag, F. 2004. Spatial scaling of net primary productivity using subpixel information. *Remote Sensing of Environment* 93:246–258.

Wang, G., Grtner, G. Z., and Anderson, A. B. 2004a. Up-scaling methods based on variability-weighted and simulation for inferring spatial information cross scales. *International Journal of Remote Sensing* 25(22):4961–4979.

Wang, G., Grtner, G. Z., and Anderson, A. B. 2004b. Spatial variability based algorithms for scaling up spatial data and uncertainties. *IEEE Transactions on Geoscience and Remote Sensing* 42(9):2004–2015.

Wang, G., Grtner, G. Z., Howard, H. R., and Anderson, A. B. 2008. Optimal spatial resolution for collection of ground data and multi-sensor image mapping of a soil erosion cover factor. *Journal of Environmental Management* 88:1088–1098.

Wang, G., Oyana, T., Zhang, M., Adu-Prah, S., Zeng, S., Lin, H., and Se, J. 2009. Mapping and spatial uncertainty analysis of forest vegetation carbon by combining national forest inventory data and satellite images. *Forest Ecology and Management* 258(7):1275–1283.

Wang, G., Zhang, M., Gertner, G. Z., Oyana, T., McRoberts, R. E., and Ge, H. 2011. Uncertainties of mapping forest carbon due to plot locations using national forest inventory plot and remotely sensed data. *Scandinavia Journal of Forest Research* 26:360–373.

Wu, J. 1999. Hierarchy and scaling: Extrapolating information along a scaling ladder. *Canadian Journal of Remote Sensing* 25:367–380.

Wu, J., and Qi, Y. 2000. Dealing with scale in landscape analysis: An overview. *Geographical Information Sciences* 6:1–5.

Zhang, M., and Wang, G. 2008. The forest biomass dynamics of Zhejiang Province. *Acta Ecologica Sinica* 28(11):5665–5674.

Zhang, M., Wang, G., Zhou, G., Ge, H., Xu, L., and Zhou, Y. 2009. Mapping of forest carbon by combining forest inventory data and satellite images with co-simulation based up-scaling method. *Acta Ecologica Sinica* 29(6):2919–2928.

Zhang, Y. 1999. Measuring gravities and pulpwood rates of poplar species. *Journal of Liaoning Forestry Science & Technology* 2:50–51.

7

ESTIMATING GRASSLAND CHLOROPHYLL CONTENT FROM LEAF TO LANDSCAPE LEVEL: BRIDGING THE GAP IN SPATIAL SCALES

Yuhong He

7.1 INTRODUCTION

Approximately one-fifth of Earth's land surface is covered by grasslands, both natural and disturbed (He and Mui, 2010). Grasslands worldwide have experienced signifi-cant degradation and extinction due to urbanization, overgrazing, land use conversion (i.e. crop and pastoral lands), climate change, and mining. As an example, the Canadian tallgrass prairies once covered approximately $1000 \, km^2$ of southern Ontario. However, less than 1% of its historic range in southern Ontario remains. It is thus imperative that research is done in the way of conservation, management, and restoration efforts for the endangered grassland prairie ecosystems.

Quantifying vegetation biochemical properties could advance our understanding of vegetation physiological processes and provide insight into detection and mon-itoring of vegetation conditions under the context of environmental stressors (Evans, 1989; Yoder and Pettigrew-Crosby, 1995; Wu et al., 2008). One of the primary vegetation biochemical properties, chlorophyll (Chl a + b), is responsible for photo-synthesis and has been used to characterize plant physiological status and health under stress (i.e., Blackburn, 1998; Zarco-Tejada et al., 2002), investigate plant

Scale Issues in Remote Sensing, First Edition. Edited by Qihao Weng.
© 2014 John Wiley & Sons, Inc. Published 2014 by John Wiley & Sons, Inc.

photosynthetic capacity (i.e., Gitelson et al., 2006), and estimate vegetation productivity (i.e., Peng and Gitelson, 2011).

Methods for predicting and estimating Chl a + b at the leaf scale are based on light–foliar interactions that are recorded by remote sensing data and have been in development for several decades. Spectral data in the visible and near-infrared regions of the spectrum were originally used to develop a number of spectral indices [e.g., the normalized difference vegetation index (NDVI)] (Rouse et al., 1974) for estimating vegetation chlorophyll content. However, these indices have been found to be insensitive to medium and high chlorophyll concentrations (Gitelson and Merzlyak, 1994). Recent studies have developed several new spectral indices based on other visible wavelength bands instead of the red wavelengths around the region of 670–680 nm. For example, studies have found that the region near 700 nm is highly sensitive to chlorophyll concentrations (e.g., Gitelson and Merzlyak, 1996; Gitelson et al., 2005; Ciganda et al., 2009), and according to that finding, the narrow-band-based indices simple ratio ($SR_{[700,750]}$) (Gitelson and Merzlyak, 1996) and $NDVI_{[705,750]}$ (Datt, 1999) were found to be well correlated with total chlorophyll content of different types of leaves. Further, Sims and Gamon (2002) examined hundreds of leaves of nonrelated plant species and proved that reflectance in the spectral channel around 700 nm was the most sensitive indicator of chlorophyll and that indices $SR_{[700,750]}$ and $NDVI_{[705,750]}$ could be used as a measure of chlorophyll content. In the tall grassland, a previous study further demonstrated that the index $SR_{[700,750]}$ is better than $NDVI_{[705,750]}$ in estimating vegetation pigment (Wong, 2012).

At the leaf scale, strong relationships have been reported between spectral indices and chlorophyll measurements in many leaf types originating from a wide range of ecosystems. However, the predictive capability of spectral indices in estimating chlorophyll content remains uncertain at the canopy and landscape scales. At the canopy scale, remote sensing estimation of pigments has been performed through different empirical methods (Johnson et al., 1994; Curran et al., 1997; Daughtry et al., 2000; Zarco-Tejada et al., 1999, 2000, 2002, 2004). One method is to correlate leaf-level pigment content directly to the canopy-level reflectance spectral data measured in the field or by airborne or satellite sensors. This method, however, may not produce reliable results as canopy pigment composition depends strongly on plant species as well as on the canopy structure (Zhang et al., 2008). The other commonly used method includes correlating the canopy-integrated pigment content to the canopy-level reflectance spectral data (Yoder and Pettigrew-Crosby, 1995; Daughtry et al., 2000). The canopy-integrated pigment content is obtained by multiplying the leaf pigment content by the corresponding canopy biophysical parameters such as leaf area index (LAI) or biomass. For example, Jago et al. (1999) defined canopy chlorophyll content as chlorophyll concentration × biomass within an area covered by a pixel. Similarly, Ciganda et al. (2009) estimated total chlorophyll in maize canopies using LAI × leaf chlorophyll content. The canopy-integrated approach markedly improved current techniques proposed for pigment quantification in the canopy. However, the major assumption of the canopy-integrated method is that all leaves in the plant have the same pigment content. Consequently, the method might be successful when only one plant species homogeneously covers each pixel of the hyperspectral image. For a homogeneous canopy such as heterogeneous grasslands (Zhang et al., 2008; He and Mui, 2010), the method is not suitable because other factors such as leaf age, light gradients, and leaf angle distribution would contribute to pigment variation within such a canopy. Using species percentage cover data, the author recently proposed an alternative method that could scale vegetation chlorophyll content from the leaf to canopy or higher level for heterogeneous grasslands (Wong and He, 2013). Understanding grassland chlorophyll properties are critical not only at the leaf and canopy level but also at the landscape level because a study at this scale could provide a better understanding of ecosystem nutritional and health status. Further, landscape scale pigment estimation could provide information on potential regional or global responses of vegetation to climate change.

The objective of this study was thus to bridge the gap in spatial scales through estimating grassland chlorophyll content from leaf to landscape level. Specifically, this study uses a spectral index to estimate grassland Chl a + b content at the leaf, canopy, and landscape levels using ground and satellite remote sensing data. At the leaf level, this study examined the relationship between leaf Chl a + b content data and lab-derived hyperspectral reflectance data. The leaf-level Chl a + b content was then scaled to the canopy level through a newly proposed simple method described in the methodology section. The derived canopy-level Chl a + b was then correlated to field hyperspectral data and space remote sensing data. The canopy-level Chl a + b was further scaled to the landscape level to correlate with both field and satellite remote sensing data.

7.2   STUDY AREA

This study was conducted at the Koffler Scientific Reserve at Jokers Hill, a 350-ha field station owned by the University of Toronto and situated 50 km north of Toronto, Ontario (44.03′ N, 79.29′ W; http://www.ksr.utoronto.ca). The study sites were established in undisturbed old fields dominated by temperate tall grass prairies. Some common species found in the study plots are common milkweed (*Asclepias syriaca*), bird vetch (*Vicia cracca* L.), Canada goldenrod (*Solidago*), bog goldenrod (*Euthamia graminifolia* (L.) Salisb.), Queen Anne's lace (*Daucus carota* L.), rye grass (*Lolium perenne* L.), red fescue (*Festuca rubra* L.), and Canada thistle (*Cirsium arvense*).

7.3   DATA AND METHODS

7.3.1   Field Data Collection

Field data collection was conducted July 1 to 15, 2011, which is the maximum growing season. This date collection period was chosen because chlorophyll is more stable at the time window, while in the early or late growing season chlorophyll content in vegetation could be highly variable. Sixteen sites distributed over upland, sloped, and valley areas were randomly chosen in the undisturbed old field in Koffler Science Reserve. Each study site was formed by two 50-m transects that were oriented perpendicular to one another (forming a cross). Each transect was made up of 10 plots 50 cm × 50 cm with a 5-m distance from one plot to another. In each plot, a set of grassland composition, biophysical, biochemical, and ground hyperspectral data was measured using various instruments, field visual determination, and lab-based approaches. The details are discussed below.

Samples were collected for the three dominant species in each chosen plot to be used for lab-based spectral reflectance measurements and chlorophyll extraction. Leaf spectral reflectance was measured for each species in a controlled laboratory environment using an analytical spectral device (ASD) FieldSpec 3 Max Portable spectroradiometer with a fiber optic. The illumination source was an ASD Pro Lamp (14.5 V, 50 W) that is adapted for indoor lab diffuse reflectance measurements over the region 350–2500 nm. The wavelength range for the collected reflectance data was 350–2500 nm with a spectral sample of 1.4 nm in the 350–1050-nm range and 2 nm in the 1000–2500-nm range. For optimal spectral measurements, the fiber-optic sensor was pointed opposite to the angle of the light source, which is approximately 45° pointing downward, with the leaf sample being placed underneath the point of highest light intensity. Five replicate measurements were taken for each leaf sample and were averaged to suppress the measurement noise.

Canopy spectral reflectance was measured at each plot using the same instrument that was used for lab-based leaf spectral reflectance measurement. The fiber-optic sensor, with a field view of 25°, was hand held approximately 1 m above and perpendicular to the ground. Calibration was conducted frequently with a calibrated diffuse white reference panel provided by ASD Inc. during field sampling to minimize measurement noise. To minimize atmospheric perturbations, spectral measurements were taken during cloudless days between 10:00 AM and 2:00 PM. Canopy composition data were collected using a 0.5 m × 0.5 m frame. The frame was lightly tossed onto a random spot around the plot. Then the percentage top layer cover (grass, forbs, shrubs, and standing dead), lower layer cover (litter, moss, lichen, rock, bare ground), and species cover were visually determined within the frame and recorded. Photographs were taken over each frame for more careful scrutiny back in the laboratory.

Samples of the three dominant species were collected from every other plot for species chlorophyll extraction. The samples were collected from every other plot because field observation indicated that the difference in vegetation species and their coverage within 5 m (the distance between two neighboring plots) are marginal. The collected vegetation samples were kept on ice in a dark thermal cooler and then transferred immediately to a −20°C freezer to prevent leaf pigments from deteriorating. In order to extract chlorophyll, each leaf sample was first placed in a microtube containing a solution of 100% acetone and 30 mg of sodium bicarbonate. The tubes were then placed into a mini–bead beater in order to vigorously break the leaf cells and separate the chlorophyll content. Next, the homogenized solution was centrifuged to separate the fibrous plant material from the supernatant (the chlorophyll content suspended within the acetone). The process was repeated three times in order to completely extract as much chlorophyll pigment as possible from the leaf material. The absorbance at 447, 645, and 662 nm of each finished sample was measured using a spectrophotometer (Thermo Scientific, Detroit, MI). The chlorophyll concentration of each sample was then calculated using the fresh sample weight, sample extraction volume, and absorption coefficients as reported by Lichtenthaler (1987). The Chl a + b concentration is given in micrograms per milligram. The chlorophyll concentration data for the same species were then averaged and converted to the chlorophyll content of the species (in grams per square meter).

7.3.2   QuickBird Image Acquisition and Preprocessing

In this study, a multispectral QuickBird image with a spatial resolution of 2.4 m was acquired on July 5, 2010, from the Digital Globe archival collection. The spectral bandwidth of QuickBird includes the blue (430–545 nm), green (466–620 nm), red (590–710 nm), and near-infrared (NIR) (715–918 nm) regions of the spectrum. The image was geometrically and radiometrically corrected by the distributor. Atmospheric correction was conducted using the ATCOR-2 algorithm in PCI Geomatica 10 using weather conditions obtained from Environment Canada's National Climate Data.

7.3.3   Spectral Indices

The reflectance values in the wavelength regions of 700 and 750 nm were combined to calculate the red edge index (i.e., R750/R700) since it was found to be well correlated with total chlorophyll content of different types of leaves (Datt, 1999; Gamon and Surfus, 1999; Sims and Gamon, 2002; Wong, 2012). These calculations were performed using the hyperspectral reflectance gathered in the lab and field. For the QuickBird image acquired at the spaceborne level, the broadbands red (band 3) and NIR (band 4) were used to calculate SR [R, NIR].

7.3.4   Scaling Up from Leaf to Canopy and Landscape Levels

Although hyperspectral data at the canopy level have been used to detect changes in canopy chlorophyll content in dense vegetation (e.g., crops and forests) (Daughtry et al., 2000; Yoder and Pettigrew-Crosby, 1995; Zhang et al., 2008; Wu et al., 2008), it has proven to be difficult to achieve the same outcome on grassland ecosystems characterized by heterogeneous vegetation (He and Mui, 2010). Therefore, the leaf-level chlorophyll measurements were scaled to the canopy level to produce a canopy-integrated chlorophyll content. Specifically, the chlorophyll content in the middle plot at each arm in each site (a total of four plots in each site) was calculated by multiplying leaf Chl a + b content for each dominant species with its percentage cover and summing the resulting values for all species considered. Only one plot in each arm was selected to remove the autocorrelation between plots within each arm. Furthermore, the canopy Chl a + b content (Chl_{canopy}, in grams per square meter) is averaged from the Chl a + b contents of all the plots in a site. This canopy-integrated approach was used in this study as it accounts for species heterogeneity and nongreen components, such as dead litter and bare soil, which have a disproportionately strong effect on canopy reflectance variability (He and Mui, 2010). In order to scale canopy chlorophyll contents to the landscape level, canopy chlorophyll data were averaged from plots to represent landscape-level chlorophyll contents ($Chl_{landscape}$, in grams per square meter).

7.3.5 Relating Remote Sensing Data to Chlorophyll Data

The relationship between grassland chlorophyll content and ground and satellite remote sensing data at all scales were investigated using linear regression analysis, where the regression was fit with the least-squares method. The coefficient of determination (R^2) was used to determine how much of the variation in vegetation chlorophyll content can be explained by remote sensing data.

7.4 RESULTS

7.4.1 Chlorophyll Content and Laboratory Remote Sensing Data at Leaf Scale

Chlorophyll data were extracted in the laboratory based upon field samples collected for dominant tall grass species. The results of the eight most common species are shown in Table 7.1. Samples of the bird vetch held the highest chlorophyll content of Chl a + b, 26.4 g/m², with a standard deviation of 6.3 g/m², while the red fescue held the lowest value, 18.4 g/m², with a low standard deviation of 0.6 g/m².

The lab-based spectral reflectance of selected five types of tallgrass species is shown in Figure 7.1. The spectral signature for each species is typical of green vegetation (i.e., a peak in reflectance in the green wavelength, a decrease in red, and a sharp rise in the NIR region). Although differences between species were marginal in most of the wavelengths, some variation was observed in the green and NIR regions.

TABLE 7.1 Descriptive Statistics of Chl a + b Content (g/m²) for Selected Species

Species	Numbers of Samples	Mean ± STD	Maximun	Minimum
Bird vetch	62	26.4 ± 6.3	48.2	12.6
Rye grass	146	16.1 ± 6.9	46.5	3.4
Canada goldenrod	68	10.3 ± 3.4	21.8	1.1
Common milkweed	30	10.9 ± 4.0	23.5	4.6
Queen anne's lace	10	13.8 ± 3.4	18.9	11.5
Bog goldenrod	16	11.5 ± 5.2	20.1	5.2
Canada thistle	20	9.8 ± 2.9	18.4	4.0
Red fescue	77	8.0 ± 2.9	18.4	0.6

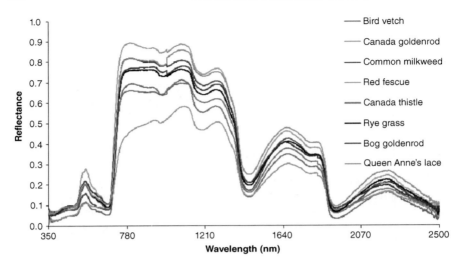

FIGURE 7.1 Lab-measured leaf spectral reflectance for selected dominant species. (See the color version of this figure in Color Plates section.)

Significant relationships ($p < 0.01$) were found between R750/R700 and vegetation chlorophyll content at the leaf level (Figure 7.2). Sixty-four percent of Chl a + b variation was explained by SR[700, 750].

7.4.2 Chlorophyll Content and Remote Sensing Data at Canopy Scale

The canopy-level chlorophyll content was calculated and found to have a range from 2.97 to 23.5 g/m² for the study plots (Figure 7.3). The field hyperspectral data–derived

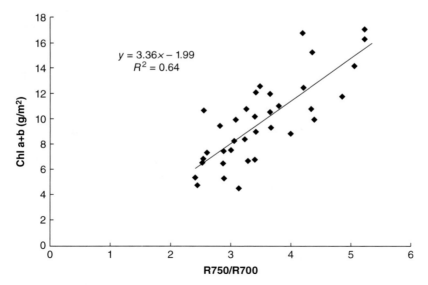

FIGURE 7.2 Relationship between leaf-level spectral reflectance and chlorophyll content.

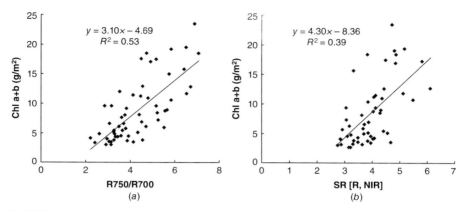

FIGURE 7.3 Relationship between (*a*) ground R750/R700 and canopy chlorophyll content and (*b*) QuickBird SR[*R*, NIR] and canopy chlorophyll content.

spectral index R750/R700 has a range of 2.7–6.1, while the QuickBird-derived spectral index SR[*R*, NIR] has a range of 2.2–7.7.

The significant relationship between field hyperspectral data-derived spectral index R750/R700 and chlorophyll content at the canopy level are shown in Figure 7.3. Overall, the simple ratio index R750/R700 explained 53% of variation for canopy Chl a + b. The relationship between QuickBird-derived spectral index SR[*R*, NIR] and canopy Chl a + b is also significant but with a much lower R^2 (39%), indicating QuickBird data captured much less variation in canopy chlorophyll.

7.4.3 Chlorophyll Content and Remote Sensing Data at Landscape Scale

A significant correlation was found between field hyperspectral data–derived SR and chlorophyll content at the landscape level (Figure 7.4*a*). The red edge index R750/R700 again explained a greater variation for chlorophyll content ($R^2 = 0.75$).

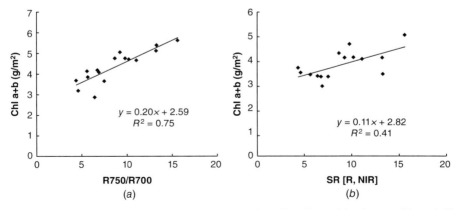

FIGURE 7.4 Relationship between (*a*) ground SR[700,750] and landscape chlorophyll content and (*b*) QuickBird SR[*R*, NIR] and landscape chlorophyll content.

The relationships between QuickBird-derived SR[R, NIR] and chlorophyll content at the landscape level were also significant with slightly lower R^2 (0.41) (Figure 7.4b).

7.5   CONCLUSIONS AND DISCUSSION

With species percentage cover data, this study calculated Chl a + b content for heterogeneous grasslands at the canopy and landscape levels. This canopy integration approach is simple and shown to be capable of estimating vegetation chlorophyll content beyond leaf level. By relating the chlorophyll contents at leaf, canopy, and landscape scales with remote sensing data, we found it is possible to estimate chlorophyll contents at three levels using both ground and space remote sensing data.

At the species level, 64% of the variation in Chl a + b was explained by lab-based remote sensing–derived spectral index. Comparing with other similar studies, we found that these R^2 values were generally lower. For example, Gitelson and Merzlyak (1997) estimated chlorophyll content on higher plant leaves over a range of species (i.e., *Acer*, *Castanea*, *Vitis vinifera*, and *Fagus*) using R750/R700 and found the R^2 values ranged from 90 to 98%. The relatively lower R^2 values found in this study could be explained by the fact that the leaf samples were taken in the maximum growing season (July 1–15). In this season, vegetation is uniformly green so that species have a similar level of chlorophyll content. However, spectral data may vary due to differences in leaf attributes such as size and shape and thickness. Consequently, those species with a similar level of chlorophyll content may not have the same spectral indices values, leading to a lower overall R^2.

At the canopy level, 53% of the variation in chlorophyll content was explained by field remote sensing–derived spectral index, while only 39% of the variation was explained by satellite remote sensing. At the landscape level, 75% of the variation in vegetation chlorophyll was explained by field remote sensing data, and 41% of the variation was explained by satellite remote sensing data. In comparison with R^2 values at the canopy level, the lower R^2 values between satellite remote sensing and chlorophyll data at canopy and landscape levels could be explained by two reasons:

1. The different estimation capability between narrow-band index (calculated from ground hyperspectral data) and broad-band index (calculated from QuickBird) for quantifying chlorophyll. The narrow-band spectral indices have long been demonstrated to be better predictors in comparison with broad-band spectral indices for quantifying chlorophyll content.
2. Imperfect atmospheric and geometric corrections for the QuickBird image.

Although atmospheric effect was corrected, the spectral index derived from pixels could still be affected by cloud/haze interference. As for the geometric correction, the field plots were paired to the nearest pixel possible in a 2.4-m-resolution QuickBird image for chlorophyll content estimation at the canopy (0.5 m × 0.5 m) and landscape (50 m × 50 m); therefore, a slight displacement in georeferencing may still exist. In spite of these differences, the significant results at the canopy and landscape scale imply that it is possible to accurately estimate chlorophyll contents using both ground hyperspectral and satellite multispectral remote sensing data. The recent availability of spaceborne hyperspectral data such as CHRIS/PROBA (the Compact High Resolution Imaging Spectrometer/Project for On Board Autonomy) and the development of more effective narrow-band chlorophyll indices provide new opportunity to estimate the biochemical properties of vegetation at the landscape or even higher level.

As scales increased from leaf to landscape, more and more new spectral features are included in the remote sensing data. For example, at the leaf scale, leaf Chl a + b variation and anatomical differences such as cell size, air spaces, and cell wall thickness can all contribute to spectral differences. At the canopy level, besides leaf characteristics, canopy compositions including standing dead, litter, and bare soil are incorporated into the spectral data. At the landscape level, a complex array of ground and atmospheric features and sensing techniques, including illumination and viewing geometry, affect the imagery data. One would thus assume that relationships between vegetation chlorophyll properties and spectral indices would become weaker and weaker as scale increased from species to landscape level. This is true when the chlorophyll is scaled from the leaf to the canopy level, and the R^2 value is decreased from 64% of leaf level to 53% of canopy level. Unexpectedly, this study found that the relationships at the landscape level became stronger than that in the landscape and level levels. The unexpected result may be explained by a few factors: (1) The less regression samples (only 16 sites in the landscape level) with large variation in the sampling sites compared to much more regression samples at the canopy levels with less variation in the samples. (2) When scaling the Chl a + b data from leaf to canopy level, the variation in species percentage cover may contribute to higher variation in canopy Chl a + b data. Further, when scaling the Chl a + b data from the canopy to the landscape level, an effect of averaging could potentially exist. Cumulatively, these factors probably contributed to the higher coefficients of determinations at the landscape level. These factors also represent inherent limitations to using the approach as a scaling tool, unless the sampling points at three scales are controlled and variation in the samples is taken into account in future studies and applications.

Chlorophyll content in leaf keeps changing through the entire growing season, while our analysis is only based on one set of field data from July 1 to July 15, 2011. The regression models established in this study may only be suitable for estimating chlorophyll content in the maximum growing season. In the early or late growing season, other factors in the field could cause a different relationship between chlorophyll content and spectral indices. A time-series analysis is thus needed to provide more insights on the relationships between chlorophyll content and spectral indices over the growing season.

ACKNOWLEDGMENT

The support of the NSERC Discovery Grant RGPIN-386183 to Dr. Yuhong He is gratefully acknowledged. Sections 7.2, 7.3.4, and 7.5, and Figures 7.2 and 7.4 are used with the permission of the Canadian Aeronautics and Space.

REFERENCES

Blackburn, G. A. 1998. Quantifying chlorophylls and carotenoids at leaf and canopy scales: An evaluation of some hyperspectral approaches. *Remote Sensing of Environment* 66:273–285.

Ciganda, V., Gitelson, A., and Schepers, J. 2009. Non-destructive determination of maize leaf and canopy chlorophyll content. *Journal of Plant Physiology* 166:157–167.

Curran, P. J., Kupiec, J. A., and Smith, G. M. 1997. Remote sensing the biochemical composition of a slash pine canopy. *IEEE Transactions on Geoscience and Remote Sensing* 35:415–420.

Datt, B. 1999. Remote sensing of water content in eucalyptus leaves. *Australian Journal of Botany* 47:909–923.

Daughtry, C. S. T., Walthall, C. L., Kim, M. S., De Colstoun, E. B., and McMurtrey, III, J. E. 2000. Estimating corn leaf chlorophyll concentration from leaf and canopy reflectance. *Remote Sensing of Environment* 74:229–239.

Evans, J. R. 1989. Photosynthesis and nitrogen relationships in leaves of C3 plants. *Oecologia* 78:9–19.

Gamon, J. A., and Surfus, J. S. 1999. Assessing leaf pigment content and activity with a reflectometer. *New Phytologist* 143:105–117.

Gitelson, A. A., and Merzlyak, M. N. 1994. Spectral reflectance changes associated with autumn senescence of *Aesculus hippocastanum L.* and *Acer platanoides L.* leaves. Spectral features and relation to chlorophyll estimation. *Journal of Plant Physiology* 143:286–292.

Gitelson, A. A., and Merzlyak, M. N. 1997. Remote estimation of chlorophyll content in higher plant leaves. *International Journal of Remote Sensing* 18:2691–2697.

Gitelson, A. A., Viña, A., Rundquist, D. C., Ciganda, V., and Arkebauer, T. J. 2005. Remote estimate of canopy chlorophyll content in crops. *Geophysical Research Letters* 32:1–4.

Gitelson, A. A., Viña, A., Verma, S. B., Rundquist, D. C., Arkebauer, T. J., Keydan, G., Leavitt, B., Ciganda, V., Burba, G. G., and Suyker, A. E. 2006. Relationship between gross primary production and chlorophyll content in crops: Implications for the synoptic monitoring of vegetation productivity. *Journal of Geophysical Research* 111:D08S11.

Gitelson, A. G., Yacobi, Y. Z., Karnieli, A., and Kress, N. 1996. Reflectance spectra of polluted marine waters in Haifa Bay, Southeastern Mediterranean. Features and application for remote estimation of chlorophyll concentration. *Israel Journal of Earth Sciences* 45:127–136.

He, Y., and Mui, A. 2010. Scaling up semi-arid grassland biochemical content from the leaf to the canopy level: Challenges and opportunities. *Sensors* 10:11072–11087.

Jago, R. A., Cutler, M. E. J., and Curran, P. J. 1999. Estimating canopy chlorophyll concentration from field and airborne spectra. *Remote Sensing of Environment* 68:217–224.

Johnson, L. F., Hlavka, C. A., and Peterson, D. L. 1994. Multivariate analysis of AVIRIS data for canopy biochemical estimation along the oregon transect. *Remote Sensing of Environment* 47:216–230.

Lichtenthaler, H. K. 1987. Chlorophyll and carotenoids: Pigments of photosynthetic biomembranes. *Methods in Enzymology* 148:331–382.

Peng, Y., and Gitelson, A. A. 2011. Application of chlorophyll-related vegetation indices for remote estimation of maize productivity. *Agricultural and Forest Meteorology* 151:1267–1276.

Rouse, J. W., Haas, R. H., Schell, J. A., Deering, D. W., and Harlan, J. C. 1974. Monitoring the vernal advancement of retrogradation of natural vegetation. National Aeronautics and Space Administration, Goddard Space Flight Center (NASA/GSFC), Type III, Final Report.

Sims, D. A., and Gamon, J. A. 2002. Relationships between leaf pigment content and spectral reflectance across a wide range of species, leaf structures and developmental stages. *Remote Sensing of Environment* 81:337–354.

Wong, K. 2012. Remote Sensing of Tall Grasslands: Estimating Vegetation Biochemical Contents at Multiple Spatial Scales and Investigating Vegetation Temporal Response to Climate Conditions. Ph.D. Dissertation, University of Toronto.

Wong, K., and He, Y. 2013. Estimating grassland chlorophyll content using remote sensing data at the species, canopy, and landscape scales. *Canadian Journal of Remote Sensing* 39:1–12.

Wu, C., Niu, Z., Tang, Q., and Huang, W. 2008. Estimating chlorophyll content from hyperspectral vegetation indices: Modeling and validation. *Agricultural and Forest Meteorology* 148:1230–1241.

Yoder, B. J., and Pettigrew-Crosby, R. E. 1995. Predicting nitrogen and chlorophyll content and concentrations from reflectance spectra (400–2500 nm) at leaf and canopy scales. *Remote Sensing of Environment* 53:199–211.

Zarco-Tejada, P. J., Miller, J. R., Mohammed, G. H., Noland, T. L., and Sampson, P. H. 1999. Canopy optical indices from infinite reflectance and canopy reflectance models for forest condition monitoring: Application to hyperspectral CASI data. Paper presented at the International Geoscience and Remote Sensing Symposium (IGARSS) 3:1878–1881.

Zarco-Tejada, P. J., Miller, J. R., Mohammed, G. H., Noland, T. L., and Sampson, P. H. 2000. Chlorophyll fluorescence effects on vegetation apparent reflectance: II. Laboratory and airborne canopy-level measurements with hyperspectral data. *Remote Sensing of Environment* 74:596–608.

Zarco-Tejada, P. J., Miller, J. R., Mohammed, G. H., Noland, T. L., and Sampson, P. H. 2002. Vegetation stress detection through chlorophyll a + b estimation and fluorescence effects on hyperspectral imagery. *Journal of Environmental Quality* 31:1433–1441.

Zarco-Tejada, P. J., Miller, J. R., Morales, A., Berjón, A., and Agüera, J. 2004. Hyperspectral indices and model simulation for chlorophyll estimation in open-canopy tree crops. *Remote Sensing of Environment* 90:463–476.

Zhang, Y., Chen, J. M., Miller, J. R., and Noland, T. L. 2008. Leaf chlorophyll content retrieval from airborne hyperspectral remote sensing imagery. *Remote Sensing of Environment* 112:3234–3247.

PART III

SCALE AND LAND SURFACE PROCESSES

8

VISUALIZING SCALE-DOMAIN MANIFOLDS: A MULTISCALE GEO-OBJECT-BASED APPROACH

Geoffrey J. Hay

8.1 INTRODUCTION

. . . Scale affects everything . . .

Understanding, modeling, managing, and forecasting landscape-based pattern–process interactions through multiple temporal and spatial scales represent key research activities in landscape ecology (Wu and Hobbs, 2002; Wu, 2007). To achieve these activities, ecologists and others are increasingly relying upon and integrating advances in areas outside traditional ecology, such as geoinformatics, remote sensing, complexity theory, computer vision, and pattern recognition (Hay et al., 1997, 2001; Burnett and Blaschke, 2003; Hay and Marceau, 2004; Castilla et al., 2008; Steiniger and Hay, 2009; Wu and Li, 2009; Powers et al., 2012). The necessity for doing so is clear: Landscapes are complex systems composed of a large number of heterogeneous components that interact in a nonlinear way, are hierarchically structured, and are scale dependent (Waldrop, 1992; Nicolis and Prigogine, 1989; Kay and Regier, 2000; Hay et al., 2002a,b; Wu and Marceau, 2002). Remote sensing technologies represent the only source of large-area data, and geoinformatics, computer vision, pattern recognition, and complexity theory provide useful concepts, methods, models, tools, and insight for defining, visualizing, querying, and managing these data (Hay et al., 2003). Furthermore, since the advent of NASA's mission to planet Earth with its plethora of Earth-orbiting sensors and their daily generation of

Scale Issues in Remote Sensing, First Edition. Edited by Qihao Weng.
© 2014 John Wiley & Sons, Inc. Published 2014 by John Wiley & Sons, Inc.

terabytes of different multispatial, multispectral, multiradiometric, and multitemporal resolution digital data, the necessity for automated analytical methods is paramount (Hay et al., 2005).

Critical to these research activities is the need to understand the concept of *scale* in its many guises, work within its limitations, and benefit from its potential. Though scale is considered one of the central concepts in many disciplines which study human–environment systems, a wide diversity of scale definitions across numerous disciplines (Gibson et al., 2000) can confusingly result in inappropriate sampling designs, incorrect results, and misleading conclusions (O'Neill and King, 1998; Wheatley and Johnson, 2009; Wu and Li, 2009). As an example, Wikipedia[1] provides 60 different meanings of scale, most of which are related to measurement units. Regardless of which definition is used, scale conceptually represents *the* window of perception (Hay et al., 2002a), a continuum through which entities, patterns, and processes can be perceived and linked together (Marceau, 1999). Change the spatial/ temporal characteristics of the window (e.g., size, shape, viewing time), and the corresponding view is altered. Thus, knowing how this view is altered and the effects upon the scene such alterations produce are critical to the validity and utility of decisions based on this view. Scale affects everything. Thus its mastery holds the promise of powerful insight and understanding. But what scale(s) should we choose for our analysis and how should we assess information *within* and *between* scales? Where is within and between?

From an ecological perspective, it is well recognized that the same ecological processes may reveal different patterns if observed at different scales (Peterson, 2000; Wheatley and Johnson, 2009). Thus, if we study a system at an "inappropriate" scale, we may miss the actual system dynamics and may instead identify patterns that are *artifacts of scale* (Wiens, 1989), like looking at the world through the wrong pair of glasses. Furthermore, patterns observed across scales will form the bases of hypotheses exploring underlying processes (Swihart et al., 2002). Ideally what we need are adaptive methods that are able to automatically query the varying sized, shaped, and spatially distributed components of a landscape and "tell us" the correct scale(s) and locations over which to conduct analysis (Hay et al., 2001; Hay and Marceau, 2004; Hay et al., 2005; Drăguţ et al., 2010).

In an effort to address these challenges, we pose the multipart question "What does a scale domain look like; where is it located, and how can we visualize it?" In response, we introduce and describe a novel geo-object-based framework that integrates high-resolution remote sensing imagery, hierarchy theory, and scale space (SS) for automatically visualizing and modeling dominant (i.e., most salient) landscape structures through multiple scales and within uniquely defined scale domains. We also report on three main goals:

(i) We describe a three-tier hierarchical methodology for automatically delineating and assessing the dominant structural components within 200 different multiscale representations of a complex agro-forested landscape.

---

[1] http://en.wikipedia.org/wiki/Scale, last accessed May 24, 2012.

(ii) By considering *scale-space events* as *critical domain thresholds*, we describe and apply a new scale-domain topology that facilitates improved querying and analysis of this complex multiscale scene.

(iii) We spatially model and visualize the hierarchical structure of dominant geo-objects within a scene as *scale-domain manifolds*—which we suggest represents a multiscale extension to the *hierarchical scaling ladder* as defined in the *hierarchical patch dynamics paradigm*.

To better evaluate these results, we provide a brief background on hierarchical theories, geo-object-based hierarchies, multiscale analysis, and image objects. This is followed by a brief description of the methods underlying linear SS and blob feature detection, which is accompanied by a discussion of the visualization results.

8.1.1 Hierarchical Theories

Ecologists have long recognized that many "natural" processes produce clusters of entities that emerge (and interact) at a specific range of spatial, spectral, and temporal scales. These clusters result in visually *distinct* spatial patterns [typically referred to as *patches—a relatively homogeneous area that differs from its surrounding*[2] (Forman, 1995)] which are typically generated by a small set of self-organizing principles (Allen and Starr 1982; Waldrop, 1992). Therefore, one way to understand, explain, and forecast the effects of *natural processes* is to examine these *natural patterns* at their corresponding *natural scales* of emergence (Wessman, 1992; Rosin, 1998; Levin, 1999; Hay et al., 2001). To assist in this task, the conceptual framework of *hierarchy theory* has been adopted in this chapter, as it builds upon this idea of natural scales. Here the term *scale* refers to the spatial dimensions (i.e., grain and extent) at which entities, patterns, and processes can be observed and measured. *Grain* refers to the smallest distinguishable component (i.e., spatial resolution), while *extent* refers to the entire *area or scene* under analysis. For related definitions of scale see Quattrochi and Goodchild (1996), Marceau (1999), Marceau and Hay (1999), Wu and Li (2005), Manson (2006), and Wu and Li (2009).

Hierarchy theory was developed in the framework of general system's theory, mathematics, and philosophy in the 1960s and 1970s (Wu and Loucks, 1995) and is generally regarded as being introduced into ecology by Allen and Starr (1982). However, it should be noted that early work by Watt (1947), Whittaker (1953), and others embraces ideas that are implicitly hierarchical (Urban et al., 1987). Hierarchy theory attempts to analyze the effect of scale on the organization of complex systems (Simon, 1973). It does not assume that a system is hierarchically structured but rather partitions "the world" into hierarchical levels to simplify the analysis of cross-scale interactions (Allen and Starr, 1982; Ahl and Allen, 1996; Peterson, 2000). Here, "complex systems" are characterized by a large number of components that interact in a nonlinear way and exhibit adaptive properties through time (Kay, 1991).

---

[2] This is also noted as such in Wikipedia (http://en.wikipedia.org/wiki/Landscape_ecology), last accessed April 21, 2013.

Within hierarchy theory, a hierarchically organized system can be visualized as a (three-tiered) nested system in which levels exhibiting progressively slower behavior are at the top (level +1), while those reflecting successively faster behavior are seen as a lower level in the hierarchy (level −1). The level of interest is referred to as the *focal level* (level 0) and it exists between the other two. From an ecological perspective, hierarchy theory states that complex ecological systems, such as landscapes, are composed of loosely coupled levels (*scale domains*) which operate within a distinct range of temporal and spatial scales. Interactions tend to be stronger and more frequent *within* a level than *among* levels, and *scale thresholds* separate domains by representing relatively sharp transitions or critical locations where shifts occur in the relative importance of variables influencing a process (Allen and Starr, 1982; Meentemeyer, 1989; Wiens, 1989). These shifts then manifest themselves as changes in spatial patterns. This important response enables the perception and description of complex systems by decomposing them into their fundamental parts and interpreting their interactions (Simon, 1962).

Building on these ideas, the *hierarchical patch dynamics paradigm* (HPDP) represents an organizational framework that attempts to integrate the vertical components of hierarchy theory and the horizontal components of patch dynamics theory (PDT) to express the relationships among pattern, process, and scale explicitly within the context of landscapes (Wu and Loucks, 1995). In the HPDP, the hierarchical structure of a system is conceptualized as a "scaling ladder," where each rung corresponds to a threshold within the scale continuum (Wu, 1999). Reference is made to extrapolating information along a hierarchical scaling ladder by varying grain or extent or both; however, limited information is provided on how to do so or where the ladder or its rungs should exist within the landscape or over what spatial resolution(s). Conceptually, one would scale (i.e., transfer information) between dominant land-scape patches (i.e., internally homogeneous image objects); however, defining meaningful patches at different scales, the "correct" method(s) for scaling between them is not trivial (Hay et al., 1997; Hay and Marceau, 1998; Wu and Li, 2005) and remains an active body of research in numerous domains.

In more recent HPDP work, Wu et al. (2003) discuss the concept of *ecosystem functional types* (EFTs) and describe scaling between these entities. It is suggested that EFTs provide a way of clustering a large number of local ecosystems into a smaller number of categories that each have similar functional properties in terms of biogeochemical cycling (Reynolds and Wu, 1999). Thus, EFTs effectively represent multiscale patches within the HPDP. One of the challenges of this approach is the necessity for detailed information regarding biogeochemical cycling, which is seldom ubiquitously available.

In an effort to automate patch delineation from remote sensing imagery, new ideas and methods are being developed and implemented by the GEOBIA (Geographic Object-Based Image Analysis) community (Hay and Castilla, 2008; Dey et al., 2010; Blaschke, 2010). Based on an integration of earlier definitions (Hay and Castilla, 2008; Hay and Blaschke, 2010), and the closing panel discussion of GEOBIA 2012,[3] GEOBIA may simply be described as "a sub-discipline of Geoinformatics devoted to

---

[3] http://www.inpe.br/geobia2012/keynote.php.

developing automated methods to partition remote sensing imagery into meaningful image-objects (i.e., patches), and assessing their characteristics through scale." (Hay and Castilla, 2008, p. 79).

8.1.2 Geographic Object-Based Hierarchies

Hierarchies are granular partitions (i.e., parts composed of parts), and thus from a geographical or landscape perspective, they can be divided into either *taxonomies* (representing partitions of the conceptual domain) or *zonations* (partitions resulting from projecting the taxonomy onto the territory) (Castilla, 2003). This important distinction is compatible with the *object-oriented* (OO) paradigm as defined in computer science. In this paradigm, two kinds of hierarchical structures are recognized, each of which describes different types of relationships between classes (Graham, 2001). A *class* refers to a group of objects with similar attributes and behaviors, and an *object* is any physical or conceptual entity. The first type of OO hierarchy describes the *aggregation relationship* by asking if a *focal* object (e.g., tree) is *a part of* a higher level object (e.g., forest stand), or reciprocally if it is *composed of a* lower level object (e.g., bole, branches, and leaves). True responses to both questions typically result in a *nested* hierarchy (Figure 8.1).

The second type of OO hierarchy describes the *generalization–specialization relationship* and asks whether an object at the lower levels is *a kind of* focal object. This results in generalization: For example, a pine is a kind of tree. Conversely, the upper level class *can be* the object at the focal level. For example, vegetation can be a tree, grass, or flowers, which represents specialization. True responses typically result in both *nested* and *unseated* hierarchies. For example, in Figure 8.2, *pine* and *tree* are each a kind of class above, while *vegetation* and *tree* can be the class beneath, and thus they are hierarchically nested. Similarly, *maple* and *flowers* are a kind of vegetation; however, both cannot be a kind of tree. Thus, flowers are unseated with regards to tree but nested with regards to vegetation. The important point to appreciate is that while both types of hierarchies may contain the same focal object (i.e., tree), their

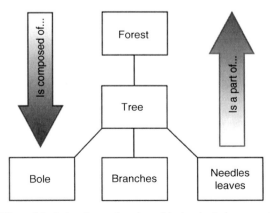

FIGURE 8.1 Hierarchical structures showing object-oriented aggregation relationship.

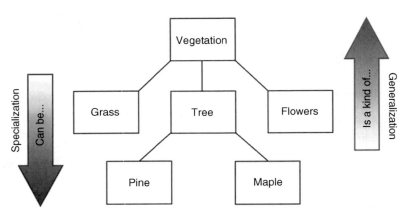

FIGURE 8.2   Hierarchical structures showing object-oriented generalization/specialization relationship.

relationships are completely different. Thus the questions that can be posed and the information resulting from each type of hierarchy will be very different. This is one critical reason why it is important to understand how patterns and processes are/are not (hierarchically) connected through scale.

As Cousins (1993) notes, recognizing the existence of these different types of hierarchies (nested and unseated) and the warnings against mixing them have not been fully understood or heeded across a broad range of disciplines. Similarly, Wu (1999, p. 4) states that "levels in the traditional hierarchy of ecological organization (i.e., individual-population-community-ecosystem[4]-landscape-biome-biosphere) are definitional and do not necessarily meet scalar criteria." Rowe (2001) also remarks that "the biological hierarchy of cell-organ-organism-ecosystem has been imprudently extrapolated to include psychological and social/cultural phenomena".

8.1.3   Multiscale Analysis

Over the last four decades, a number of computational techniques have been developed that incorporate concepts and theory from computer vision and machine learning to generate multiscale representations (Jähne, 1999) and detect important features in images (Starck et al., 1998; Dey et al., 2010). These include edge detectors (Canny, 1986), mathematical morphology (Haralick et al., 1987), watersheds (Beucher and Lantuéjoul, 1979), image texture analysis (Hay et al., 1996), spectral unmixing (Settle and Drake, 1993), neural nets (Foody, 1999), Bayesian networks (Robert, 2001), fuzzy logic (Wang, 1990), and multiscale techniques such as fractals (Mandelbrot, 1967), quadtrees (Klinger, 1972), pyramids (Klinger and Dyer, 1976), and wavelets (Daubechies, 1988). However, the results of these techniques often fall

---

[4] Cousins (1993) notes that while the concept of "ecosystem" is a subjectively determined aggregate with boundaries given by an observer, it is possible to define an ecological object which substitutes for ecosystem in a hierarchy of functional objects (pp. 77–78).

short when compared with those of human interpreters (Hay et al., 2003). This is in part because these techniques (traditionally) do not take into consideration or incorporate the concept of *object* within their analysis (Castilla and Hay, 2008). Yet this is fundamental to human cognition (Biederman, 1987).

Recently, this limitation has been more widely recognized, resulting in new object-based technologies being developed and applied to landscape analysis (Hay et al., 2001, 2005; Hay and Marceau, 2004; Blaschke, 2010; Chen et al., 2012; Syed et al., 2012) supporting the development of a growing GEOBIA community.[5] In 2000, eCognition[6] became the first commercially available software for multiscale object-based segmentation (Baatz and Schäpe, 2000). Here, segmentation follows a proprietary bottom-up region merging technique starting with one-pixel objects, which are iteratively merged into larger objects based on a user-defined (spectral) scale parameter.[7] A key limitation of this method is that there is no intuitive relationship between the scale parameter in eCognition (which is unitless) and spatial measures (i.e., area) specific to the objects of interest composing a scene. This limitation has recently been overcome by Castilla (2003) and Hay et al. (2005) with the development of *size-constrained region merging* (SCRM), where scene delineation is based on either manually or automatically [derived from object-based image statistics (Hay et al., 2005)] defining the minimum, maximum, and average size of all scene objects. More recent commercial software such as ENVI FX[8] also included unitless scale and aggregation parameters but with the advantage that as the user changes these parameters (along a Graphical User Interface (GUI) sliding threshold bar), the scene is segmented in real time allowing for the scale/aggregation levels to correspond to the boundaries of the scene objects of interest, also called image objects).

8.1.4 Image Objects

While remote sensing images are capable of providing high-resolution spatial, spectral, radiometric, and temporal data relevant for landscape analysis, these data are not composed of discrete landscape objects such as trees, forests, rivers, and cities; instead their fundamental primitive(s) are (typically millions of) pixels. Fortunately, *image objects* can be defined within a remote sensing image (Castilla and Hay, 2008). In simple terms, image objects are groups of pixels with meaning in the real world. More specifically, image objects are individually resolvable entities located within a digital image that are *perceptually* generated from high-resolution pixel groups (Hay et al., 2001), which are sufficiently internally "homogeneous" so as to be distinguishable from their surroundings. High resolution (H-res) corresponds to the situation where a single real-world object (i.e., a tree crown) is visually modeled

---

[5] International conferences have taken place biannually since OBIA, 2006 (Salzburg, Austria); GEOBIA 2008 (Calgary Alberta Canada), GEOBIA 2010 (Ghent, Belgium); and GEOBIA 2012 (Rio de Janerio, Brazil).
[6] http://www.ecognition.com/products, last accessed May 24, 2012.
[7] It also includes a user-defined shape heterogeneity parameter.
[8] ENVI FX Feature Extraction (http://www.exelisvis.com/language/en-us/productsservices/envi/enviex .aspx), last accessed May 24, 2012.

within a digital scene by many individual pixels; whereas low resolution (L-res) implies that a single pixel represents the integrated signal of many smaller real-world objects (e.g., leaves and branches within a single 30-m pixel) (Woodcock and Strahler, 1987). Because a H-res scene typically models complex land cover composed of varying sized, shaped, and spatially distributed image objects, there is often no single "optimal" scale for analysis; rather there are *many optimal scales* specific to the image objects that exist/emerge within a scene (Hay et al., 1994, 1997, 2002a,b; Marceau et al., 1994; Chen et al., 2011; Powers et al., 2012). Therefore, we support the idea that multiscale analysis (including the generation of scale hierarchies) should be guided by the innate spatial resolution of the salient landscape objects composing a scene. To achieve this, we describe the concept of linear scale-space and blob feature detection applied to a high-resolution remote sensing image and integrate these methods with ideas from hierarchical theories.

8.2 METHODS

The following methods were developed to explore the following questions: What does a scale domain look like, where is it located, and how can we visualize it?

Multiscale analysis requires two primary components: (i) the generation of a multiscale representation and (ii) a (multiscale) object delineation tool. To satisfy these requirements, we describe the use of *linear scale-space* and *blob-feature detection* (SS) [as defined by Lindeberg (1994) and adapted by Hay et al. (2002a,b)] for generating a multiscale representation of a complex agroforest landscape (Hall et al., 2004) and for automatically delineating dominant multiscale components from this representation.

8.2.1 Study Site and Data Set

Building on the work of Hay et al. (2002a), the data used in this study are a panchromatic (PAN) 500×500-pixel subimage of an IKONOS-2 (Geo) scene acquired in September 2001 (Figure 8.3a). IKONOS-2 provides 11-bit multispectral data in the red, green, blue, and near-infrared channels at 4.0 m spatial resolution and an 11-bit panchromatic (PAN) channel at 1.0 m resolution. Since the PAN channel covers a significant portion of the wavelengths represented by the four multispectral channels, a well-known 4.0-km^2 subarea of the 1.0-m PAN image was selected and resampled to 4.0 m. This resampled spatial resolution represents a trade-off between a fine enough grain and a reasonable spatial extent with which to evaluate object evolution over multiple scales. Resampling was conducted using *object-specific upscaling*, which is considered a robust (object-based) upscaling technique (Hay et al., 1997, 2001, 2003, 2004, 2005). Based on the computational demands required by SS processing, all data were linearly resampled to 8 bits (prior to analysis).

Geographically, this study area (Figure 8.3b) represents a portion of the highly fragmented agroforested landscape typical of the Haut Saint-Laurent region of southwestern Québec, Canada (Bouchard and Domon, 1997). The vegetation in

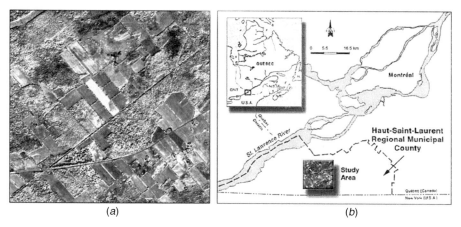

(a)   (b)

FIGURE 8.3 IKONOS subimage and study site map: (*a*) 500 × 500-pixel 4.0-m pan-chromatic image; (*b*) map location of image.

this area is dominated by a beech-maple climax forest (*Fagus grandifolia* Ehrh–*Acer saccharum* Marsh.) situated on uncultivated moraine islets, with cereal crops grown in the rich lowland marine clay deposits of the Champlain Sea (Meilleur et al., 1994). In Figure 8.3*a*, two thin dark-toned roads diagonally divide the scene into three main sections, while numerous hedge rows (narrow bright textured diagonal linear features) separate smooth-toned agricultural fields and textured forested areas. At the top left of the image, a (wide, smoothly textured) portion of an electrical utility corridor is visible, while near the image center, a bright-toned cornfield appears illuminated when compared to its neighboring darker toned alfalfa fields.

8.2.2 Linear Scale Space

Scale Space (SS) represents an uncommitted framework for early visual operations that was developed by the computer vision community to automatically analyze real-world structures at multiple scales (*t*), specifically when there is no a priori information about these structures or the appropriate scale(s) for their analysis (Lindeberg, 1994). The term *uncommitted framework* refers to observations made by a front-end vision system (i.e., an initial-stage measuring device) such as the retina or a camera that involves "no knowledge" and "no preference" for anything. In this text, we emphasize linear SS, though we note that a significant body of work exists on nonlinear SS. Interested readers are referred to a general definition on Wikipedia[9] as well as to more recent, related literature (Tzotsos et al., 2011; Doxani et al., 2012).

8.2.2.1 Gaussian Operators   When scale information is unknown within a scene, the only reasonable approach for an uncommitted vision system is to represent the

---

[9] http://en.wikipedia.org/wiki/Scale_space, last accessed April 21, 2013.

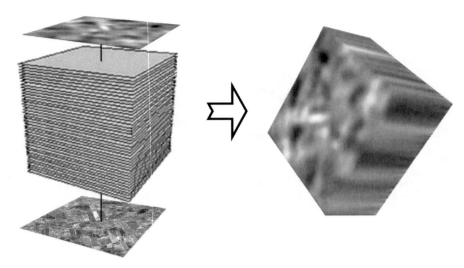

FIGURE 8.4   Linear SS stack or SS cube. The smallest scale (original image t_0) is on the bottom and the largest scale (t_{200}, most smoothed) is on the top. At the right side of the stack, the diffusive patterns of patches/objects through scale are visible.

input data at all (i.e., multiple) scales. Thus, the basic premise underlying SS is that a multiscale representation of a signal (such as a remotely sensed image) is an ordered set of derived signals showing structures at coarser scales that constitute "simplifications" of corresponding structures at finer scales. Simplification results from applying Gaussian filters to an initial image (i.e., the signal) at a range of increasing kernel sizes resulting in a "scale-space primal sketch" or "stack" of progressively "smoothed" image layers, where each new layer represents convolution at an increased scale (Figure 8.4). More explicitly, each smoothed layer is created by convolving the nth-order *derivative of a Gaussian* function with the original image, where the scale of each derived signal is defined by selecting a different (incremented) standard deviation for the derivative of a Gaussian function (at each new iteration).

The use of Gaussian operators is both fundamental and essential for SS analysis. An in-depth discussion of this topic is beyond the scope of this chapter; however, we note that Gaussian kernels (and all partial derivates) are solutions of the *linear isotropic diffusion equation*, and thus the exact behavior of Gaussian smoothing is well known. Most importantly, this means recognizing that no "artifacts" are produced in the "scaled" images as a result of the Gaussian function. In addition, Gaussian kernels also exhibit similarity with biological visual operators, and they satisfy the axioms for an uncommitted vision system, which includes linearity (i.e., no a priori knowledge is required), and no preference for location, orientation, and scale (Weickert et al., 1997).

In this work we have only used the zero-order derivative and applied it 200 times with a scale increment of one. This results in a SS stack of 200 increasingly smoothed

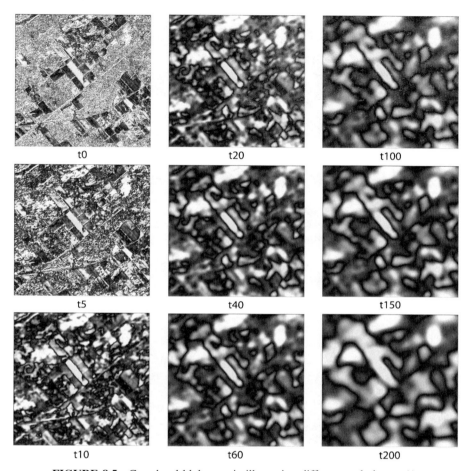

FIGURE 8.5 Gray-level blob mosaic illustrating different scale layers (*t*).

images, which models the evolution of the original image through scale. Each hierarchical layer in a stack represents convolution at a fixed scale, with the smallest scale (i.e., finest object details) at the bottom and the largest scale (i.e., coarsest object features) at the top (Figure 8.4). The main features that arise at each scale within a stack are smooth regions, which are brighter or darker than the background and which stand out from their surroundings. These regions are referred to as "gray-level blobs" (Lindeberg, 1993) (Figure 8.5).

8.2.3 Blob–Feature Detection

The second component of multiscale analysis involves applying an algorithm(s) to automatically delineate relevant structures within the image. Typically, edge, ridge, and corner detection have been applied to the SS primal sketch. While the related concepts and applications of edge detection in ecology are both useful and interesting, they also represent a relatively well-defined body of knowledge (Forman, 1995). However, the notion of detecting blobs and their association with *multiscale object-based* frameworks (Hay et al., 2002a; Burnett and Blaschke, 2003; Syed et al., 2011) and the *hierarchical patch dynamics paradigm* (Wu, 1999) represent a more recent body of multiscale research where exciting opportunities and new discoveries still exist. It is this avenue of research that we will report on here.

To delineate dominant features within the SS stack, we apply *blob–feature detection*. Its primary objective is to link structures (i.e., gray-level blobs) at different scales in SS, to higher order objects called SS blobs and to extract significant features based on their appearance and persistence through scales. When blobs are evaluated as a volumetric structure within a stack, it becomes apparent that some structures visually persist through scale, while others disappear (Figure 8.6). Consequently, an important premise of SS is that bloblike structures which "persist" in SS are likely candidates to correspond to significant structures in the image and thus in the real-world scene.

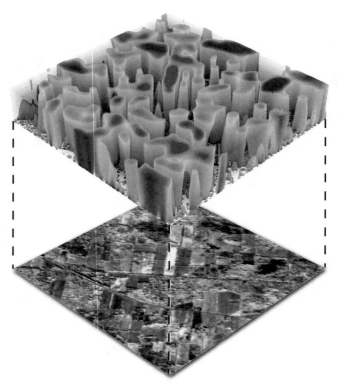

FIGURE 8.6   Gray-level stack with opacity filters applied to illustrate persistence of blob structures through scale with original image on bottom to provide context.

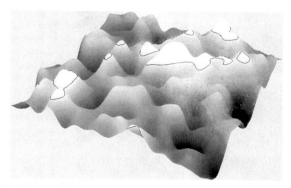

FIGURE 8.7   Scale layer (t_{150}) modeled as 2.5D topographic surface. Contours represent an instance of flood subsidence when two or more peaks become connected. These connected boundaries become the region of support for delineated blobs.

8.2.3.1 *Gray-Level Blob Delineation*   Operationalizing this concept of SS persistence is not trivial and involves a number of components, which we will only briefly describe. For more detailed information see Lindeberg (1994) and Hay et al. (2002a). In simple terms, gray-level blob delineation may be explained with a watershed analogy. In this context, gray-level blobs at each scale (t) are treated as objects with extent both in two-dimensional (2D) space (x, y) and in the gray level (z axis). Thus, dark values topologically represent valleys, while bright values represent peaks. By working with one image layer in the SS stack at a time, we consider each scale as a flooded 3D gray-level landscape. As the water level gradually sinks, peaks appear. At some instance, two different peaks become connected (Figure 8.7).

The corresponding "connected" elevation level is called the "base level" of the blob. These are used for delimiting the 2D spatial extent or "region of support" of each blob, which is defined as a *binary blob* (Figure 8.8).

8.2.3.2 *Binary Blobs*   The actual technique for defining binary blobs involves convolving the original image with the Laplacian of the Gaussian at different standard deviations. This results in 200 new derivate images. The Laplacian is a second-derivative operator that is invariant to rotation. Thus it is insensitive to directional discontinuities (points, lines, edges). We use the same standard deviation values (of the Gaussian operators) for generating binary blobs that were used to create the gray-level blobs. Zero crossings (also called *thresholds*) are defined in each new derivative image, which result in binary blobs. Once all the 2D binary blobs (x, y, t) are defined, they are combined within a new stack to create *3D hyperblobs* (Figure 8.9). Essentially, binary blobs represent a discretization (or *objectification*) of corresponding gray-scale blobs.

From a landscape ecology perspective, binary blobs spatially correspond to *patches*, that is, spatially discrete landscape components. Thus, the opportunity exists to generate landscape metrics for the binary blobs at each scale; however, we caution the reader that doing so would represent only a partial (linear) analysis of landscape

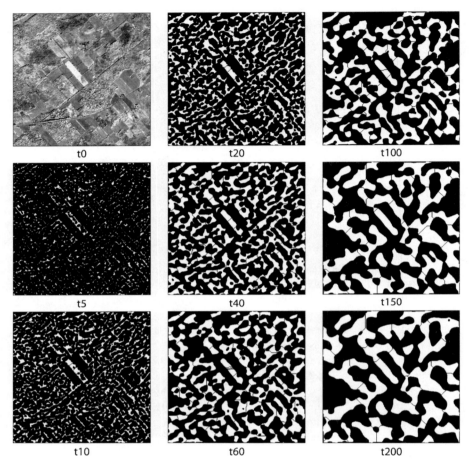

FIGURE 8.8   Binary blob mosaic illustrating same scale layers (*t*) as Figure 8.5.

components and the landscape is intrinsically nonlinear. In addition, it should be noted that many metrics do not scale well (Wu and Li, 2005) and a lack of formalization among metrics significantly limits their utility (Bogaert, 2003; Li and Wu, 2004).

8.2.3.3   Scale-Space Events, Lifetimes, and Normalization   Within a single hyper-blob, there are four primary types of visible structures referred to as *bifurcations* or *SS events* (Figure 8.10). These consist of the following:

- *Creations*—One (or more) new blob appears.
- *Annihilations*—One (or more) blob disappears.
- *Merges*—Two (or more) blobs merge as one.
- *Splits*—One blob splits into two (or more).

FIGURE 8.9 Hyperblob stack composed of 2D binary blobs. For illustrated purposes only, each binary layer has been assigned a value equal to its scale. Thus dark values (i.e., smallest scales) are on the bottom, while the brightest value (i.e., largest scale) is at the top.

The ability to define these SS events represents a critical component of SS analysis, as scales between bifurcations are linked together forming the *lifetime* (Lt_n) and topological structure of individual SS blobs (Figure 8.11). The lifetime defines the persistence—or evolution—of a structure through scale (see Section 8.3.1.2 for a discussion on SS lifetimes), and the topology provides a way to contextually query SS components.

As blob behavior is strongly dependent upon image structure, it is possible that an *expected* image behavior may exist (Lindeberg, 1994). Thus statistics are extracted from a large number of stacks that describe how random noise blobs can be expected to behave in scale space. These statistics are then used to generate a normalized (4D) SS volume for each SS blob. In our processing we generated 100 individual stacks resulting from 100 different random (white-noise) images the same size as the original

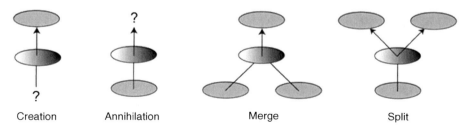

FIGURE 8.10 Four generic blob events. The gradient shaded disks represent the focal blob, i.e., the blob under analysis. The ? indicates that no blobs exist at this scale. We note that "creation" and "merge" describe events leading to the focal level, while "split" and "annihilation" describe events from the focal level.

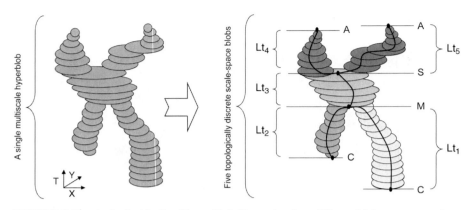

FIGURE 8.11   A single, idealized hyperblob illustrating four different blob events: creations (C), merges (M), splits (S), and annihilations (A). The number of scales between SS events represents the lifetime (Lt_n) of a SS blob. Five different Lt_n are illustrated.

500×500 pixel image. Each random SS stack was composed of 200 layers with a scale increment of one. We note that the data required to conduct this normalization procedure are an important reason (and limitation) for performing analysis on a relatively small scene (500×500 pixels). For more in-depth information regarding this normalization procedure see Hay et al. (2002a).

8.2.3.4 SS Blob Ranking and Vectors   Once the lifetime of each SS blob is defined, the binary blobs that correspond to a specific lifetime are used as a mask to extract the integrated *normalized (4D) volume* (x, y, z, Lt_n) of each individual SS blob from a normalized gray-level stack. These resulting normalized volumes are then ranked from the highest to the lowest, and a user-determined number of significant SS blobs are generated, from which the scale (t) representing the maximum 3D gray-level blob volume (x, y, z) of each hyperblob is extracted. From these layers the associated 2D spatial support (i.e., binary blob) is identified, converted from a raster to a vector polygon, and then related back to the corresponding structures in the image for further examination (Figure 8.12). Thus, 4D SS blobs are simplified to 3D gray-level blobs, which are further simplified to their 2D support region [i.e., modeled as a geographic information system (GIS) polygon] and then to their corresponding real-world structure in the original image. For a detailed nonmathematical description of SS and blob feature detection, see Hay et al. (2002a).

8.3   DISCUSSION

8.3.1   Integrating Hierarch Theory and Scale Space

8.3.1.1 Integrating Hierarchy Theory to Reduce SS Processing Requirements
An important limitation of SS is that within an SS cube a significant amount of "redundant" data result in large stack sizes. Based on the original gray-scale image

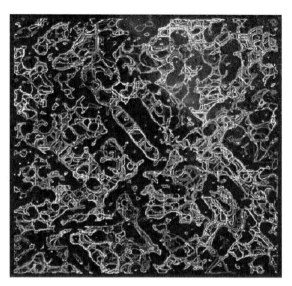

FIGURE 8.12   Ranked blobs converted to individually queriable polygons. Note how the different (colored) polygons overlay each other making analysis nontrivial. Compare with Figure 8.3a. (See the color version of this figure in the Color Plates section.)

size of 249 kB, we generated 101^{10} (additional) stacks, with sizes ranging from 200 to 980 MB. To reduce the random access memory (RAM) requirements when defining SS blob topology, we integrated a three-tier hierarchical approach in the programming language IDL[11] (interactive data language) to parallel-process multidimensional array structures. Thus, instead of loading the entire stack into memory and trying to assess the varied structure of potentially thousands of individual SS blobs per stack, we only need to load three scales of a SS stack into memory at a time. From a hierarchy theory perspective, we evaluate the blob locations at the "focal" scale (see Figure 8.10) and establish links with blobs in the scale above and with those below. We then shift up an additional scale in the cube while dropping the bottom scale, always keeping only three scales in memory at one time. We then repeat this procedure until the last scale has been processed. This simple yet elegant application of hierarchy theory provides the potential to automatically evaluate much larger landscape areas than the 4-km^2 scene defined here. We also suggest that when this application is combined with image tiling, and with running multiple instances of the software (on different stacks) at the same time (on multicore CPUs), the ability to assess significantly larger scenes will be further improved. However, the implementation of these ideas is not the focus of this chapter.

---

[10] The original SS stack and 100 random-noise SS stacks.
[11] http://en.wikipedia.org/wiki/IDL_(programming_language), last accessed May 24, 2012.

8.3.1.2   *From SS Lifetimes to Scale Domain-Level Topology*   The result of this
processing is a large number of ranked (blobs) polygons defined from three dimensions (x, y, t) that physically exist within the same area (x, y). As a result they visually
obscure each other when overlaid on the same study area (Figure 8.12), and querying
each polygon is not trivial. To solve this, we need an improved topology for each
ranked blob. This is achieved by considering SS events as critical thresholds within a
hyperblob, where fundamentally different geometric structures exist in both scale and
the landscape. Thus, from an ecological perspective, the lifetime of a SS blob (i.e., the
scales between two events) may be considered as a *level* within a specific scale
domain. And the domain can be defined by the first and last scales within a stack.
To define these levels, each hyperblob is topologically defined as a unique
multiscale entity, and its corresponding SS events are isolated (as defined in Section
8.2.2.3, Figure 8.11). Then, the location of the first SS event of each hyperblob is
geometrically defined regardless of its event type (i.e., creation, annihilation, etc.) and
at what scale it exists within the stack. Then the second, third, and nth events of
each hyperblob are isolated until the last possible event is defined. These event values
are then considered as domain-level attributes and are assigned to their corresponding
ranked blobs.

This process can be visualized as generating unique domain levels each of which
pass through a different set of newly defined events (Figure 8.13). For example, one
level passes through all the first events of each hyperblob, another through all second
events, and so on, until the last level has been defined as passing through the last set of
events of their corresponding hyperblobs. Then the "ranked" 2D blobs located within
each newly defined domain level (the dark disks in Figure 8.13) are provided a new

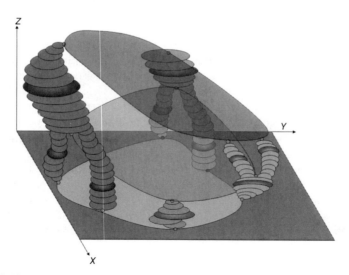

FIGURE 8.13   Conceptual model of three domain-level manifolds (layers) passing through
SS events (red dots). Dark-toned disks represent the highest ranked blobs located within each of
the three levels. (See the color version of this figure in the Color Plates section.)

corresponding domain-level attribute. This method of topologically defining a domain-level attribute provides a unique way to query, partition, and evaluate the ranked blobs and their related real-world structures, as many ranked blobs can and do exist within the full horizontal areal extent covered by a domain level, but no more than one ranked blob can exist within the same vertical space defined between two levels (Figure 8.13). Thus the problem of overlapping ranked blobs (Figure 8.12) is resolved because blobs can be queried and evaluated by the unique levels in which they exist.

8.3.1.3 Four New Shared-Event Classes Defined Within the ($500x$ by $500y$ by $200t$) SS cube we distinguish 10371 single events of the four known event types (Table 8.1). However, we also identified four new classes of shared events (composed of 1429 shares) that to our knowledge have never been defined before (Table 8.2). We surmise that the reason for these shared events has to do with our choice of scale increment between successive layers within the stack and/or noise in the original image.

In our processing, we have defined 200 individual scales, where the scale increment is unity. Thus, at each layer in the SS stack, the standard deviation of the Gaussian filter is increased by unity ($STD + 1$). However, this is an arbitrarily chosen value. We could have chosen fractional components as has been done by others (ter Haar Romeny et al., 2000), but in doing so, we would still have to arbitrarily select this value. This is because in theory a scale continuum is continuous; however, digitally we are limited to a discrete (i.e., binary) computational environment.

These new shared events include creations and annihilations (C&A), creations and splits (C&S), merges and annihilations (M&A), and merges and splits (M&S)

TABLE 8.1 Single-Event Types and Quantities

Event Types	Number	Percent
Creations (C)	4,516	42.1
Annihilations (A)	1,528	14.2
Merges (M)	3,309	30.9
Splits (S)	1,378	12.8
Total	**10,731**	**100**

TABLE 8.2 Shared-Event Types and Quantities

Event Types	Number	Percent
C & A	949	66.4
C & S	63	4.4
M & A	26	1.8
M & S	391	27.4
Total	**1429**	**100**

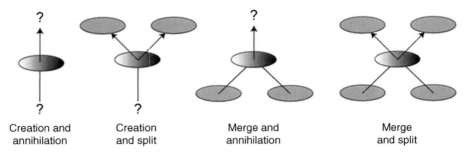

| Creation and | Creation | Merge and | Merge |
| annihilation | and split | annihilation | and split |

FIGURE 8.14   Four new dual-instant blob events. The ? represents no blobs found at these scales.

(Figure 8.14). Conceptually, events are dimensionless points, but computationally they represent at a minimum a single pixel in SS (x, y, t). More commonly, they are blob areas composed of many pixels. In an ideal computational system, a C&A would never exist, because there would be (infinitely) many scales between the two shared events that would allow for each individual event to be isolated at a different t. However, for practical reasons, the resolution of our system has been defined as a whole-number increment of STD $+ 1$. Since each C&A only exists at a single scale, these events have a correspondingly short lifetime; consequently their persistence is low, as is the probability of them being ecologically meaningful and/or spatially dominant in the landscape. Thus, it is more likely that C&A events are the result of image noise even though great effort was taken to normalize the data (see Section 8.2.3.3). This supposition corresponds with our findings that 84.5% (802/949) of all C&A events are defined within the first 11 of 200 scales, after which their numbers dramatically diminish at higher t. Image noise—typically small random variations in the tone of the data—is integrated into the stronger background signal of the matrix through scale, which is the same pattern we see manifest in these data.

8.3.2   SS Events and Domain Thresholds: What Does a Scale Domain Look Like and Where Do We Go from Here?

What does a scale domain look like? To the best of our knowledge, no one has spatially modeled a scale domain based on remote sensing data for a single instant in time. Wu (1999) has proposed the hierarchical scaling ladder in relation to the HPDP, but this is a conceptual model. Similarly, time–space scales have been created illustrating zooplankton biomass variability as in the Stommel diagram (Haury et al., 1978), and boreal forest structures in relation to disturbance and atmospheric possesses, as described by Holling (1992), but none of these explicitly illustrate a single instance, and none of them actually define in detail what a scale domain may look like or the area over which they "exist." In the process of conceptualizing *scale-domain manifolds* and methods to visualize them, it became apparent that while the domain-level concept works for topologically discriminating ranked blobs (Section 8.3.1.2), the assumption that a single scale domain exists between the first and last

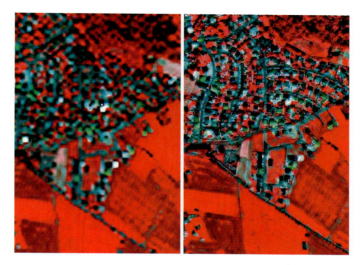

FIGURE 2.12 Multispectral IKONOS image in false-color CIR display (band 4: red, band 3: green, band 2: blue) reduced to 10 m (left) and fusion result with panchromatic 1-m IKONOS image (right). The fusion process produces excellent pan-sharpening and spectral preservation results and is evaluable as a 1-m multispectral image.

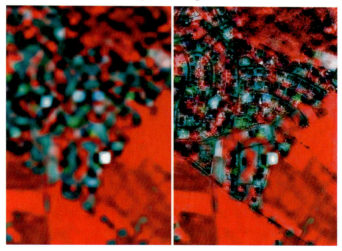

FIGURE 2.13 Multispectral IKONOS image in false-color CIR display (band 4: red, band 3: green, band 2: blue) reduced to 20 m (left) and fusion result with panchromatic 1-m IKONOS image (right). The fusion process produces a multispectral image of improved spatial resolution but some spectral changes begin to occur which would make a multispectral analysis questionable. Also the fused image begins to look like a "colorized black-and-white photo."

Scale Issues in Remote Sensing, First Edition. Edited by Qihao Weng.
© 2014 John Wiley & Sons, Inc. Published 2014 by John Wiley & Sons, Inc.

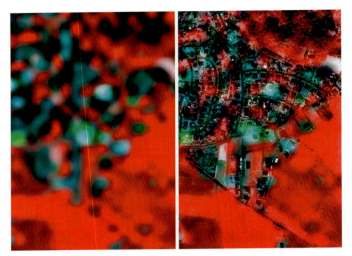

FIGURE 2.14 Multispectral IKONOS image in false-color CIR display (band 4: red, band 3: green, band 2: blue) reduced to 30 m (left) and fusion result with panchromatic 1-m IKONOS image (right). The fusion process no longer produces a single image. The effects of "color overlay" on a panchromatic image make this image very hard to interpret.

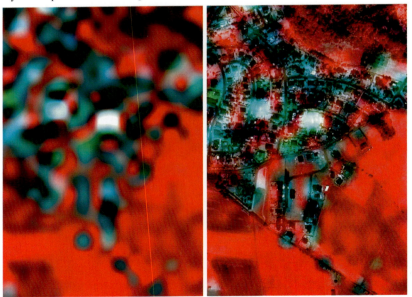

FIGURE 2.15 Multispectral IKONOS image in false-color CIR display (band 4: red, band 3: green, band 2: blue) reduced to 32 m (left) and fusion result with panchromatic 1-m IKONOS image (right). The fused image shows no significant differences from the fused 30-m image in Figure 2.14.

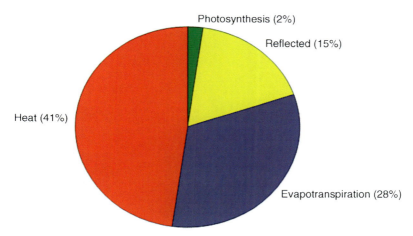

FIGURE 3.2 Partitioning of surface energy fluxes in Hubbard Brook (Bormann and Likens, 1979; Gosz et al., 1978; Kay, 1978).

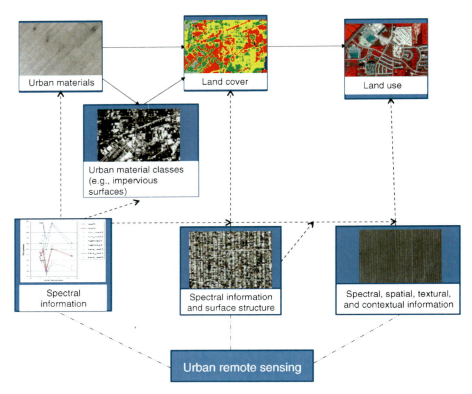

FIGURE 4.1 Relationship among remote sensing of urban materials, land cover, and land use (after Weng and Lu, 2009).

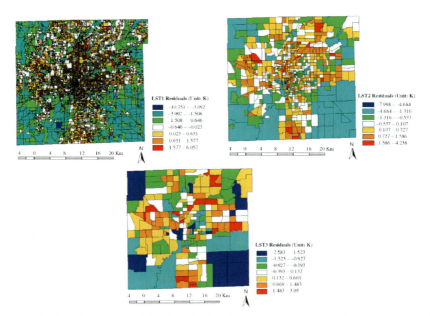

FIGURE 4.2 Distribution of LST residuals at black, black group, and census tract level.

FIGURE 5.1 Photographs showing effects of leaf beetles on *Tamarix* stands at Lower Dolores River in 2009 (a); Humbolt River in 2007 (b); Walker River in 2006 (c); and Upper Colorado River site in 2007 (d).

May 1, 2012 Left and Right Views, Virgin River Tower Site

June 1, 2012

July 1, 2012

August 1, 2012

September 1, 2012

FIGURE 5.2 Series of wide-angle phenocam images showing progression of defoliation of *Tamarix* by leaf beetles on Virgin River. Two cameras mounted on the same tower capture different views of the floodplain.

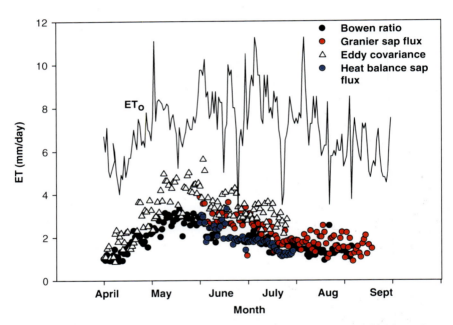

FIGURE 5.6 Comparison of *Tamarix* ET estimates in 2009 at stressed site (Diablo) on Lower Colorado River, CA.

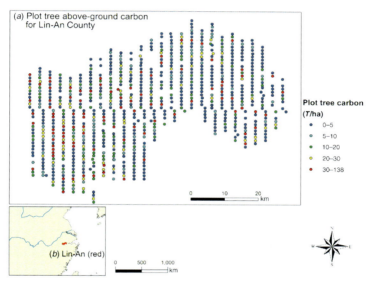

FIGURE 6.1 (*a*) Locations of forest inventory sample plots and their values of above-ground forest carbon and (*b*) location of study area.

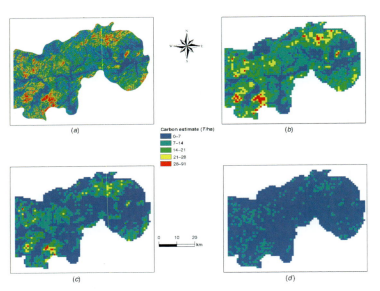

FIGURE 6.4 (*a*) Estimation map of above-ground forest carbon at spatial resolution (SR) of 30 m × 30 m using sequential Gaussian cosimulation; (*b*) above-ground forest carbon map at spatial resolution of 990 × 990 m by scaling up sample plot and image data from spatial resolution of 30 m × 30 m using PSCPS upscaling method; (*c*) above-ground forest carbon map at spatial resolution of 990 × 990 m by scaling up sample plot and image data from spatial resolution of 30 m × 30 m using PSCBS; and (*d*) difference map of forest carbon estimates between two methods.

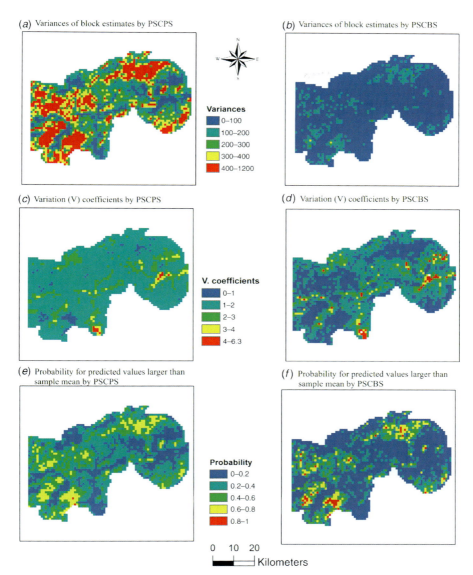

FIGURE 6.5 Variances of block predicted values for above-ground forest carbon at spatial resolution of 990 m × 990 m by (*a*) PSCPS and (*b*) PSCBS, respectively; their coefficients of variation by (*c*) PSCPS and (*d*) PSCBS, respectively; and probability maps for block predicted values larger than a given threshold—sample mean by (*e*) PSCPS and (*f*) PSCBS, respectively.

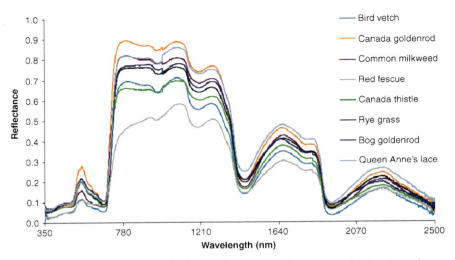

FIGURE 7.1 Lab-measured leaf spectral reflectance for selected dominant species.

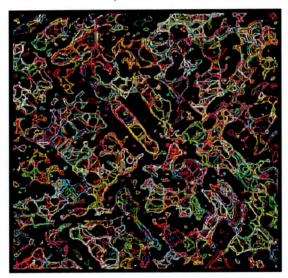

FIGURE 8.12 Ranked blobs converted to individually queriable polygons. Note how the different colored polygons overlay each other making analysis nontrivial. Compare with Figure 8.3*a*.

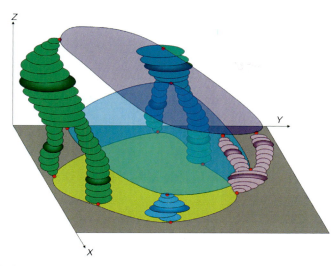

FIGURE 8.13 Conceptual model of three domain-level manifolds (layers) passing through SS events (red dots). Dark-toned disks represent the highest ranked blobs located within each of the three levels.

FIGURE 8.15 Colorized model of five scale-domain manifolds based on annihilation events that have been extracted from 200-layer SS stack (shown in Figure 8.4). The first scale models 1221 annihilation events, the second 152, the third 64, the fourth 28, and the final scale models 10 annihilation events. The original study site (panchromatic image 500xy) is shown on the bottom to provide context.

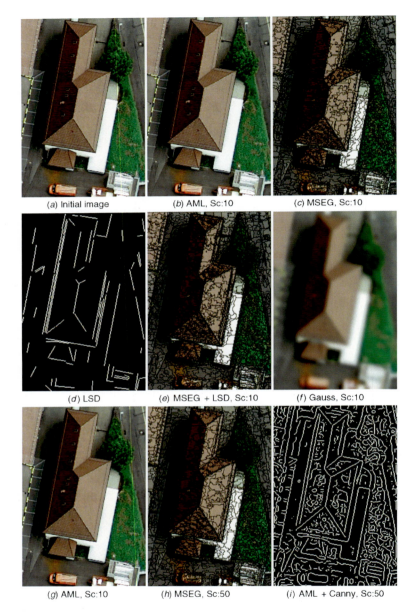

(a) Initial image   (b) AML, Sc:10   (c) MSEG, Sc:10

(d) LSD   (e) MSEG + LSD, Sc:10   (f) Gauss, Sc:10

(g) AML, Sc:10   (h) MSEG, Sc:50   (i) AML + Canny, Sc:50

FIGURE 9.1 Comparison of region merging segmentation results using scale-space representations and advanced edge features. Four scales from various steps of the proposed methodology are presented. (a,f,k,p) Initial remote sensing aerial image with spatial resolution of 5 cm along with three Gaussian scales; (b,g,l,q) scale-space representation at various selected scales; (c,h,m,r) initial image objects using MSEG algorithm applied to scale-space representation; (d,i,n,s) advanced edge and line features used in the following step; (e,j,o,t) results of edge-enhanced MSEG algorithm proposed in this research.

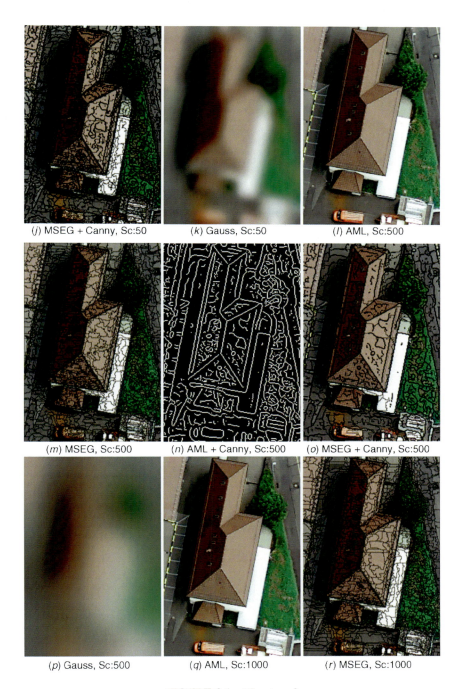

(*j*) MSEG + Canny, Sc:50   (*k*) Gauss, Sc:50   (*l*) AML, Sc:500

(*m*) MSEG, Sc:500   (*n*) AML + Canny, Sc:500   (*o*) MSEG + Canny, Sc:500

(*p*) Gauss, Sc:500   (*q*) AML, Sc:1000   (*r*) MSEG, Sc:1000

FIGURE 9.1   (*Continued*)

(s) AML + Canny, Sc:1000   (t) MSEG + LSD, Sc:1000

FIGURE 9.1 (*Continued*)

FIGURE 9.2 Comparison of various segmentation algorithms on a DMC aerial multispectral image with 5 cm pixel- size (copyright Intergraph Corp.): (*a*) standard MSEG with scale parameter 100; (*b*) mean-Shift segmantation with default parameters; (*c*) multiresolution segmentation (eCognition) with default parameters; (*d*) edge-constrained MSEG without merging edge objects in last pass for demonstration purposes; (*e*) edge-constrained MSEG with Canny edge features used; (*f*) edge-constrained MSEG with LSD edge features used.

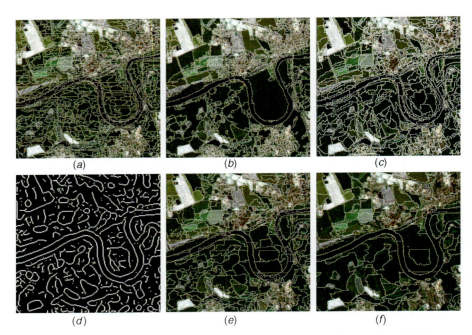

FIGURE 9.4 Comparison of various segmentation algorithms on Landsat TM data set (Dessau, Germany); (*a*) standard MSEG results with scale parameter 100; (*b*) mean-shift segmantation with default parameters; (*c*) multiresolution segmentation (eCognition) with default parameters (scale 10, shape 0.1); (*d*) Canny edge detection applied on AML scale-space representation; (*e*) edge-constrained MSEG with Canny edge features used and scale parameter 100; (*f*) edge-constrained MSEG with Canny edge features used and scale parameter 400.

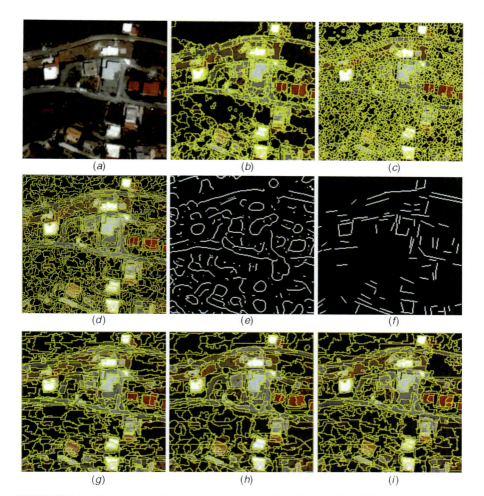

FIGURE 9.5 Comparison of various segmentation algorithms on QuickBird data set (Eastern Attika, Greece): (*a*) original image; (*b*) mean-shift segmentation with default parameters; (*c*) watershed segmentation with default parameters; (*d*) multiresolution segmentation (eCognition) with default parameters; (*e*) canny edge detection applied on AML scale-space representation; (*f*) LSD line features extracted from original image; (*g*) standard MSEG results with scale parameter 100; (*h*) edge-constrained MSEG with Canny edge features used and scale parameter 100; (*i*) edge-constrained MSEG with LSD line features used and scale parameter 100.

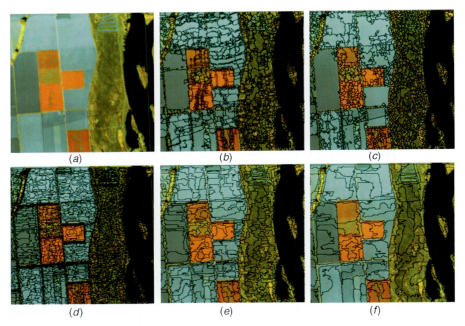

FIGURE 9.6 Comparison of various segmentation algorithms on CASI hyperspectral data set (copyright Remote Sensing Laboratory, NTUA) with 95 spectral bands (Axios River, Thessaloniki, Greece): (*a*) original image; (*b*) mean-shift segmentation with default parameters; (*c*) watershed segmentation with default parameters; (*d*) standard MSEG results with scale parameter 900; (*e*) standard MSEG on AML scale-space representation and scale parameter 900; (*f*) edge-constrained MSEG with Canny edge features used and scale parameter 900.

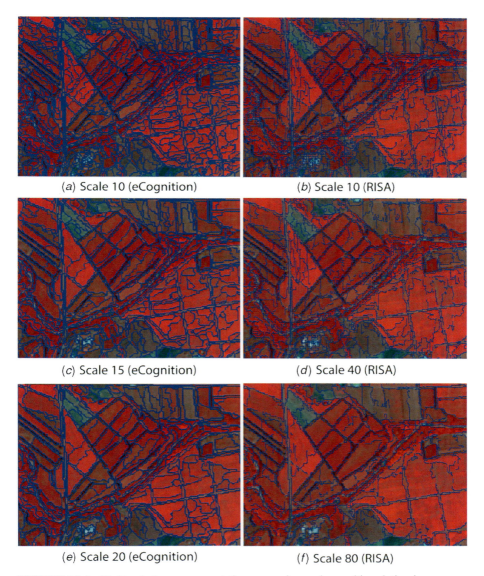

(a) Scale 10 (eCognition) (b) Scale 10 (RISA)

(c) Scale 15 (eCognition) (d) Scale 40 (RISA)

(e) Scale 20 (eCognition) (f) Scale 80 (RISA)

FIGURE 10.4 Multiscale image segmentation comparison using multiresolution image segmentation algorithm in eCognition and RISA from SPOT 5 data. (Adapted from Wang et al., 2010.)

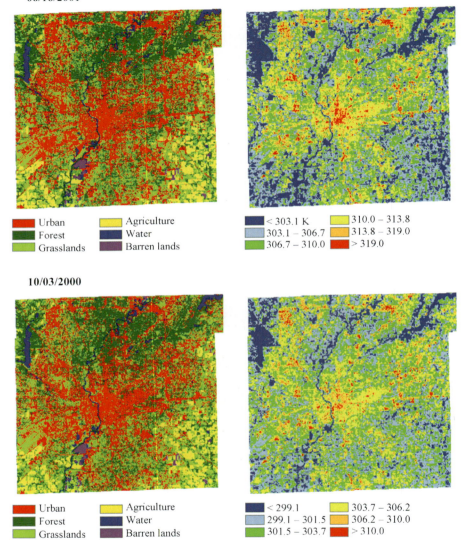

FIGURE 11.2 LULC (left) and LST (right) maps for the city of Indianapolis (image dates: June 16, 2001 and October 3, 2000).

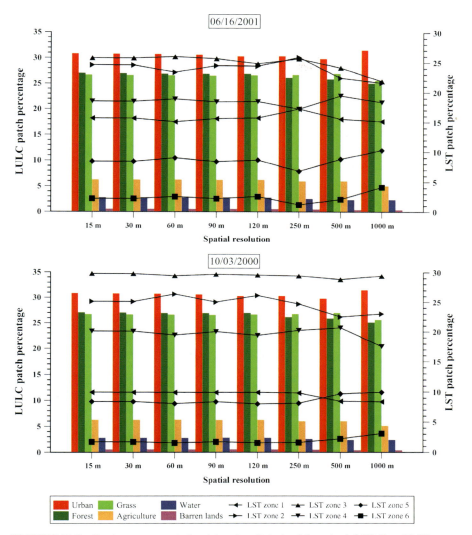

FIGURE 11.3 Patch percentage index (class-level) derived from both LULC and LST maps for two image dates: June 16, 2001 and October 3, 2000.

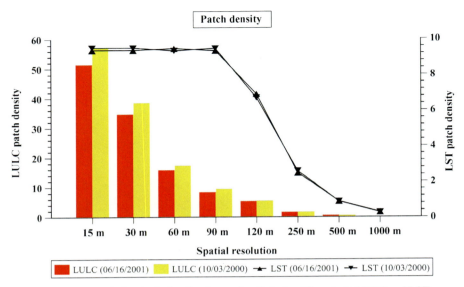

FIGURE 11.4 Patch density index (landscape-level) derived from both LULC and LST maps for two image dates: June 16, 2001 and October 3, 2000.

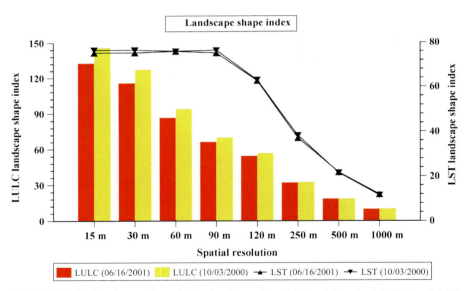

FIGURE 11.5 Landscape shape index (landscape-level) derived from both LULC and LST maps for two image dates: June 16, 2001 and October 3, 2000.

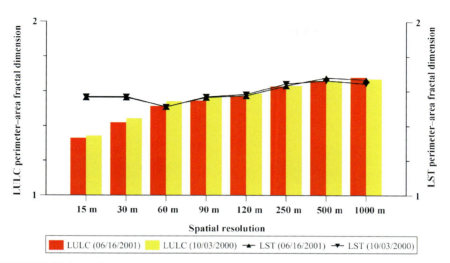

FIGURE 11.6 Perimeter–area fractal dimension index (landscape-level) derived from both LULC and LST maps for two image dates: June 16, 2001 and October 3, 2000.

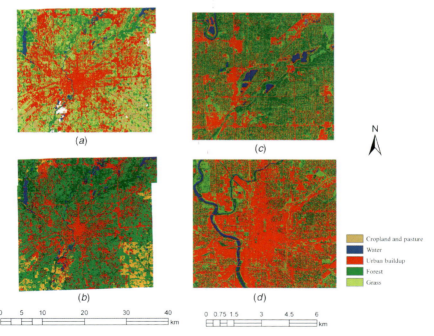

FIGURE 12.5 Sample land use and land cover maps: ETM + 00 (*a*), ASTER01 (*b*), IKONO01 subscene (*c*), and IKONO03 subscene (*d*).

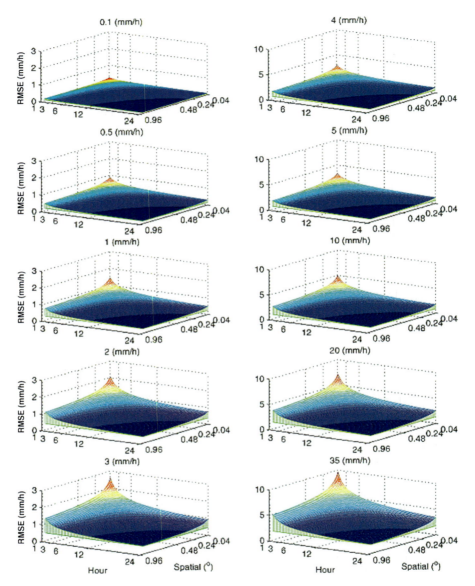

FIGURE 13.4 Plot of the error distribution with respect to space scale L, time scale T, and rain intensity R using the optimal parameters.

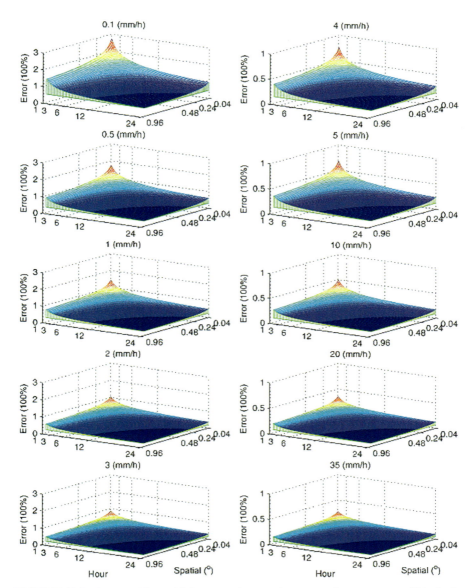

FIGURE 13.5 Same as Figure 13.3, expect the error is expressed as percentage (%) of rain rates.

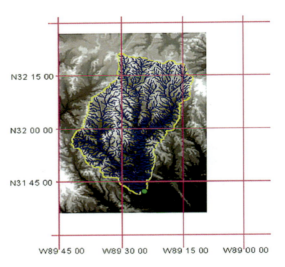

Leaf River Near Collins, Mississippi
USGS # 02472000
Basin Area : 1949 km²
~ 44 x 44 km²

FIGURE 13.6 Study area for streamflow uncertainty estimation associated with rainfall uncertainty.

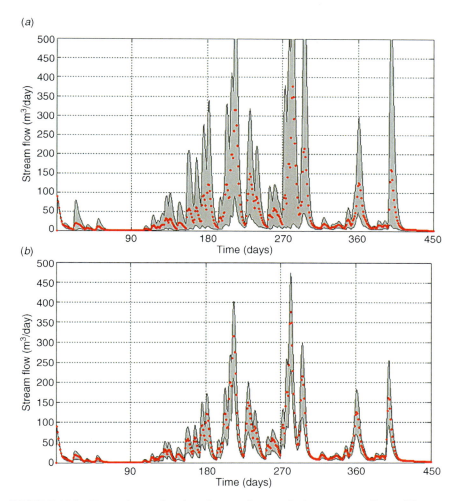

FIGURE 13.8 Uncertainty estimation of stream flow prediction corresponding to 95 percentile confidence using rainfall error propagation: (a) fixed rainfall error 50% and (b) scale-dependent rainfall error according to Equation (13.1). The red dots are mean of the 100 ensemble streamflow sequences simulated from satellite rainfall estimates.

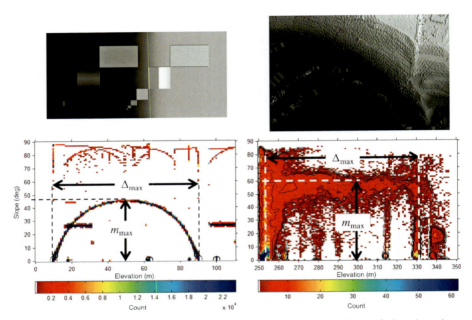

FIGURE 14.5 Measurement of terrain parameters from visualizations of elevation–slope histograms. Top: input DSM; bottom; elevation–slope histograms as images.

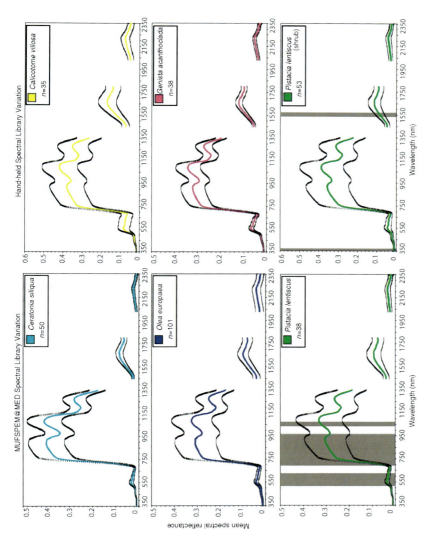

FIGURE 15.4 Mean reflectance spectrum of common Mediterranean vegetation obtained with aeral lift mobile platform MUFSPEM@MED (left) and hand held (right). Spectra are flanked by standard deviation of mean (black lines) at 95% confidence level. Wavelengths where the variance between three plants compared in each spectral library was equal are shaded gray on the lower graphs. Spectral regions around 1400, 1940, and 2400 nm were removed due to atmospheric noise. Figure adopted from Manevski et al. (2012).

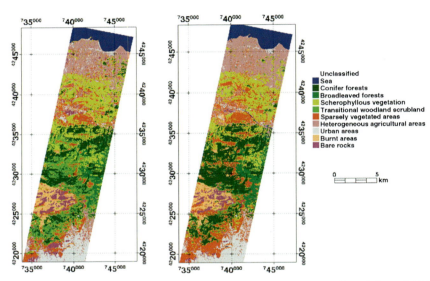

FIGURE 15.5 Hyperion pixel-based classification using SVM classifier (left) and ANNs (right). Adopted from Petropoulos et al. (2012a).

layers of the SS stack may be invalid. This is because the corresponding number of layers (200) has been arbitrarily designated (without reference to anything ecological), which is an obvious limitation of this method. It could have just as easily been 20 or 2222 as 200. However, the concept that interactions tend to be stronger and more frequent *within* a domain than *among* domains has provided critical insight into how to remedy this assumption.

Conceptually, all event types that are generated from a single hyperblob exist between a creation and annihilation. Thus corresponding creations, splits, and merges are all members within the same scale domain. Only annihilations represent a discrete thresholding of the structure(s) above and below. Therefore, by integrating the concept of hierarchical thresholds with geostatistics and 3D visualization techniques, we show that annihilation events can be visually modeled as scale-domain manifolds (Figure 8.15). To visualize and model these domain structures, we first define all annihilation events and then locate the center pixel of each annihilation blob in a similar manner as previously defined for domain levels (Section 8.3.1.2). That is, the center location (x, y) of the first annihilation event of each hyperblob within the entire stack is defined as a unique set of domain points regardless of the scale (t) that exists within the stack. Then the locations of the second, third, and nth annihilation event of all other hyperblobs are similarly isolated until the last possible event is defined. Delaunay triangulation is then

FIGURE 8.15 Colorized model of five scale-domain manifolds based on annihilation events that have been extracted from a 200-layer SS stack (shown in Figure 8.4). The first scale models 1221 annihilation events, the second 152, the third 64, the fourth 28, and the final scale models 10 annihilation events. The original study site (panchromatic image $500xy$) is shown on the bottom to provide context. (See the color version of this figure in Color Plates section.)

applied independently to each set of domain points, and the resulting surfaces are independently interpolated with a polynomial function to generate a visually smooth, topographically variable manifold/surface.

We suggest that the resulting manifold structures (Figure 8.15) correspond to a *modified* "scaling ladder" as conceptualized by Wu (1999), except that in this case, instead of a single ladder that represents the entire landscape at "constant scales" (i.e., the spaces between the rungs of the ladder are constant), we have essentially generated many geo-object-based ladders, with variable spacing between rungs. Conceptually, the base of each ladder corresponds to the center (x, y) of the spatially dominant structures (i.e., patches/image objects) in the scene, while the height of the rungs is defined by the location of the annihilation points at unique scales (t) within the stack. These annihilation points signify the spatial extent (i.e., range of influence) of the initial object through scale. Consequently, coarser scale domains will include fewer larger objects, each composed of an aggregation of smaller scale objects. Thus the resulting manifold will be smoother at higher scales (as illustrated in Figure 8.15).

Conceptually, if a manifold were stretched between the center point of all the first rungs of the many ladders, then another were stretched between rungs 2, 3, . . . (similar to that described in Section 8.3.1.2), this would result in a multiscale hierarchical scaling manifold, where the space between each manifold represents unique scale domains of different size and shape that are defined by the *persistence* of the objects they model. We note that while we have modeled "hard" domain manifolds (i.e., actual threshold-based structures), this is for illustration purposes only. Conceptually, these structures would have a diffusive (i.e., Gaussian) composition that corresponds to the notion of the "near decomposability of hierarchical structures," as outlined in hierarchy theory. Additionally an important result these domain manifolds visually demonstrate is that there is no single scale of analysis for assessing the different sized, shaped, and spatially distributed components (i.e., patches) within a scene but rather that object-based scales are required (Hay and Marceau, 2004; Hay et al., 2005).

8.4 CONCLUSION

To better understand how landscape components interact through scale, we need appropriate theory and methods for generating multiscale representations of a scene, techniques to automatically define landscape components (i.e., patches or image objects) of interest in remote sensing imagery, and object topology that facilitates the ability to link and query these image objects within appropriate hierarchical structures. This is necessary as both nested and unseated hierarchies may contain the same focal object, but their relationships are completely different; consequently the information resulting from each type of hierarchy will be very different, which in turn will affect the methods used for scaling, the questions that can be posed, and the conclusions that can be drawn (see Section 8.1.2).

What does a scale domain look like and where is it located? In this chapter we describe a novel integration of scale space and hierarchy theory for automatically defining and visually modeling salient landscape structures at multiple scales. Our results support the idea that multiscale analysis (including the establishment of scale hierarchies) should be guided by the innate spatial resolution of the (salient) geo-objects composing a scene. Scale space originates from the computer vision community, where it was developed to analyze real-world structures with no a priori information about the scene being assessed. Its basic premise is that a multiscale representation of a signal (such as a remote sensing image of a landscape) is an ordered set of derived signals showing structures at coarser scales that constitute simplifications of corresponding structures at finer scales. To define scale domains we first describe methods that define and link structures at different scales in scale space to higher order objects, called scale-space blobs, from which we extract significant features based on their appearance and persistence over all scales. This is based on the idea that bloblike structures which persist in scale space are likely candidates to correspond to significant structures in the image and thus in the landscape.

By integrating concepts from scale space and hierarchy theory:

(i) We describe a three-tier hierarchical methodology for automatically delineating and assessing the dominant structural components within 200 different multiscale representations of a complex agroforested landscape (with reduced processing demands).

(ii) We implement a new multiscale geo-object topology and define four new "species" of dual-instance blob events (Section 8.3.1.3).

(iii) We visually model five different "landscape scale domains" based on the novel idea of using SS annihilation events as critical domain thresholds.

(iv) We further suggest that the resulting manifolds may be considered a multiscale extension to the hierarchical scaling ladder as defined in the *hierarchical patch dynamics paradigm*.

Though beyond the scope of this chapter, we provide ideas for how to implement these methods over larger scenes and suggest that these domain structures may represent critical landscape scale thresholds. Thus they may be used as templates to define the grain and extent over which scale-dependent ecological models could be developed and applied and the limits over/between which landscape data can be appropriately scaled.

ACKNOWLEDGMENTS

Dr. Hay completed this chapter while on sabbatical in Brisbane, Australia, at the Centre for Spatial Environmental Research, School of Geography Planning and Environmental Management, The University of Queensland, and expresses thanks to Dr. Stuart Phinn and his team for the collaboration and ideas that have been shared. Research support has been provided by an NSERC Discovery Grant and Tecterra funding. Appreciation is expressed to Dr. Jingle Wu and Dr. Guillermo Castilla for insightful comments on earlier versions of this manuscript. Data were provided with funds from a team grant from FCAR, Gouvernement du Québec, awarded to Dr. André Bouchard and Dr. Danielle Marceau.

REFERENCES

Ahl, V., and Allen, T. F. H. 1996. *Hierarchy Theory: A Vision, Vocabulary, and Epistemology.* New York: Columbia University Press.

Allen, T. F. H., and Starr, T. B. 1982. *Hierarchy Perspective for Ecological Complexity.* Chicago: University of Chicago Press.

Baatz, M., and Schäpe, A. 2000. Multiresolution segmentation: An optimization approach for high quality multiscale image segmentation. In J. Strobl and T. Blaschke (Eds.), *Angewandte Geogr. Informationsverarbeitung XII.* Heidelberg: Wichmann, pp. 12–23.

Beucher, S., and Lantuéjoul, C. 1979. Use of watersheds in contour detection. Int. Workshop on Image Processing, Real-time Edge and Motion Detection/Estimation, Rennes, France, pp. 17–21, September, CCETT/IRISA (Centre commun d'études de télédiffusion et télécommunications/Institut de recherche en informatique et systèmes aléatoires), Report No. 132, 2.1–2.12.

Biederman, I. 1987. Recognition-by-components: A theory of human image understanding. *Psychological Review* 94(2):115–147.

Blaschke, T. 2010. Object based image analysis for remote sensing. *ISPRS International Journal of Photogrammetry and Remote Sensing* 65(1):2–16.

Bogaert, J. 2003. Lack of agreement on fragmentation metrics blurs correspondence between fragmentation experiments and predicted effects. *Conservation Ecology* 7(1):r6; available at: http://www.consercol.org/vol7/iss1/resp6.

Bouchard, A., and Domon, G. 1997. The transformation of the natural landscapes of the Haut-Saint-Laurent (Québec) and its implication on future resources management. *Landscape and Urban Planning* 37(1–2):99–107.

Burnett, C., and Blaschke, T. 2003. A multi-scale segmentation/object relationship modelling methodology for landscape analysis. *Ecological Modelling* 168(3):233–249.

Canny, J. 1986. A computational approach to edge detection. *IEEE Transactions on Pattern Analysis and Machine Intelligence* 8(6):679–697.

Castilla, G. 2003. Object-oriented Analysis of Remote Sensing Images for Land Cover Mapping: Conceptual Foundations and a Segmentation Method to Derive a Baseline Partition for Classification. Ph.D. Dissertation, Polytechnic University of Madrid; http://www.montes.upm.es/Servicios/biblioteca/tesis/GCastillaTD_Montes.pdf.

Castilla, G., and Hay, G. J. 2008. Image-objects and geo-objects. In T. Blaschke, S. Lang, and G. J. Hay (Eds.), *Object-Based Image Analysis. Spatial Concepts for Knowledge-Driven Remote Sensing Applications.* Berlin: Springer, pp. 91–110.

Castilla, G., Hay, G. J., and Ruiz, J. R. 2008. Size-constrained Region Merging (SCRM): An automated delineation tool for assisted photointerpretation. *Photogrammetric Engineering & Remote Sensing* 74(4):409–419.

Chen, G., Hay, G. J., Carvalho, L. M. T., and Wulder, M. 2012. Object Based Change Detection. *International Journal of Remote Sensing* 33(14):4434–4457.

Chen, G., Hay, G. J., and St-Onge, B. 2011. A GEOBIA framework to estimate forest parameters from lidar transects, Quickbird imagery and machine learning: A case study in Quebec, Canada. *International Journal of Applied Earth Observation and Geoinformation* 15:28–37.

Cousins, S. H. 1993. Hierarchy in ecology: Its relevance to landscape ecology and geographic information systems. In R. H. Young, D. R. Green, and S. H. Cousins (Eds.), *Landscape Ecology and GIS*. New York: Taylor & Francis, pp. 75–86.

Daubechies, I. 1988. Orthonormal bases of compactly supported wavelets. *Communications on Pure and Applied Mathematics* 41:906–966.

Dey, V., Zhang, Y., and Zhong, M. 2010. A review on image segmentation techniques with remote sensing perspective. ISPRS TC VII Symposium Vienna, Austria, July 5, Vol. XXXVIII, Part 7A, pp. 31–42.

Doxani, G., Karantzalos, K., and Tsakiri-Strati, A. 2012. Monitoring urban changes based on scale space filtering and object-oriented classification. *International Journal of Applied Earth Observations and Geoinformation* 15:38–48.

Drăguţ, L., Tiede, D., and Levick, S. R. 2010. ESP: A tool to estimate scale parameter for multiresolution image segmentation of remotely sensed data. *International Journal of Geographical Information Science* 24(6):859–871.

Foody, G. 1999. Image classification with a neural network: From completely-crisp to fully-fuzzy situations. In P. M. Atkinson and N. J. Tate (Eds.), *Advances in Remote Sensing and GIS Analysis*. Chichester: Wiley, pp. 17–37.

Forman, R. T. T. 1995. *Land Mosaics: The Ecology of Landscapes and Regions*. Cambridge, UK: Cambridge University Press.

Gibson, C. C., Ostrom, E., and Ahn, T. K. 2000. The concept of scale and the human dimensions of global change: A survey. *Ecological Economics* 32:217–239.

Graham, I. 2001. *Object-Oriented Methods. Principles and Practices*, 3rd ed. Reading, MA: Addison-Wesley.

Hall, O., Hay, G. J., Bouchard, A., and Marceau, D. J. 2004. Detecting dominant landscape objects through multiple scales: An integration of object-specific methods and watershed segmentation. *Landscape Ecology* 19(1):59–76.

Haralick, R. M., Sternberg, S. R., and Zhuang, X. 1987. Image analysis using mathematical morphology. *IEEE Transactions on Pattern Analysis and Machine Intelligence* 9(4):532–550.

Haury, L. R., McGowan, J. A., and Wiebe, P. H. 1978. Patters and processes in the time-space scales of plankton distributons. In J. H. Steele (Ed.), *Spatial Pattern in Plankton Communities*. New York: Plenum, pp. 277–327.

Hay, G. J., and Blaschke, T. 2010. Forward: Special issue on geographic object-based image analysis (GEOBIA). *Photogrammetric Engineering and Remote Sensing* 76(2): 121–122.

Hay, G. J., Blaschke, T., Marceau, D. J., and Bouchard, A. 2003. A comparison of three image-object methods for the multiscale analysis of landscape structure. *ISPRS Journal of Photogrammetry and Remote Sensing* 57(5–6):327–345.

Hay, G. J., and Castilla, G. 2008. Geographic Object-Based Image Analysis (GEOBIA): A new name for a new discipline. In T. Blaschke, S. Lang, and G. J. Hay (Eds.), *Object-Based Image Analysis. Spatial Concepts for Knowledge-Driven Remote Sensing Applications*. Berlin: Springer, pp. 75–89.

Hay, G. J., Castilla, G., Wulder, M. A., and Ruiz, J. R. 2005. An automated object-based approach for the multiscale image segmentation of forest scenes. *International Journal of Applied Earth Observation and Geoinformation* 7:339–359.

Hay, G. J., Dube, P., Bouchard, A., and Marceau, D. J. 2002a. A scale-space primer for exploring and quantifying complex landscapes. *Ecological Modelling* 153(1–2):27–49.

Hay, G. J., and Marceau, D. J. 1998. Are image-objects the key for upscaling remotely sensed data? *Proceedings of Modelling of Complex Systems, July 12–17, New Orleans*, pp. 88–92.

Hay, G. J., and Marceau, D. J. 2004. Multiscale Object-Specific Analysis (MOSA): An integrative approach for multiscale landscape analysis. In S. M. de Jong and F. D. van der Meer (Eds.), *Remote Sensing Image Analysis: Including the Spatial Domain*. Book series: Remote Sensing and Digital Image Processing, Volume 5. Chapter 3. Dordrecht: Kluwer Academic.

Hay, G. J., Marceau, D. J., and Bouchard, A. 2002b. Modelling multiscale landscape structure within a hierarchical scale-space framework. *International Archives of Photogrammetry and Remote Sensing* 34(4):532–536.

Hay, G. J., Marceau, D. J., Dubé, P., and Bouchard, A. 2001. A multiscale framework for landscape analysis: Object-specific analysis and upscaling. *Landscape Ecology* 16(6):471–490.

Hay, G. J., and Niemann, K. O. 1994. Visualizing 3-D texture: A three dimensional structural approach to model forest texture. *Canadian Journal of Remote Sensing* 20(2):90–101.

Hay, G. J., Niemann, K. O., and Goodenough, D. G. 1997. Spatial thresholds, image-objects and upscaling: A multi-scale evaluation. *Remote Sensing of Environment* 62(1):1–19.

Hay, G. J., Niemann, K. O., and McLean, G. 1996. An object-specific image-texture analysis of H-resolution forest imagery. *Remote Sensing of Environment* 55(2):108–122.

Holling, C. S. 1992. Cross-scale morphology, geometry, and dynamics of ecosystems. *Ecological Monographs* 62(4):447–502.

Jähne, B. 1999. A multiresolution signal representation. In B. Jähne, H. Haußecker, and P. Geißler (Eds.), *Handbook on Computer Vision and Applications*. Boston: Academic, pp. 67–90.

Kay, J. 1991. A Non-equilibrium thermodynamic framework for discussing ecosystem integrity. *Environmental Management* 15(4):483–495.

Kay, J. J., and Regier, H. 2000. Uncertainty, complexity, and ecological integrity: Insights from an ecosystem approach. In P. Crabbe, A. Holland, L. Ryszkowski, and L. Westra (Eds.), *Implementing Ecological Integrity: Restoring Regional and Global Environmental and Human Health*. Dordrecht: Kluwer, NATO Science Series, Environmental Security, pp. 121–156.

Klinger, A. 1972. Patterns and search statistics. In J. Rustagi (Ed.), *Optimizing Methods in Statistics*. New York: Academic Press, pp. 303–339.

Klinger, A., and Dyer, C. R. 1976. Experiments in picture representation using regular decomposition. *Comp. Graphics Image Process* 5:68.

Levin, S. A. 1999. *Fragile Dominions: Complexity and the Commons*. Reading, MA: Perseus Books.

Li, H., and Wu, J. 2004. Use and misuse of landscape indices. *Landscape Ecology* 19:389–399.

Lindeberg, T. 1993. Detecting salient blob-like image structure and their scales with a scale-space primal sketch: A method for focus of attention. *International Journal of Computer Vision* 11(3):283–318.

Lindeberg, T. 1994. *Scale-Space Theory in Computer Vision*. Dordrecht: Kluwer Academic.

Mandelbrot, B. 1967. The fractal geometry of nature. *Science* 156:636–642.

Marceau, D. J. 1999. The scale issue in the social and natural sciences. *Canadian Journal of Remote Sensing* 25(4):347–356.

Marceau, D. J., and Hay, G. J. 1999. Contributions of remote sensing to the scale issue. *Canadian Journal of Remote Sensing* 25(4):357–366.

Marceau, D. J., Howarth, P. J., and Gratton, D. J. 1994. Remote sensing and the measurement of geographical entities in a forested environment. Part 1: The scale and spatial aggregation problem. *Remote Sensing of Environment* 49(2):93–104.

Meentemeyer, V. 1989. Geographical perspectives of space, time, and scale. *Landscape Ecology* 3:163–173.

Meilleur, A., Bouchard, A., and Bergeron, Y. 1994. The relation between geomorphology and forest community types of the Haut-Saint-Laurent, Quebec. *Vegetatio* 111:173–192.

Nicolis, G., and Prigogine, I. 1989. *Exploring Complexity: An Introduction*. New York: W H Freeman.

O'Neill, R. V., and King, A. W. 1998. Homage to St. Michael; Or, why are there so many books on scale? In *Ecological Scale Theory and Applications*. New York: Columbia University Press, pp. 3–15.

Peterson, G. 2000. Scaling ecological dynamics: Self-organization, hierarchical structure, and ecological resilience. *Climate Change* 44(3):291–309.

Powers, R., Hay, G. J., and Chen, G. 2012. How wetland type and area differ through scale: A case study of Alberta's Boreal Plains. *Remote Sensing of Environment* 117:135–145.

Quattrochi, D. A., and Goodchild, M. F. (Eds.) 1996. *Scale in Remote Sensing and GIS*. Boca Raton, FL: Lewis Publishers.

Reynolds, J. R., and Wu, J. 1999. Do landscape structural and functional units exist? In J. D. Tenhunen and P. Kabat (Eds.), *Integrating Hydrology, Ecosystem Dynamics and Biogeochemistry in Complex Landscapes*. New York: Wiley, pp. 275–298.

Robert, C. P. 2001. *The Bayesian Choice: From Decision-Theoretic Foundations to Computational Implementation*. New York: Springer.

Rosin, P. L. 1992. Representing curves at their natural scales. *Pattern Recognition* 25 (11):1315–1325.

Rowe, J. S. 2001. Transcending this poor Earth—á la Ken Wilber. *Trumpeter* 17:1; available at http://trumpeter.athabascau.ca/index.php/trumpet/article/view/138/161, last accessed October 20, 2013.

Settle, J. J., and Drake, N. A. 1993. Linear mixing and the estimation of ground cover proportions. *International Journal of Remote Sensing* 14(6):1159–1177.

Simon, H. A. 1962. The architecture of complexity. *Proceedings of the American Philosophical Society* 106:467–482.

Simon, H. A. 1973. The Organization of Complex Systems. In H. H. Pattee (Ed.), *Hierarchy Theory: The Challenge of Complex Systems*. New York: George Braziller, pp. 1–27.

Starck, J.-L., Murtagh, F., and Bijaoui, A. 1998. *Image Processing and Data Analysis: The Multiscale Approach*, Cambridge, UK: Cambridge University Press, 315 pp.

Steiniger, S., and Hay, G. J. 2009. Free and open source geographic information tools for landscape ecology: A review. *Ecological Informatics* 4(4):183–195.

Swihart, R. K., DunningJr., G. B., and Waser, P. M. 2002. Grey matters in ecology: Dynamics of pattern, process, and scientific progress. *Bulletin of the Ecological Society of America* 83:149–155.

Syed, A. H., Saber, E., and Messinger, D. 2011. Scale-space representation of remote sensing images using an object-oriented approach. *SPIE* 8053(1):805308.

Syed, A. H., Saber, E., and Messinger, D. 2012. Encoding of topological information in multiscale remotely sensed data: Applications to segmentation and object based image analysis. *Proceedings of the 4th GEOBIA, May 7–9, Rio de Janeiro, Brazil*, pp. 102–107.

ter Haar Romeny, B. M., and Florack, L. M. J. 2000. Front-end vision, a multiscale geometry engine. Proc. First IEEE International Workshop on Biologically Motivated Computer Vision (BMCV2000), May 15–17, Seoul, Korea. In *Lecture Notes in Computer Science*, Vol. 1811. Heidelberg: Springer, pp. 297–307.

Tzotsos, A., Karantzalos, K., and Argialas, D. 2011. Object-based image analysis through nonlinear scale-space filtering. *ISPRS Journal of Photogrammetry and Remote Sensing* 66:2–16.

Urban, D. L., O'Neill, R. V., and Shugart, Jr.H. H. 1987. Landscape ecology. A hierarchical perspective can help scientists understand spatial patterns. *Bioscience* 37(2):119–127.

Waldrop, M. M. 1992. *Complexity. The Emerging Science at the Edge of Order and Chaos.* New York: Simon and Schuster.

Wang, F. 1990. Improving remote sensing image analysis through fuzzy information representation. *Photogrammetric Engineering and Remote Sensing* 56(8):1163–1169.

Watt, A. S. 1947. Pattern and process in the plant community. *Journal of Ecology* 35:1–22.

Weickert, J., Ishikawa, S., and Imiya, A. 1997. On the history of Gaussian scale-space axiomatics. In J. Sporring, M. Nielsen, L. Florack, and P. Johansen (Eds.), *Gaussian Scale-Space Theory*. Dordrecht: Kluwer, pp. 45–59.

Wessman, C. A. 1992. Spatial scales and global change: Bridging the gap from plots to GCM grid cells. *Annual Review of Ecological Systems* 23:175–200.

Wheatley, M., and Johnson, C. 2009. Factors limiting our understanding of ecological scale. *Ecological Complexity* 6:150–159.

Whittaker, R. H. 1953. A consideration of climax theory: The climax as a population and pattern. *Ecol. Monogr.* 23:41–78.

Wiens, J. A. 1989. Spatial scaling in ecology. *Functional Ecology* 3(4):385–397.

Woodcock, C. E., and Strahler, A. H. 1987. The factor of scale in remote sensing. *Remote Sensing of Environment* 21(3):311–332.

Wu, J. 2007. Scale and scaling: A cross-disciplinary perspective. In J. Wu and R. Hobbs (Eds.), *Key Topics in Landscape Ecology*. Cambridge, UK: Cambridge University Press, pp. 115–142.

Wu, H., and Li, Z.-H. 2009. Scale issues in remote sensing: A review on analysis, processing and modeling. *Sensors* 9:1768–1793.

Wu, J. 1999. Hierarchy and scaling: Extrapolating information along a scaling ladder. *Canadian Journal of Remote Sensing* 25:367–380.

Wu, J., and Hobbs, R. 2002. Key issues and research priorities in landscape ecology: An idiosyncratic synthesis. *Landscape Ecology* 17:355–365.

Wu, J., and Li, H. 2005. Concepts of scale and scaling. In J. Wu, K. B. Jones, H. Li, and O. L. Loucks (Eds.), *Scaling and Uncertainty Analysis in Ecology*. New York: Columbia University Press.

Wu, J., David, J. L., and Jenerette, G. D. 2003. Linking land use change with ecosystem processes: A hierarchical patch dynamics model. In S. Guhathakurta (Ed.), *Integrated Land Use and Environmental Models*. New York: Springer, pp. 99–119.

Wu, J., and Marceau, D. 2002. Modelling complex ecological systems: An introduction. *Ecological Modelling* 153(1–2):1–6.

Wu, J., and Loucks, O. L. 1995. From balance of nature to hierarchical patch dynamics: A paradigm shift in ecology. *Quarterly Review of Biology* 70:439–466.

9

MULTISCALE SEGMENTATION AND CLASSIFICATION OF REMOTE SENSING IMAGERY WITH ADVANCED EDGE AND SCALE-SPACE FEATURES

ANGELOS TZOTSOS, KONSTANTINOS KARANTZALOS, AND DEMETRE ARGIALAS

9.1 INTRODUCTION

The current need for automated image analysis and computer vision technological tools requires a processing scheme able to encapsulate effectively the content of remote sensing data. However, Earth's landscape structure is complex, the context varies, and so does the appearance of the images, being a combination of many different intensities, representing natural features such as vegetation, geomorphological and hydrological features, man-made objects (e.g., buildings, roads), and artifacts caused by variation in illumination of the terrain (e.g., shadows).

Furthermore, roads, infrastructure, vegetation, landforms, and other land features appear in different sizes and geographical scales in images (e.g., country road vs. interstate, tree stands vs. forest, maisonette vs. polygon building, rill vs. river). Only in a few special circumstances do the objects of interest belong to a certain scale while the remaining ones, to be discarded, belong to another. In most cases such a global scale threshold is not possible since the desired information is present at several scales (Witkin, 1983; Lindeberg, 1994; Weickert, 1998; Meyer and Maragos, 2000).

Scale-space representations and multiscale image analysis provide the framework to explore the entire image content by detecting the scale(s) at which objects or patterns appear and are most distinctly identified. Toward this end and parallel to the

Scale Issues in Remote Sensing, First Edition. Edited by Qihao Weng.
© 2014 John Wiley & Sons, Inc. Published 2014 by John Wiley & Sons, Inc.

human visual system, several multiscale low-level processes (i.e., filtering, segmentation, etc.) have been developed during which a series of representations of the same image are computed (from fine to coarse) and used for the recognition of Earth surface objects (Blaschke and Hay, 2001; Hay et al., 2002; Hall and Hay, 2003; Benz et al., 2004; Stewart et al., 2004; Jimenez et al., 2005; Karantzalos and Argialas, 2006; Duarte-Carvajalino et al., 2008; Ouma et al., 2008). The mathematical models and the manner for constructing these scale space representations are of fundamental importance.

In addition, during the last decade the way of classifying remotely sensed imagery has been changing, and instead of classifying individual pixels into discrete land cover classes, object-based classification approaches construct a hierarchical object representation of an image and the classifier is responsible for associating them with a land cover class (Blaschke, 2010). Therefore, it is not just the spectral signature of each pixel but the statistical, geometric, and topological characteristics of each object that play a key role during classification. Object-based classification is considered optimal for the analysis of very high resolution remote sensing imagery (with spatial resolution of 5 m per pixel or less), since this kind of imagery is introducing more complexity to the classification tasks, due to increased heterogeneity and the increased number of land cover classes that can be observed. The statistical, geometric, and contextual characteristics of the image primitives are considered by the object-based methods (much like in photointerpretation), in contrast to pixel-based approaches (Blaschke, 2010). Recent studies are highlighting that the determination of one or more optimal filtering scales for image segmentation is still a challenge and a multiscale object-based classification is a significantly better approach than the classical per-pixel classification procedure (Myint et al., 2011; Tzotsos et al., 2011).

In this chapter, an object-based image analysis framework was developed which integrates advanced scale space representations, edge and line feature detection, multiscale segmentation, and a kernel-based classification. The contributions of this approach are twofold:

- A generic framework able to process any remotely acquired raster data (satellite/airborne data, multispectral/hyperspectral data, and radar data of any spatial resolution) without the need of tuning any parameter (scale, color, texture, etc.).
- A new robust multiscale segmentation procedure (replacing an earlier segmentation algorithm in the framework) which is constrained by advanced edge-based features.

The chapter is structured as follows. In Section 9.2, the related work on scale-space representations, multiscale object-based analysis, and edge-based image segmentation is briefly presented. The developed object-based image analysis framework is detailed in Section 9.3, along with a description and detailed analysis of its different processing steps. Experimental results and the performed quantitative evaluation are given in Section 9.4. Finally, conclusions and perspectives for future work are in Section 9.5.

9.2   RELATED WORK

9.2.1   Multiscale Object-Based Image Analysis

Along with the gradual availability of Earth observation data with higher spatial and spectral resolution, research efforts in classifying remote sensing data have been shifting in the last decade from pixel-based approaches to object-based ones (Blaschke, 2010; Myint et al., 2011; Tzotsos et al., 2011). Assigning land cover classes to individual pixels can be intuitively proper and functional for low-resolution data. However, this is not the case for the emerging applications which arise from the continuously improving remote sensing sensors (Aplin and Smith, 2008; Blaschke et al., 2008). This is mostly because at higher resolutions it is a connected group of pixels that is likely to be associated with a land cover class and not just an individual pixel (Tzotsos et al., 2011).

In addition, the Earth surface exhibits various regular and irregular structures which are represented with a certain spatial heterogeneity in images composing their intensity, scale, and texture. Several important aspects of the Earth surface cannot be analyzed based on pixel information but can only be exploited based on contextual information and the topological relations of the objects of interest (Liu et al., 2008) through a multiscale image analysis (Blaschke and Hay, 2001; Hay et al., 2002; Hall and Hay, 2003; Benz et al., 2004; Stewart et al., 2004; Jimenez et al., 2005; Duarte-Carvajalino et al., 2008; Ouma et al., 2008; Dragut et al., 2010; Tzotsos et al., 2011). Starting with the observed spatial heterogeneity and variability, meaningful spatial aggregations (primitive objects) can be formed at certain image scales configuring a relationship between ground objects and image objects. Ground objects refer to real-world objects or areas of specific land cover class connected by complex spatial and contextual relations. Image objects, on the other hand, refer to knowledge-free areas (primitive objects) of an image that a segmentation algorithm provides based on various criteria. In order to configure a relationship between ground and image objects, a multiscale knowledge representation is needed, which is usually provided by the object-oriented paradigm. With such an object-based multiscale representation and analysis, which is based on certain hierarchically structured rules, the relationship between the different scales of the spatial entities is described.

During the last decade, a number of object-based image analysis software were developed in the form of proprietary software (eCognition, ENVI FX, ERDAS Objective) (Baatz and Schape, 2000) or in the form of free software (Orfeo Toolbox, EDISON, MSEG) (Tzotsos and Argialas, 2006; Inglada and Christophe, 2009; Christophe and Inglada, 2009), enabling the broad application on various engineering and environmental remote sensing studies (Benz et al., 2004; Zhou et al., 2009; Dragut et al., 2009; Blaschke, 2010; Mladinich et al., 2010). In all cases, the challenge was to construct an efficient scale-space object representation through certain multi-scale (region merging or other) segmentation techniques (Blaschke et al., 2004; Carleer et al., 2005; Jimenez et al., 2005; Neubert et al., 2006; Tzotsos and Argialas, 2006), which partition the image on several regions/objects, based on the spectral homogeneity in a local neighborhood. In addition to the spectral homogeneity criterion, shape parameters are used to define geometric properties that the segmentation algorithm must take into account when computing the overall homogeneity (scale parameter) of each image object during search for optimal merges.

In 2008, a texture optimization procedure was introduced for the MSEG algorithm (Tzotsos et al., 2008) integrating gray-level co-occurrence matrices and introducing an object-based cost measure for texture homogeneity as an additional parameter to the segmentation procedure. Such an integration of spatial and spectral information can produce a multiscale object representation but only through an iterative and exhaustive tuning (based on trial-and-error investigation) of certain parameters, for example, shape, scale, and texture (Baatz and Schape, 2000; Benz et al., 2004; Blaschke et al., 2004; Carleer et al., 2005; Hay et al., 2005; Tzotsos and Argialas, 2006; Ouma et al., 2008; Dragut et al., 2009; Zhou et al., 2009).

Other research efforts were based on the construction of linear scale spaces for the multiscale analysis of several landscape structures (Blaschke and Hay, 2001; Hay et al., 2002, 2003; Stewart et al., 2004) or on the construction of multiscale representations through object-specific analysis and upscaling through the computation of a number of coarse and fine scales by sampling the initial image (Hall and Hay, 2003). Furthermore, other studies employed unsupervised classification algorithms for both optical and radar data (Derrode and Mercier, 2007; Jung, 2007) or multiple hierarchical segmentation (Akcay and Aksoy, 2008).

More recent research efforts are focusing on optimizing the segmentation procedure through a data-driven thresholding approach (Martha et al., 2011) and on constructing advanced nonlinear scale-space representations for efficient supervised classification (Tzotsos et al., 2011) and change detection over urban areas (Doxani et al., 2012).

9.2.2 Scale-Space Remote Sensing Data Representations

Earth surface objects cannot be represented to a single scale but rather to many. The use of scale-space image representations is thus of fundamental importance for a number of image analysis and computer vision tasks. It dates back to the 1960s and was first introduced by Iijima (Weickert et al., 1999). In western literature many linear scale-space methods were introduced (Witkin, 1983; Koenderink, 1984; Lindeberg, 1994), and respectively many isotropic multiscale operators were developed. Either through Gaussian filtering or through isotropic multiresolution analysis (by downsampling the initial data), all linear scale-space approaches present the same important drawback: image edges are blurred and new nonsemantic objects may appear at coarse scales (Witkin, 1983; Paragios et al., 2005; Ouma et al., 2008). Under a hierarchical multiscale segmentation or an object-based classification framework, the thematic information to be extracted is directly related with the primitive image objects computed at every scale. The better these primitive objects represent real-world entities, the better they can describe the semantics of the image (Hay and Castilla, 2006; Blaschke et al., 2008; Hofmann et al., 2008; Tzotsos et al., 2011). Therefore, the selection of the appropriate approach for constructing the multiscale image and hierarchical object representation is of great importance.

Since linear scale-space approaches, by acting isotropically in the image domain, delocalize and blur image edges, nonlinear operators and nonlinear scale spaces have been studied and applied in various image processing and computer vision applications. Following the pioneering work of Perona and Malik (1990), there has been a flurry of activity in partial differential equation and anisotropic diffusion filtering techniques (Weickert, 1998). For remote sensing applications, a number of anisotropic diffusion schemes have been proposed and applied to aerial and satellite data sets (Lennon et al., 2002; Camps-Valls and Bruzzone, 2005; Karantzalos and Argialas, 2006; Duarte-Carvajalino et al., 2007; Ouma et al., 2008; Plaza et al., 2009), combined, in most cases, with pixel-based classification techniques. All their scale-space formulations, though, were based on either diffusions during which the average luminance value is preserved or geometrically driven ones formulated under a variational framework. Although these formulations may reduce the problems of isotropic filtering, they do not eliminate them completely: spurious extrema and important intensity shifts may still appear (Meyer and Maragos, 2000; Karantzalos et al., 2007; Tzotsos et al., 2011).

Therefore, another way to produce nonlinear scale-spaces is through mathematical morphology and, in particular, with morphological levelings, which have been introduced by Meyer (1998) and further studied by Matheron (1997) and Serra (2000). Morphological levelings overcome the drawback of spurious extrema or important intensity shifts and possess a number of desired properties for the construction of elegant scale-space representations. Levelings, which are a general class of self-dual morphological operators, do not displace contours through scales and are characterized by a number of desirable properties for the construction of nonlinear scale-space representations. They satisfy the following spatial and spectral properties/axioms (Meyer and Maragos, 2000; Meyer, 2004; Karantzalos et al., 2007; Tzotsos et al., 2011):

- Invariance by spatial translation
- Isotropy, invariance by rotation
- Invariance to a change of illumination
- Causality principle
- Maximum principle, excluding the extreme case where g is completely flat

In addition, levelings:

- Do not produce new extrema at larger scales
- Enlarge smooth zones
- Create new smooth zones
- Are particularly robust (strong morphological filters)
- Do not displace edges

Designing and formulating an optimal scale space framework are still active areas of research. Recent efforts include studies on certain scale-space formulation (Nilufar et al., 2012; Ouzounis et al., 2012), studies on a varying stopping time (Gilboa, 2008), and studies on the behavior on corner and other local descriptors (Zhong et al., 2009; Jiang et al., 2011; Kimmel et al., 2011; Xu et al., 2012).

9.2.3 Edge-Based Segmentation

Extracting primitives (i.e., contours, edges, lines, etc.) is a basic low-level operation in the human visual system. Along with the aim to understand and simulate the human vision, the importance of building up computational models for the perception of primitives is a major component in many applications of computer vision, such as object/pattern recognition, robot vision, remote sensing, and medical image analysis.

These primitives give important information about the geometric content of images. Most of the Earth surface objects and in particular most man-made objects are made of flat surfaces with certain geometric features. In addition, many shapes can be described roughly or in detail with edge and line primitives. Therefore, edge or line segments can be used as a low-level feature description in order to extract information from images and can serve as the basic tool to analyze and detect more complex shapes (Von Gioi et al., 2010; Papari and Petkov, 2011; Wang and Oliensis, 2010; Chia et al., 2012). In the context of scale-space representations, image primitives may support more stable and efficient representations since their description can be independent of the object size. Edge and line primitives, defined mainly by the object geometric properties, allow the robust and efficient feature comparison in various scales. The latter is of major importance due to large appearance variations of object instances belonging to the same class.

However, even the recent more sophisticated edge and line detectors cannot produce connected segments and suffer from the Earth surface complexity pictured in images, shadows, occlusions, and so on. Therefore, recent efforts are trying to merge the advantages of edge/line detection and image segmentation techniques in order to produce connected object contours/boundaries and a comprehensive object description (Pavlidis and Liow, 1990; Kermad and Chehdi, 2002; Cufi et al., 2003). Certain primitive combinations have been proposed in order to describe more efficiently object boundaries (Chia et al., 2012; Klonus et al., 2012). Another recent study proposed a region-based unsupervised segmentation and classification algorithm which included the computation of an edge strength model (Yu et al., 2012). This edge penalty model improved segmentation performance by preserving segment boundaries.

9.3 METHODOLOGY

The main objective here was to design the overall framework in order to be generic, robust, and able to process effectively a wide variety of remote sensing data, such as hyperspectral and multispectral data from ground, aerial, and space borne sensors, radar data, and digital elevation models. It is based upon the object-based image analysis (OBIA) approach, which generally includes low-, medium-, and high-level image processing subtasks:

- Preprocessing steps (geometric and radiometric corrections, filtering, scale-space image simplification, edge detection, band math expression computations, etc.)
- Image segmentation (in order to produce single-level or multilevel hierarchies of primitive objects within the image space)
- Computation of image object properties based on spectral, shape, topological, and context features
- Definition of object-oriented class hierarchy and representation of knowledge through rules or through training with samples
- Classification (learning techniques or rule-based systems to perform the classification task)
- Accuracy assessment in order to derive the quality of the resulting classification
- Vectorization steps (create the output to spatial databases and integrate the information to thematic maps).

The developed approach is integrating certain advanced computer vision and machine learning methods for implementing the above tasks. Although the proposed methodology is following the above general OBIA approach, which is also available in several proprietary and free software solutions, the originality of the developed framework lies in the novel approach that excludes the parameter tuning step, in the robustness of the multiscale hybrid (edge-based and region-based) segmentation algorithm, in the extensive use of nonlinear scale-space representations for image simplification purposes, and in the integration with kernel-based machine learning methods for classification.

Briefly the developed framework consists of the following steps: First, for every band of the initial image, a scale-space representation is generated using the anisotropic morphological leveling (AML) formulation (Karantzalos et al., 2007). The supported type of the imagery can be up to double precision and of any number of bands. A feature extraction step is then applied on the scale-space stack. For this step two algorithms were tested and are presented here: the Canny edge detector (Canny, 1986) and the line segment detector (LSD) (Von Gioi et al., 2010). A multiscale segmentation algorithm is applied afterward which is able to integrate the simplified scale-space stack along with the corresponding edge information. During this step primitive image objects are formed. The procedure starts from single pixel, and through pairwise merges bounded by edge information, several levels of image objects are produced. The multiscale object hierarchies are been constructed without any parameter tuning. In a similar way with Tzotsos et al. (2011), here the edge features are produced without tuning the edge extraction parameters (Figure 9.1). Last in the processing order comes a dual classification procedure using a support vector machine classifier. The first classification is performed on all scale-space representations and their corresponding segmentations, while the second optimal one is performed after an interim accuracy assessment.

9.3.1 Scale-Space Filtering

The first step in the developed approach is the construction of the nonlinear scale-space representation in order to elegantly simplify raw data. Anisotropic diffusion methods are used widely in computer vision applications to simulate the filtering procedures that are performed in the human vision system. Such methods provide robust simplification of images without the loss of important information such as edges that are of high importance for higher level processing algorithms. Especially in OBIA applications, where very high resolution data are usually processed, it is very important to simplify the complexity of the initial data and provide a multiscale representation since different features of the image reside in different scales.

For this preproccesing step, the AMLs (Karantzalos et al., 2007) were incorporated in the processing scheme. Anisotropic morphological levelings are a combination of morphological levelings with anisotropic markers and are employed in order to achieve better segmentation results, reduce the heterogeneity of image data, reduce oversegmentation, and create accurate image objects. Figure 9.1 shows that by creating a series of simplified data scale-space filtering leads to a multiscale segmentation (without tuning any segmentation parameters such as texture, color, shape, etc.).

Starting from the initial image and for every available band, a scale-space representation was generated using the AML formulation. Using iterative anisotropic morphological operations and increasing scales (10, 50, 100, 500, 1000) a scale-space cube (3D) representation was constructed from each initial band. The result of this step was a scale-space stack with simplified versions of the raw data. Note that during this process edge information was preserved in all scales, contrary to isotropic (e.g., Gaussian) filtering that loses edge information as scale increases (Figures 9.1$f,g,k,l,p,$ q). In the following sections further analysis of Figure 9.1 will follow.

9.3.2 Multiscale Segmentation Based on Advanced Edge Features

Edge and line features were computed for every image in the scale-space stack. Edge information was obtained from the standard Canny detector (Canny, 1986) and the recent LSD (Von Gioi et al., 2010, 2012).

The Canny edge detector was employed in order to provide primitive edge features that were integrated to the implemented region merging algorithm. Throughout this research, the variance parameter of the Canny detector was set stable to 5. In Figures 9.1e,n,s results from the application of the Canny edge detection on scale-space images is shown. The LSD is a linear-time detector giving subpixel accurate results. The LSD algorithm starts by computing the level line angle at each pixel to produce a level line field, that is, a unit vector field such that all vectors are tangent to the level line going through their base point. Then, this field is segmented into connected regions of pixels that share the same level line angle up to a certain tolerance. These connected regions are called line support regions (Von Gioi et al., 2012). Each line support region (a set of pixels) is a candidate for a line segment. The principal inertial axis of the line support region is used as the main rectangle direction.

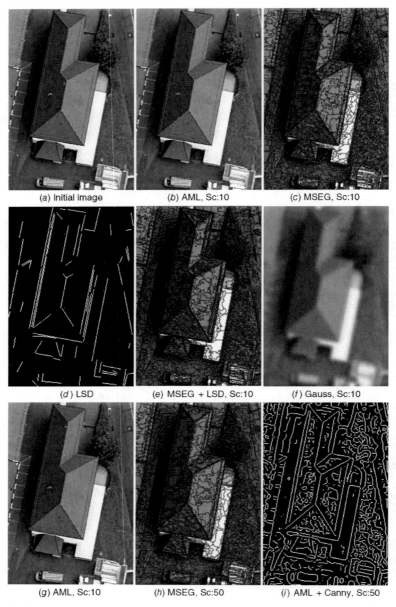

FIGURE 9.1 Comparison of region merging segmentation results using scale-space representations and advanced edge features. Four scales from various steps of the proposed methodology are presented. (*a,f,k,p*) Initial remote sensing aerial image with spatial resolution of 5 cm along with three Gaussian scales; (*b,g,l,q*) scale-space representation at various selected scales; (*c,h,m,r*) initial image objects using MSEG algorithm applied to scale-space representation; (*d,i,n,s*) advanced edge and line features used in the following step: (*e,j,o,t*) results of edge-enhanced MSEG algorithm proposed in this research. (See the color version of this figure in Color Plates section.)

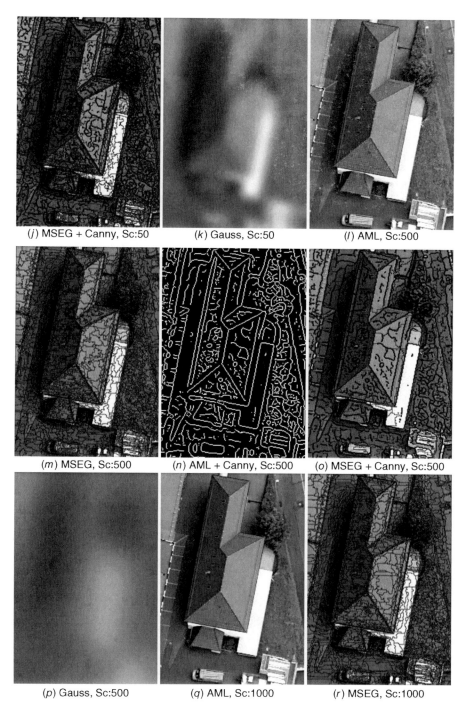

(j) MSEG + Canny, Sc:50   (k) Gauss, Sc:50   (l) AML, Sc:500

(m) MSEG, Sc:500   (n) AML + Canny, Sc:500   (o) MSEG + Canny, Sc:500

(p) Gauss, Sc:500   (q) AML, Sc:1000   (r) MSEG, Sc:1000

FIGURE 9.1 (*Continued*).

(s) AML + Canny, Sc:1000   (t) MSEG + LSD, Sc:1000

FIGURE 9.1   (*Continued*).

After examining and validating line support regions and testing that they are aligned properly, a selection of meaningful rectangles is provided as the final result. In Figure 9.1*d* the application of LSD on a very high resolution aerial scanner image is demonstrated. LSD has been designed to be automated and includes an internal filtering and simplification procedure with constant scale of 0.8.

For the multiscale segmentation procedure, an improved version of the initial MSEG algorithm (Tzotsos and Argialas, 2006) was implemented. The initial MSEG is a region-based multiscale segmentation algorithm recently developed for object-oriented image analysis. Briefly, starting from a pixel representation it creates objects through continuous pairwise object fusions executed in iterations (passes). For each pass, every object is evaluated in relation with its neighboring objects toward the optimal pair of objects adequate for fusion. In every pass, an image object can be merged only once, aiming at a balanced object growth. The MSEG algorithm defines a cost function for each object merge and then implements various optimization techniques to minimize this cost. The cost function is implemented using the measure of homogeneity (color and shape) in the same way as with other approaches (Baatz and Schape, 2000). The threshold of the allowed merging cost for the segmentation procedure is called a scale parameter, since it implicitly dictates the area growth of the image objects. Results from the application of the MSEG algorithm are shown in Figures 9.1*c,h,m,r*. Through this research, the parameters of the MSEG algorithm were set stable, the color parameter was set to 0.8, and the shape parameter was set to 0.2. The goal was to allow the simplified data (from the scale-space stack) to control the way that image segments and objects are being created and not the region merging procedure. In a previous study (Tzotsos et al., 2011) is was shown that there is no need for tuning the segmentation parameters when the approach includes a reliable edge-preserving formulation for the scale-space computation.

The MSEG algorithm was improved in order to be able to integrate edge information (as a constraint) during the segmentation procedure. The goal was to design a more robust and generic segmentation procedure that would be able to take into account advanced edge and line features. In particular, the region merging algorithm starts by selecting initialization points throughout the image using the SPE (start point estimation) module (Tzotsos and Argialas, 2006) and a queue of pixels is created in order to be able to achieve reproducibility. Then, iterative pairwise fusions occur within the image space, starting from single-pixel objects, in a way that local heterogeneity is minimized (for color and shape criteria). During this pairwise merging of image objects, the edge information is used as a boundary. Two adjacent pixels will not be merged into an object if one or both reside on top of an edge. After the first pass of the region merging procedure, image objects of one or two pixels exist, with edge pixels being constrained and not merged to each other. During the following passes edge objects (still single pixels) are not merged, thus not permitting object merging between image regions that are separated by a line or a continuous edge feature. After several passes (iterations) converging of the algorithm occurs and no more object merging is performed due to scale parameter. At this point the edge objects are still intact by the region merging procedure, thus binding the procedure into respecting edge features. Finally a last iteration of the algorithm is forced on edge objects only and a selection is made to which neighboring object they should be merged based on local heterogeneity. This step is taking advantage of the fact that both Canny and LSD features are one pixel wide, and thus edge objects are always capable of merging with nonedge objects.

A certain novelty of the developed segmentation process is that it does not use an edge penalty model for the edge compensation as presented in other approaches (Kermad and Chehdi, 2002; Cufi et al., 2003; Yu et al., 2012); rather it uses a topological constraint effective throughout the region merging procedure. The results of this enhancement is presented in Figures 9.1e,j,o,t showing very promising segmentation results. A scale parameter of value 100 was used for all tests on Figure 9.1, showing that scale-space representation effectively provides the scale of the obtained objects. More results are presented and discussed in the following sections.

9.3.3   Kernel-Based Classification

For the developed approach, an Support Vector Machine (SVM) classification scheme (Vapnik, 1998; Tzotsos, 2006; Tzotsos and Argialas, 2008; Tzotsos et al., 2011) was employed. After the multiscale segmentation which is constrained by edge information, image objects were extracted and object properties were computed forming the feature space of the classification step. For each primitive image object, spectral, shape, and spatial properties (such as mean band values, standard deviation, Gray-Level Co-occurrence Matrix (GLCM) texture features, area, perimeter, compactness, and number of neighbors) were extracted by the topological model used to handle the object topology. This model was proposed by

Lehmann (2008) but also was developed independently in MSEG (Tzotsos and Argialas, 2006). The computed properties are bound to each object by a unique identifier within the object hierarchy of the image. Some of the objects are selected as samples and their properties formed a training set for the SVM (Tzotsos and Argialas, 2008).

The SVM classifier seeks to find the optimal separating hyperplane between classes by focusing on the training data (support vectors) that are placed at the edge of the class descriptors. Training data other than support vectors are discarded. Thus, not only an optimal hyperplane is fitted but less training samples are effectively used as well (Tzotsos and Argialas, 2008). This method works very well for classes that are linearly separable. In the case that image classes are not linearly separable, the SVM maps the feature space into a higher dimensionality using kernels (Vapnik, 1998; Theodoridis and Koutroumbas, 2003) and then separates classes in that new feature space forming the support vectors.

Since the SVM classification method was initially designed for binary classification problems, a heuristic one-against-one strategy was employed for multiclass classification (Hsu and Lin, 2002). Many binary classifiers were applied for each pair of classes and for every object of the image and then a max-win operator determined the final classification of the object. An n-fold cross-validation scheme was also used in order to define the parameters needed for the training procedure. After specification of parameters, the training set was used to train the classifier. Primitive image objects were classified using this trained SVM algorithm using the one-against-one strategy. Finally a quality assessment took place using ground truth data that were not used during the training procedure.

The above classification procedure was repeated for all scales (values starting from 1 to a maximum value defined by the user) in the scale-space representation in order to determine the best classification accuracy, as proposed in Tzotsos et al. (2011). The best scale was then selected based on the best classification accuracy. After determining the best scale to perform classification, a final classification step took place to produce the optimal results.

To sum up, the initial data set was simplified and a successive series of simplified images were constructed forming a nonlinear scale space. The simplified imagery that was derived was then used to extract edge and line features using advanced methods. An edge-enhanced multiscale image segmentation algorithm was employed to provide primitive image objects from the scale-space images without the tuning of any standard parameter. Finally, a classification step was performed to complete the OBIA tasks and to evaluate the developed method.

9.4   EVALUATION AND DISCUSSION

As already stated, the overall objective of the present research was (a) to introduce a generic and robust framework able to process any kind of remote sensing data without tuning any parameters (scale, texture, color, etc.) during the computation, (b) to introduce a multiscale segmentation algorithm which is constrained by advanced edge-based features at various scales, and (c) to evaluate the developed methodology in various remote sensing data sets.

In Figure 9.1 a general overview of the developed method is presented. Starting from the initial image (Figure 9.1a), Gaussian filtering at different scales demonstrates the loss of edge information due its isotropic character (Figures 9.1f,k,p). These results are directly compared with the AML scale-space representations at equivalent scales. One can observe in Figures 9.1b,g,l,q that edge information is preserved while the initial image is simplified. For example, in the tile roof of the building the single tiles are more difficult to distinguish as scale increases. The results from the application of the standard MSEG algorithm on the simplified images using the same scale parameter value (100) are shown as well (Figures 9.1c, h,m,r). The standard MSEG algorithm performs well across object boundaries but produces over segmented results and the mean object size is increasing along with scale. The edge and line feature extraction at various scales is demonstrated in Figure 9.1 as well. The result from the application of the LSD in the original image is shown in Figure 9.1d. This is an impressive result, demonstrating that LSD is robust and works well for man-made objects, even if not all of the building sides have been detected correctly. The results of the Canny edge detector are also presented at different scales (Figures 9.1i,n,s). It can be observed that due to the simplified data through the AML scale-space computation, more clear edge features are detected which accurately describe object boundaries. Less false detections have also been detected inside homogeneous regions as, for example, in the tile roof region. Furthermore, results from the application of the improved MSEG algorithm are presented in Figures 9.1e,j,o,t. The first result (Figure 9.1e) shows how the developed algorithm is constrained by the detected LSD line features. On the homogeneous regions there is not much difference, which is normal since the same AML scale is used for both Figures 9.1c and e. The second result (Figure 9.1j) shows how the Canny edges are preserved inside the roof segments and how the improved segmentation method has been robustly constrained by edge information. This result is better than Figure 9.1h, where image objects are over segmented and arbitrarily set inside a homogeneous region of the image. The third result in Figure 9.1o shows that the combination of edge information with region merging in higher scales is outperforming the standard MSEG algorithm (Figure 9.1m) at the same AML scale and segmentation scale parameter. Moreover, the result in Figure 9.1t shows how the scale space in combination with the edge-constrained segmentation tackles the over-segmentation issue shown in Figure 9.1r.

These aforementioned results demonstrate that the developed method outperforms earlier efforts (Tzotsos and Argialas, 2008; Baatz and Schape, 2000; Tzotsos et al., 2011). In addition, in order to further validate the developed algorithm's experimental results and demonstrate its performance under several type of data sets and settings, a variety of remote sensing data have been selected with different spatial and spectral characteristics. In the following sections, the developed method was compared against previous research efforts (Tzotsos et al., 2011) and other standard OBIA implementations implemented in Orfeo Toolbox (Inglada and Christophe, 2009) and eCognition (Benz et al., 2004).

9.4.1 Very High Spatial Resolution Airborne Imagery

The developed methodology was applied to a variety of very high and ultrahigh resolution remote sensing imagery. At first a 5-cm-resolution image from a DMC airborne digital scanner was tested. These data are practically impossible to handle using traditional pixel-based classification and image analysis approaches. As shown in Figure 9.2, it is possible to segment this image into primitive objects in order to construct a feature space for OBIA classification. In this figure, a comparison of

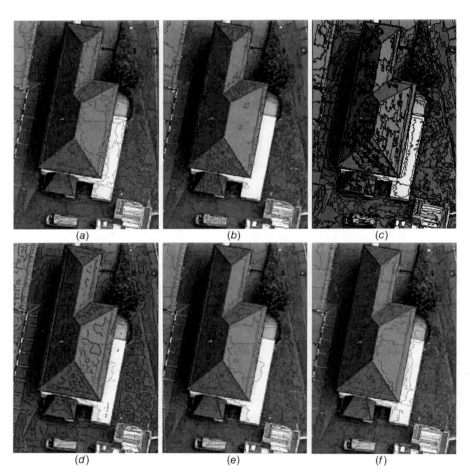

FIGURE 9.2   Comparison of various segmentation algorithms on a DMC aerial multispectral image with 5 cm pixel- size (copyright Intergraph Corp.): (*a*) standard MSEG with scale parameter 100; (*b*) mean-Shift segmantation with default parameters; (*c*) multiresolution segmentation (eCognition) with default parameters; (*d*) edge-constrained MSEG without merging edge objects in last pass for demonstration purposes; (*e*) edge-constrained MSEG with Canny edge features used; (*f*) edge-constrained MSEG with LSD edge features used. (See the color version of this figure in Color Plates section.)

various segmentation methods is performed. Initially, the standard MSEG segmentation algorithm is tested in Figure 9.2*a* at a scale of 100. The MSEG algorithm is applied on a scale-space AML representation and the result is achieved without any parameter tuning.

In Figure 9.2*d* results from the application of the developed edge-constrained MSEG algorithm are demonstrated. The edge objects are not merged to the rest of the image objects, and they remain unmerged until a final step concludes the segmentation procedure and produces the result in Figure 9.2*e*. A comparison of the developed algorithm with mean-shift algorithm (Comaniciu and Meer, 2002) (Figure 9.2*b*) shows that while in some image regions the mean shift can merge large parts of the image into one object, it fails to do so in other areas of the same texture. This behavior is not optimal and can lead to problems for classification steps since the mean object size varies. On the other hand, multiresolution segmentation as provided by eCognition (Figure 9.2*c*) manages to segment the image with a homogeneous object size, but it suffers from oversegmentation problems, especially on the roof objects. For both algorithms the default values were used to avoid parameter tuning. The developed method (Figures 9.2*e,f*) manages to obtain similar objects in size, which can be very applicable in a multiscale OBIA approach to classification. The main difference between Figure 9.2*e* and Figure 9.2*f* is that image edges are larger in number as derived from the Canny algorithm, and this leads to better results in areas that LSD (Figure 9.2*f*) has not detected any straight lines in the image. So the LSD method is not very suitable for curved or small edges.

In order to evaluate the developed method and to show the advantages of edge-constrained segmentation algorithm, a test similar to the one performed in Tzotsos et al. (2011) was deployed. For this test a very high resolution aerial scanner image was used with four spectral bands in order to perform full-scale object-based image analysis tests. The initial image was segmented using a simple MSEG algorithm without parameter tuning (default values of scale parameter 100, color 0.8, and shape 0.2 were selected). After primitive objects were obtained, a training set was given to a kernel-based classifier (SVM) to perform learning based on the feature space introduced by object spectral and shape properties. For this test four generic land cover classes were used: vegetation, tile roofs, bright roofs, and asphaltlike materials. A set of training samples/objects was introduced to the SVM and a classification was performed. Using ground truth data, a quantitative evaluation was performed and a confusion matrix is presented in Table 9.1. The accuracy of the object-based classification was 88.07% similar to the results reported for this approach in Tzotsos and Argialas (2008).

A similar approach was then followed for the same image, with the same segmentation parameters and the same training and testing samples. This time, a scale-space AML representation was used to provide anisotropic diffusion and simplification of the initial data set. After SVM classification and evaluation of results (Table 9.1) an overall accuracy of 89.29% was achieved, similar to the accuracy reported in Tzotsos et al. (2011).

Finally the developed method was applied in a similar manner to the same data. A scale-space AML method was used to simplify the initial data set. A segmentation

TABLE 9.1 Quantitative Results of Classification Accuracy for High-Spatial-Resolution Airborne Multispectral Data Set

	Vegetation	Tile Roofs	Bright Roofs	Asphalt Like
Classification Accuracy with MSEG Only				
Vegetation	15,247	0	0	2,539
Tile roofs	198	2,856	15	2,849
Bright roofs	0	1	8,362	2,064
Asphaltlike	215	34	498	35,612
Overall accuracy: 88.07%				
Classification Accuracy with AML				
Vegetation	15,523	0	0	2,263
Tile roofs	15	3,764	124	2,015
Bright roofs	0	0	8,389	2,038
Asphalt like	583	30	482	35,264
Overall accuracy: 89.29%				
Classification Accuracy with AML and Edge Enhancement				
Vegetation	15,791	34	0	1,961
Tile roofs	244	4,710	45	919
Bright roofs	0	0	8,311	2,116
Asphaltlike	141	909	475	34,834
Overall accuracy: 90.29%				

Note: The proposed OBIA methodology scored better, indicating that the enhancement of MSEG with advanced edge features along with advanced scale-space representations (AML) and the kernel classifier (SVM) outperforms earlier approaches.

step was then performed using the edge-constrained MSEG algorithm, and specifically the Canny edge features option was used. A set of primitive objects was obtained and object properties were extracted (spectral and shape features). The same training test was given to the SVM classifier and a final classification of objects was obtained. As shown in Table 9.1, the overall accuracy of the developed method outperformed the previous tests with an accuracy of 90.29%. This shows that edge features helped the segmentation procedure to obtain more meaningful objects that are capable of providing very good classification results. This procedure was repeated again with some different parameters and similar results were produced. Of course, the difference in accuracy is not wide, but it is a measure that compatible results are produced for further OBIA classification steps.

9.4.2 Radar Satellite Imagery

Experimental results include the application of the developed methodology at high-resolution SAR data (TerraSAR-X data set). The initial SAR image is shown in Figure 9.3a and the output results from the edge-constrained segmentation are compared with the ones from the mean shift and watershed (Figures 9.3b and c). The mean-shift method did not perform well and resulted in under segmentation of the shore line (Figure 9.3b). Even if the water area was successfully segmented into one

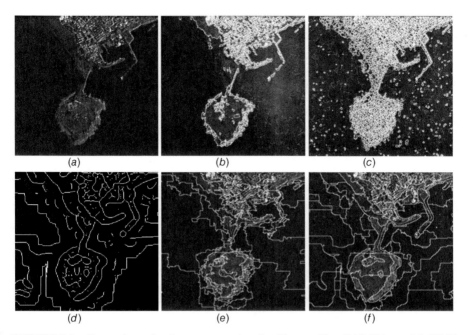

FIGURE 9.3   Comparison of various segmentation algorithms on TerraSAR-X (copyright DLR) data set (3 m ground resolution, StripMap mode, polarization HH): (*a*) initial image; (*b*) mean-shift segmantation with default parameters; (*c*) watershed segmentation with default parameters; (*d*) canny edge detection applied on AML scale-space representation; (*e*) standard MSEG results with scale parameter 400; (*f*) edge-constrained MSEG with Canny edge features used.

image object, the undersegmentation is always a very poor image segmentation performance. The applications of the watershed algorithm (Figure 9.3*c*), on the other hand, resulted in serious oversegmentation, as can be seen in Figure 9.3.

A better result was obtained with the application of the standard MSEG algorithm, although there were some problems in objects near the shore line (Figure 9.3*e*). For this reason a Canny edge feature extraction was performed (Figure 9.3*d*) and the results were imported to the edge-constrained segmentation algorithm, which outperformed all other segmentation algorithms (Figure 9.3*f*). The developed algorithm managed to obtain image objects of similar scale and, as a result of imposed edge information, the output object boundaries were more compact and approximated the image semantics more efficiently. Again no parameter tuning was performed and a default value of scale parameter 100 was used.

9.4.3   Multispectral Remote Sensing Data

The next series of tests were performed on medium- and high-spatial-resolution multispectral remote sensing data. For this, a Landsat Thematic Mapper (TM) image with pixel spatial resolution of 30 m was used as well as a QuickBird satellite image with 1 m ground resolution and four spectral bands.

For the Landsat TM imagery, the same comparison of image segmentation methods was performed and is presented in Figure 9.4. In this situation the image had a stripping noise problem, making it more difficult for the segmentation algorithms to perform well. The application of MSEG with scale parameter 100 resulted in major oversegmentation, but still the algorithm resulted in objects similar in size and scale. The stripes of the image are obvious in this segmentation result (Figure 9.4a).

The mean-shift segmentation algorithm performs much better in this specific test (Figure 9.4b), since strong simplification of the image is involved internally, making the algorithm more robust in noise presence. On the other hand, the size of the image object is variant across the image, even for the same semantic objects/areas. Multi-resolution segmentation (eCognition) produced a good result, especially on the parcel area. Some problems were observed with objects allocated on the river sides, where oversegmentation occurred (Figure 9.4c). A Canny edge detection step (Figure 9.4d) was involved and the edge-constrained segmentation was tested (Figure 9.4e). One can observe that the latter produces much better results than the standard MSEG algorithm. Object boundaries are more clear and compact, while the mean size of the

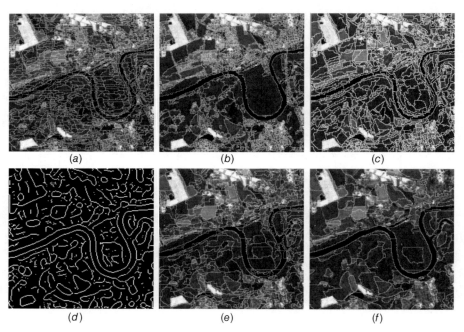

FIGURE 9.4 Comparison of various segmentation algorithms on Landsat TM data set (Dessau, Germany); (a) standard MSEG results with scale parameter 100; (b) mean-shift segmantation with default parameters; (c) multiresolution segmentation (eCognition) with default parameters (scale 10, shape 0.1); (d) Canny edge detection applied on AML scale-space representation; (e) edge-constrained MSEG with Canny edge features used and scale parameter 100; (f) edge-constrained MSEG with Canny edge features used and scale parameter 400. (See the color version of this figure in Color Plates section.)

primitive objects is approximately the same across the image. A test was also performed with an increased scale parameter (Figure 9.4*f*). The developed edge-constrained segmentation performed even better at larger scales, while the stripping problem was less apparent (Figure 9.4*f*).

For the QuickBird imagery (Figure 9.5*a*) a similar test was performed and demonstrated (Figures 9.5*e,h,f,i*) within a semiurban area. The developed algorithm

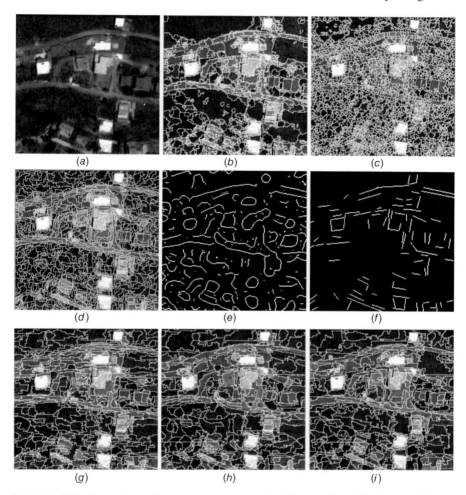

(a) (b) (c)

(d) (e) (f)

(g) (h) (i)

FIGURE 9.5 Comparison of various segmentation algorithms on QuickBird data set (Eastern Attika, Greece): (*a*) original image; (*b*) mean-shift segmentation with default parameters; (*c*) watershed segmentation with default parameters; (*d*) multiresolution segmentation (eCognition) with default parameters; (*e*) canny edge detection applied on AML scale-space representation; (*f*) LSD line features extracted from original image; (*g*) standard MSEG results with scale parameter 100; (*h*) edge-constrained MSEG with Canny edge features used and scale parameter 100; (*i*) edge-constrained MSEG with LSD line features used and scale parameter 100. (See the color version of this figure in Color Plates section.)

was able to detect building objects (Figures 9.5*h,i*) and in particular when it was constrained by the LSD features (the scale parameter for the simplification was 400). Again similar problems occurred with the mean-shift algorithm (Figure 9.5*b*) obtaining objects at different scales (i.e., larger objects in low contrast areas of the image). On the other hand, the watershed algorithm produced an oversegmentation (Figure 9.5*c*) but kept all image objects on the same scale. The multiresolution segmentation (eCognition) algorithm provided objects of the same scale (Figure 9.5*d*), with less oversegmentation problems than the watershed but did not outperform mean-shift and enhanced MSEG algorithms. Both enhanced MSEG and mean-shift algorithms had good results in building objects with the developed method having a small advantage in preserving the edges of the image semantics.

9.4.4 Hyperspectral Remote Sensing Data

The developed segmentation algorithm was also tested with hyperspectral remote sensing data obtained by a CASI aerial scanner (Figure 9.6). The spectral resolution of the data set was 95 bands and the spatial resolution was 5 m. Again, the mean-shift and watershed algorithms were tested to compare with the developed method, but in this

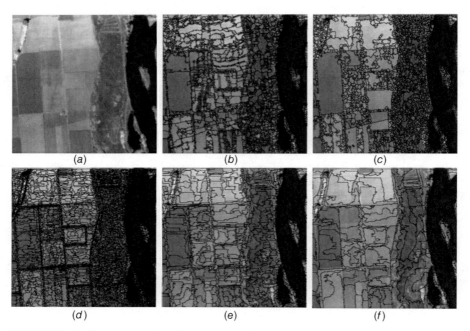

FIGURE 9.6 Comparison of various segmentation algorithms on CASI hyperspectral data set (copyright Remote Sensing Laboratory, NTUA) with 95 spectral bands (Axios River, Thessaloniki, Greece): (*a*) original image; (*b*) mean-shift segmentation with default parameters; (*c*) watershed segmentation with default parameters; (*d*) standard MSEG results with scale parameter 900; (*e*) standard MSEG on AML scale-space representation and scale parameter 900; (*f*) edge-constrained MSEG with Canny edge features used and scale parameter 900. (See the color version of this figure in Color Plates section.)

specific test, it was impossible for those algorithms to be applied to the full spectral resolution of the hyperspectral data set, since both algorithms are not designed to work on a large number of image bands. For this test, a subset of bands were used to derive the mean-shift and watershed results (Figures 9.6*b*,*c*). Both those algorithms produced similar results for this test. A mixture of small- and large-scaled objects were obtained at the same time, with the watershed being more accurate in this case, providing better results in edge areas (Figure 9.6*c*).

On the other hand, MSEG is designed to be applied to images of any spectral resolution, up to 65,535 bands. It is demonstrated in Figure 9.6*d* that the simple MSEG algorithm is performing very well given that the default scale parameter is easily reached (since it is a heterogeneity threshold) with a big number of bands contributing to object heterogeneity. Therefore, this is causing oversegmentation (Figure 9.6*d*) that cannot be considered a major problem; rather it is an effect caused by the nature of this data set. After application of a strong simplification of the AML filtering, the results were improved (Figure 9.6*e*). In all cases of MSEG application it is shown that MSEG respects the scale of the image objects in a better way that the other algorithms tested. This is very crucial for OBIA applications, especially when multiscale approaches are necessary. Finally, the enhanced MSEG algorithm was tested in Figure 9.6*f*, and the results were obviously better that other approaches shown here. The edge information was preserved (Figure 9.6*f*) and the size of the image objects still is similar for all objects on this scale.

9.5 CONCLUSIONS AND FUTURE PERSPECTIVES

A new object-based image analysis framework was proposed and developed in this research based on advanced edge features incorporated in a multiscale region merging algorithm. Advanced scale-space representations were used in order to avoid tuning of segmentation and feature extraction parameters, and a kernel-based classification was implemented to complete the OBIA framework. The developed image segmentation algorithm was shown to work on any type of remote sensing data, outperforming some widely used segmentation algorithms in some cases. The improvement of the MSEG segmantation results was demonstrated, and the edge enhancements were shown to make the algorithm robust and generic for multiscale OBIA applications. The performed qualitative and quantitative evaluation reported that the developed algorithm outperformed previous efforts, regarding both the construction of the object representations and the classification results.

In terms of performance, the proposed improvement of the MSEG algorithm had a major impact on execution times. For the very high resolution airborne image (Figure 9.1*a*), the execution time for the original MSEG algorithm was 7–8 sec on a 3-GHz PC with 4 GB of RAM running a GNU/Linux operating system. With the edge-based optimization, the number of possible object merges dropped significantly per segmentation pass, and the execution time was around 4–5 sec. Still, there is room for speed optimization improvements since for the same image the execution times were 2–3 sec for Multiresolution segmentation (eCognition) and 3–4 sec for the mean shift (Orfeo Toolbox).

An advantage of the developed OBIA framework is that it is composed of free (as in freedom) software. For the implementation of this research a number of free and open-source libraries were used (MSEG, cvAML, libSVM, OrfeoToolbox, and GDAL). The coding was performed in the C++ and Python programming languages. A disadvantage of the developed OBIA framework is that a knowledge-based classification solution is not yet integrated but is currently being worked on.

Some of the topics for further research and development are extension of the developed OBIA framework to integrate knowledge-based classification, solutions for object-specific extraction tasks based on scale-space shape priors, and adaptation of the developed methodology to specific remote sensing applications.

ACKNOWLEDGMENTS

The authors would like to thank the anonymous reviewers for their constructive comments and suggestions.

REFERENCES

Akcay, H., and Aksoy, S. 2008. Automatic detection of geospatial objects using multiple hierarchical segmentations. *IEEE Transactions on Geoscience and Remote Sensing* 46:2097–2111.

Aplin, P., and Smith, G. 2008. Advances in object-based image classification. *International Archive of Photogrammetry, Remote Sensing and Spatial Information Sciences* 37(Part B7): 725–728.

Baatz, M., and Schape, A. 2000. Multiresolution segmentation an optimization approach for high quality multi-scale image segmentation. In J. Strobl, et al. (Eds.), *Angewandte Geographische Infor-mationsverarbeitung XII.* Heidelberg: Wichmann, pp. 12–23.

Benz, U., Hofmann, P., Willhauck, G., Lingenfelder, I., and Heynen, M. 2004. Multi-resolution, object-oriented fuzzy analysis of remote sensing data for gis ready information. *ISPRS Journal of Photogrammetry and Remote Sensing* 58(3–4):239–258.

Blaschke, T. 2010. Object based image analysis for remote sensing. *ISPRS Journal of Photogrammetry and Remote Sensing* 65(1):2–16.

Blaschke, T., Burnett, C., and Pekkarinen, A. 2004. Image segmentation methods for object-based analysis and classification. In S. M.de Jong and F. D.van der Meer (Eds.), *Remote Sensing and Digital Image Analysis: Including the Spatial Domain.* Dordrecht: Kluwer Academic, pp. 211–236.

Blaschke, T., and Hay, G. 2001. Object-oriented image analysis and scale-space: Theory and methods for modeling and evaluating multi–scale landscape structure. *Interanational Archive of Photogrammetry and Remote Sensing* 34(Part4/W5): 22–29.

Blaschke, T., Lang, S., and Hay, G. 2008. *Object Based Image Analysis—Spatial Concepts for Knowledge Driven Remote Sensing Applications.* New York: Springer.

Camps-Valls, G., and Bruzzone, L. 2005. Kernel-based methods for hyperspectral image classification. *IEEE Transactions on Geoscience and Remote Sensing* 43(6):1351–1362.

Canny, J. 1986. A computational approach to edge detection. *IEEE Transactions on Pattern Analysis and Machine Intelligence* 6:679–698.

Carleer, A., Debeir, O., and Wolff, E. 2005. Assessment of very high spatial resolution satellite image segmentations. *Photogrammetric Engineering and Remote Sensing* 71(11): 1285–1294.

Chia, A.Y.S., Rajan, D., Leung, M. K., and Rahardja, S. 2012. Object recognition by discriminative combinations of line segments, ellipses, and appearance features. *IEEE Transactions on Pattern Analysis and Machine Intelligence* 34:1758–1772.

Christophe, E., and Inglada, J. 2009. Open source remote sensing: Increasing the usability of cutting-edge algorithms. *IEEE Geoscience and Remote Sensing Newsletter*, 2009, 9–15.

Comaniciu, D., and Meer, P. 2002. Mean shift: A robust approach toward feature space analysis. *IEEE Transactions on Pattern Analysis and Machine Intelligence* 24:603–619.

Cufi, X., Munoz, X., Freixenet, J., and Marti, J. 2003. A review of image segmentation techniques integrating region and boundary information. *Advances in Imaging and Electron Physics* 120:1–39.

Derrode, S., and Mercier, G. 2007. Unsupervised multiscale oil slick segmentation from sar images using a vector hmc model. *Pattern Recognition* 40:1135–1147.

Doxani, G., Karantzalos, K., and Tsakiri-Strati, M. 2012. Monitoring urban changes based on scale-space filtering and object-oriented classification. *International Journal of Applied Earth Observation and Geoinformation* 15:38–48.

Dragut, L., Schauppenlehner, T., Muhar, A., Strobl, J., and Blaschke, T. 2009. Optimization of scale and parametrization for terrain segmentation: An application to soil-landscape modeling. *Computers & Geosciences* 35(9):1875–1883.

Dragut, L., Tiede, D., and Levick, S. R. 2010. ESP: A tool to estimate scale parameter for multiresolution image segmentation of remotely sensed data. *International Journal of Geographical Information Science* 24:859–871.

Duarte-Carvajalino, J., Castillo, P., and Velez Reyes, M. 2007. Comparative study of semi-implicit schemes for nonlinear diffusion in hyperspectral imagery. *IEEE Transactions on Image Processing* 16(5):1303–1314.

Duarte-Carvajalino, J., Sapiro, G., Velez Reyes, M., and Castillo, P. 2008. Multiscale representation and segmentation of hyperspectral imagery using geometric partial differential equations and algebraic multigrid. *IEEE Transactions on Geoscience and Remote Sensing* 46(8):2418–2434.

Gilboa, G. 2008. Nonlinear scale space with spatially varying stopping time. *IEEE Transactions on Pattern Analysis and Machine Intelligence* 30:2175–2187.

Hall, O., and Hay, G. J. 2003. A multiscale object-specific approach to digital change detection. *International Journal of Applied Earth Observation and Geoinformation* 4(4):311–327.

Hay, G., Blaschke, T., Marceau, D., and Bouchard, A. 2003. A comparison of three image-object methods for the multiscale analysis of landscape structure. *ISPRS Journal of Photogrammetry and Remote Sensing* 57(5–6):327–345.

Hay, G., and Castilla, G. 2006. Object-based image analysis: Strengths, weaknesses, opportunities and threats. *Interanational Archive of Photogrammetry, Remote Sensing and Spatial Information Sciences* 36(4/C42): on CD–ROM.

Hay, G., Castilla, G., Wulder, M., and Ruiz, J. 2005. An automated object-based approach for the multiscale image segmentation of forest scenes. *International Journal of Applied Earth Observation and Geoinformation* 7(4):339–359.

Hay, G. J., Dub, P., Bouchard, A., and Marceau, D.J. 2002. A scale-space primer for exploring and quantifying complex landscapes. *Ecological Modelling* 153(1–2):27–49.

Hofmann, P., Strobl, J., and Blaschke, T. 2008. A method for adapting global image segmentation methods to images of different resolutions. In *International Conference on Geographic Object-Based Image Analysis*. ISPRS Volume 38; on CD-ROM.

Hsu, C. W., and Lin, C. J. 2002. A comparison of methods for multiclass support vector machines. *IEEE Transactions on Neural Networks* 13(2):415–425.

Inglada, J., and Christophe, E. 2009. The orfeo toolbox remote sensing image processing software. In Geoscience and Remote Sensing Symposium, 2009 IEEE International, IGARSS 2009, pp. IV–733.

Jiang, H., Yu, S., and Martin, D. 2011. Linear scale and rotation invariant matching. *IEEE Transactions on Pattern Analysis and Machine Intelligence* 33:1339–1355.

Jimenez, L. O., Rivera-Medina, J. L., Rodriguez-Diaz, E., Arzuaga-Cruz, E., and Ramirez-Velez, M. 2005. Integration of spatial and spectral information by means of unsupervised extraction and classification for homogenous objects applied to multispectral and hyperspectral data. *IEEE Transactions on Geoscience and Remote Sensing* 43(4):844–851.

Jung, C. 2007. Unsupervised multiscale segmentation of color images. *Pattern Recognition Letters* 28:523–533.

Karantzalos, K., and Argialas, D. 2006. Improving edge detection and watershed segmentation with anisotropic diffusion and morphological levelings. *International Journal of Remote Sensing* 27(24):5427–5434.

Karantzalos, K., Argialas, D., and Paragios, N. 2007. Comparing morphological levelings constrained by different markers. In G. Banon, et al. (Eds.), *Mathematical Morphology and Its Applications to Signal and Image Processing*, MCT/INPE, Brazil, pp. 113–124.

Kermad, C., and Chehdi, K. 2002. Automatic image segmentation system through iterative edge-region co-operation. *Image and Vision Computing* 20:541–555.

Kimmel, R., Zhang, C., Bronstein, A., and Bronstein, M. 2011. Are mser features really interesting? *IEEE Transactions on Pattern Analysis and Machine Intelligence* 33:2316–2320.

Klonus, S., Tomowski, D., Ehlers, M., Reinartz, P., and Michel, U. 2012. Combined edge segment texture analysis for the detection of damaged buildings in crisis areas. *IEEE Journal of Selected Topics in Applied Earth Observations and Remote Sensing* 5:1118–1128.

Koenderink, J. 1984. The structure of images. *Biological Cybernetics* 50(5):363–370.

Lehmann, G. 2008. Label object representation and manipulation with itk. *Insight Journal*, 9:1–34.

Lennon, M., Mercier, G., and Hubert-Moy, L. 2002. Classification of hyperspectral images with nonlinear filtering and support vector machines. In IEEE International Geoscience and Remote Sensing Symposium, IGARSS'02, Dec 2002, Toronto, Canada, Vol 3, pp. 1670–1672.

Lindeberg, T. 1994. *Scale-Space Theory in Computer Vision*. Dordrecht: Kluwer Academic.

Liu, Y., Guo, Q., and Kelly, M. 2008. A framework of region-based spatial relations for non-overlapping features and its application in object based image analysis. *ISPRS Journal of Photogrammetry and Remote Sensing* 63(4):461–475.

Martha, T., Kerle, N., van Westen, C., Jetten, V., and Kumar, K. 2011. Segment optimization and data-driven thresholding for knowledge-based landslide detection by object-based image analysis. *IEEE Transactions on Geoscience and Remote Sensing* 49:4928–4943.

Matheron, G. 1997. *Les Nivellements*. Technical Report. Centre de Morphologie Mathematique, Paris, France.

Meyer, F. 2004. Levelings, image simplification filters for segmentation. *International Journal of Mathematical Imaging and Vision* 20:59–72.

Meyer, F., and Maragos, P. 2000. Nonlinear scale-space representation with morphological levelings. *Journal of Visual Communication and Image Representation* 11:245–265.

Meyer, Y. 1998. The levelings. In *Mathematical Morphology and Its Applications to Image and Signal Processing*. Dordrecht: Kluwer Academic.

Mladinich, C., 2010. An evaluation of object-oriented image analysis techniques to identify motorized vehicle effects in semi-arid to arid ecosystems of the american west. *GIScience and Remote Sensing* 47:53–77.

Myint, S. W., Gober, P., Brazel, A., Grossman-Clarke, S., and Weng, Q. 2011. Per-pixel vs. object-based classification of urban land cover extraction using high spatial resolution imagery. *Remote Sensing of Environment* 115:1145–1161.

Neubert, M., Herold, H., and Meinel, G. 2006. Evaluation of remote sensing image segmentation quality—further results and concepts. Interanational Archive of Photogrammetry, Remote Sensing and Spatial Information Sciences 36(4/C42).

Nilufar, S., Ray, N., and Zhang, H. 2012. Object detection with dog scale-space: A multiple kernel learning approach. *IEEE Transactions on Image Processing* 21:3744–3756.

Ouma, Y., Josaphat, S., and Tateishi, R. 2008. Multiscale remote sensing data segmentation and postsegmentation change detection based on logical modeling: Theoretical exposition and experimental results for forestland cover change analysis. *Computers & Geosciences* 34(7):715–737.

Ouzounis, G., Pesaresi, M., and Soille, P. 2012. Differential area profiles: Decomposition properties and efficient computation. *IEEE Transactions on Pattern Analysis and Machine Intelligence* 34:1533–1548.

Papari, G., and Petkov, N. 2011. Edge and line oriented contour detection: State of the art. *Image and Vision Computing* 29:79–103.

Paragios, N., Chen, Y., and Faugeras, O. 2005. *Handbook of Mathematical Models of Computer Vision*. New York, USA, Springer.

Pavlidis, T., and Liow, Y. 1990. Integrating region growing and edge detection. *IEEE Transactions on Pattern Analysis and Machine Intelligence* 12:225–233.

Perona, P., and Malik, J. 1990. Scale space and edge detection using anisotropic diffusion. *IEEE Transactions on Pattern Analysis and Machine Intelligence* 12(7):629–639.

Plaza, A., Benediktsson, J. A., Boardman, J., Brazile, J., Bruzzone, L., Camps-Valls, G., Chanussot, J., Fauvel, M., Gamba, P., Gualtieri, A., Marconcini, M., Tilton, J., and Trianni, G. 2009. Recent advances in techniques for hyperspectral image processing. *Remote Sensing of Environment* 113(S1):S110–S122.

Serra, J. 2000. Connections for sets and functions. *Fundamentae Informatica* 41:147–186.

Stewart, S., Hay, G., Rosin, P., and Wynn, T. J. 2004. Multiscale structure in sedimentary basins. *Journal of Basin Research* 16(2):183–197.

Theodoridis, S., and Koutroumbas, K. 2003. *Pattern Recognition*. New York: Elsevier Academic Press.

Tzotsos, A. 2006. A support vector machine approach for object based image analysis. In Proceedings of 1st International Conference on Object-based Image Analysis, OBIA 2006, Salzburg, Austria, July 4–5.

Tzotsos, A., and Argialas, D. 2006. Mseg: A generic region-based multi-scale image segmentation algorithm for remote sensing imagery. In Proceedings of ASPRS 2006 Annual Conference, Reno, Nevada, May 1–5, ASPRS (on CD-ROM).

Tzotsos, A., and Argialas, D. 2008. Support vector machine classification for object-based image analysis. In T. Blaschke, S. Lang, and G. Hay (Eds.), *Object Based Image Analysis— Spatial Concepts for Knowledge Driven Remote Sensing Applications*. New York: Springer, pp. 663–679.

Tzotsos, A., Iosifidis, C., and Argialas, D. 2008. A hybrid texture-based and region-based multi-scale image segmentation algorithm. In T. Blaschke, S. Lang, and G. Hay (Eds.), *Object Based Image Analysis—Spatial Concepts for Knowledge Driven Remote Sensing Applications*. New York: Springer, pp. 221–237.

Tzotsos, A., Karantzalos, K., and Argialas, D. 2011. Object-based image analysis through nonlinear scale-space filtering. *ISPRS Journal of Photogrammetry and Remote Sensing* 66:2–16.

Vapnik, V. 1998. *Statistical Learning Theory*. New York: Wiley.

Von Gioi, R., Jakubowicz, J., Morel, J., and Randall, G. 2010. LSD: A fast line segment detector with a false detection control. *IEEE Transactions on Pattern Analysis and Machine Intelligence*, 32:722–732.

Von Gioi, R., Jakubowicz, J., Morel, J., and Randall, G. 2012. LSD: A Line Segment Detector. Image Processing On Line.

Wang, H., and Oliensis, J. 2010. Generalizing edge detection to contour detection for image segmentation. *Computer Vision and Image Understanding* 114:731–744.

Weickert, J. 1998. Anisotropic Diffusion in Image Processing. ECMI Series. Stuttgart: Teubner.

Weickert, J., Ishikawa, S., and Imiya, A. 1999. Linear scale-space has first been proposed in Japan. *Journal of Mathematical Imaging and Vision* 10(3):237–252.

Witkin, A. 1983. Scale-space filtering. In International Joint Conference on Artificial Intelligence, Karlsruhe, 1983, pp. 1019–1021.

Xu, Y., Huang, S., Ji, H., and Fermller, C. 2012. Scale-space texture description on sift-like textons. *Computer Vision and Image Understanding* 116:999–1013.

Yu, P., Qin, A., and Clausi, D. 2012. Unsupervised polarimetric sar image segmentation and classification using region growing with edge penalty. *IEEE Transactions on Geoscience and Remote Sensing* 50:1302–1317.

Zhong, B., Ma, K. K., and Liao, W. 2009. Scale-space behavior of planar-curve corners. *IEEE Transactions on Pattern Analysis and Machine Intelligence* 31:1517–1524.

Zhou, W., Huang, G., Troy, A., and Cadenasso, M. 2009. Object-based land cover classification of shaded areas in high spatial resolution imagery of urban areas: A comparison study. *Remote Sensing of Environment* 113(8):1769–1777.

10

OPTIMUM SCALE IN OBJECT-BASED IMAGE ANALYSIS

Jungho Im, Lindi J. Quackenbush, Manqi Li, and Fang Fang

10.1 INTRODUCTION

Object-based image analysis (OBIA) or geographic OBIA (GEOBIA) has become popular in the remote sensing literature since the introduction of high-spatial-resolution ($\leq$1-m) satellite remote sensing data such as IKONOS and QuickBird. Traditional pixel-based analysis often does not work well with such high-resolution data due to high-frequency components and horizontal layover caused by off-nadir look angles (Im et al., 2008b). Blaschke (2010) comprehensively reviewed over 820 articles dealing with object-based image analysis for remote sensing and provides an overview of object-based approaches that are commonly used in the remote sensing literature. The increased popularity of object-based image analysis represents a significant trend in the remote sensing context as the number of publications in the field continues to increase.

OBIA applications are broad and typically include image segmentation, edge detection, feature extraction, classification, and change detection (Blaschke, 2010; Stow, 2010). An image segmentation procedure is the first step commonly applied in OBIA. This procedure divides the image into meaningful homogeneous and nonintersecting regions based on spectral and/or spatial properties (Benz et al., 2004). Such homogeneous regions can be hierarchically organized as image objects (i.e., segments) (Blaschke, 2005). There are numerous algorithms developed for image segmentation, which Cheng et al. (2001) groups into four types: (1) histogram thresholding (Hofmann et al., 1998) that works for a single image with several thresholds; (2) image feature space clustering (Hall et al., 1992), which

Scale Issues in Remote Sensing, First Edition. Edited by Qihao Weng.
© 2014 John Wiley & Sons, Inc. Published 2014 by John Wiley & Sons, Inc.

extends histogram thresholding into multidimensions (i.e., when more than one feature is used); (3) region-based approaches (Definiens Imaging, 2005) that include region growing, region splitting, region merging, or their combinations; and (4) edge-based methods (Shih and Cheng, 2005) that first detect edges within an image. The most widely used image segmentation algorithm is perhaps the multiresolution algorithm used in eCognition software. This algorithm is a bottom-up, region-merging algorithm based on the fractal net evolution approach (FNEA) (Baatz and Schäpe, 2000; Baatz et al., 2004). An overview of different segmentation techniques and related issues can be found in Baatz and Schäpe (2000) and Schöpfer et al. (2008).

OBIA might be more appropriate than pixel-based analysis in many applications, especially when high-spatial-resolution data were used. One reason is that OBIA can utilize not only spectral information but also other information such as the shape, texture, and contextual relationships of image objects, which are called object-based metrics. Another feature of OBIA in the remote sensing classification literature is that OBIA-based classification results are relatively free from salt-and-pepper noise, which is a common problem in pixel-based analysis (Lillesand et al., 2008). Numerous studies have compared OBIA with traditional pixel-based approaches and many found that OBIA generally produced higher accuracy than pixel-based methods (Im et al., 2008b; Jensen et al., 2006; Myint et al., 2011; Vieira et al., 2012; Whiteside and Ahmad, 2005).

Although OBIA has already gained much popularity in remote sensing applications, there are several issues that need further exploration. Successful OBIA requires that image segmentation produces image objects of high quality (Baatz and Schäpe, 2000). As new or improved image segmentation methods continue to be introduced, the quantitative assessment of segmented results has gained great interest. Many studies examined the quality of image objects simply based on visual inspection of the image objects or accuracy of a subsequent analysis such as classification (Ouyang et al., 2011). However, some studies argue that the quantitative assessment of image objects in terms of real-world objects is crucial for successful classification or feature extraction (Ke et al., 2010; Möller et al., 2007; Wang et al., 2010). The methods that assess segmentation results typically compare segmented objects with the corresponding reference objects and measure similarity between them. However, there is no single method that provides a standardized way to quantitatively assess image segmentation results.

A second issue in OBIA that requires attention is object-based change detection. Unlike OBIA based on single-date imagery, object-based change detection requires additional considerations because the geometries of different dates of remote sensing data can vary. Object-based change detection approaches are often broadly grouped into (1) post-object-based classification comparison and (2) multitemporal image object change classification (Stow, 2010). Chen et al. (2012) added another group— image object change detection—that directly compares image objects between multiple dates. While post-object-based classification comparison may be the most widely used object-based change detection method, its performance is dependent on the accuracy of the object-based classifications. Regardless of the approach to be used, object-based change detection typically results in errors around real-world objects such as buildings due to misregistration, different look angles, or difference in shadow between dates (Im et al., 2008b; Stow, 2010).

Accuracy assessment of OBIA is another hot issue Clinton et al. (2010). Some studies use point data to assess OBIA results, while others use polygons (i.e., objects) as reference data. The same consideration of whether points or objects should be used is also applicable to training data for supervised classification. While accuracy assessment of OBIA is a somewhat controversial issue, there is a trend to develop a standardized framework. Details about accuracy assessment of OBIA are found in Radoux et al. (2010) and Lein (2012).

A fourth topic of consideration is optimum scale in OBIA. Image segmentation algorithms influence the geometry of image objects, and thus it is crucial to understand the parameters that are used in image segmentation algorithms and to identify how those parameters affect image objects in terms of shape and size. A successful segmentation would result in image objects similar to real-world objects. When it comes to object size, most segmentation algorithms use some type of scale parameter to control overall size of image objects. Many studies use a multiscale approach to deal with the inherently different spectral and spatial characteristics of disparate features such as buildings, roads, and forests within a scene. Some studies have tried to identify optimum scales in OBIA. Selecting optimum scale is often challenging as there is no standardized method to identify the optimality; many studies simply adopt visual interpretation of objects or the accuracy of subsequent analyses such as classification as selection criteria. In addition, scales in most segmentation algorithms are arbitrary values and provide only a relative comparison between the scales when the same segmentation algorithm is used to the same input data. There should be standardized guidelines regarding scales to facilitate the generalization of OBIA in remote sensing applications, to enable efficient comparison of different OBIA approaches, and to select optimum scales. The main objectives of this chapter are to provide a review of recent publications (since 2008) in OBIA focusing on scale optimization and discuss the related trends.

10.2   BRIEF OVERVIEW

We focused on peer-reviewed papers on OBIA published since 2008. The journals we searched for papers include *Remote Sensing of Environment* (19), *Photogrammetric Engineering and Remote Sensing* (15), *International Journal of Applied Earth Observation and Geoinformation* (8), *ISPRS Journal of Photogrammetry and Remote Sensing* (8), *IEEE Transactions on Geoscience and Remote Sensing* (3), *International Journal of Remote Sensing* (4), and others (19). We reviewed a total of 76 papers, and Figures 10.1–10.3 provide a brief overview of the selected papers for review. The number of publications has generally increased over the past five years (Figure 10.1). Most of the selected studies utilized the commercial software tool eCognition for performing the image segmentation that is a key step in OBIA

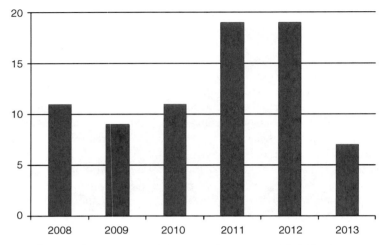

FIGURE 10.1 Number of the selected papers used for this chapter.

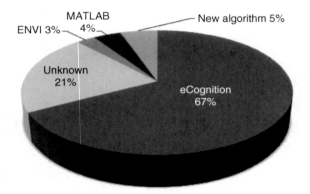

FIGURE 10.2 Image segmentation tools used in OBIA in selected papers.

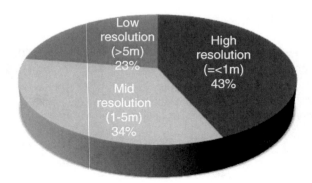

FIGURE 10.3 Spatial resolutions of data used in selected papers.

(Figure 10.2). As expected, OBIA has been more commonly applied to high-spatial-resolution data (Figure 10.3). In Section 10.3, we review these papers focusing on what scales were used in image segmentation and how optimum scale was determined.

10.3   RECENT STUDIES IN OBIA

10.3.1   Application Oriented

Land cover classification is perhaps the application field that most frequently applies OBIA. Many studies investigated land cover and land use information in urban/suburban areas using OBIA approaches (Durieux et al., 2008; Im et al., 2008a; Lizarazo and Barros, 2010; Meng et al., 2012; Stow et al., 2010; Tiede et al., 2010; Zhou et al., 2009). The heterogeneous nature of urban areas typically requires careful selection of a scale parameter, since this directly influences the size of image objects. Multiscale approaches are commonly adopted to extract urban features that vary in terms of color and shape. For example, Jacquin et al. (2008) introduced a hybrid OBIA approach to map urban sprawl in a periurban environment using SPOT images. Different scale values were used to extract urban objects at regional and local scales. They suggested a hybrid OBIA approach that combines multiscales to deal with the spatial and spectral heterogeneity of the urban surface. Huang and Qi (2010) also applied a multiscale segmentation to IKONOS data to extract land cover information using different scales for different land cover types.

As generally agreed, the higher the spatial resolution of an image, the better the expected segmentation (Holt et al., 2009; Karantzalos and Argialas, 2009; Wilschut et al., 2013). However, higher spatial resolution does not necessarily yield higher classification accuracy. For example, Lian and Chen (2011) used OBIA to extract urban features such as buildings, roads, water bodies, and vegetation from different resolution images (i.e., two SPOT data at 10 and 2.5 m, ASTER at 15 m, and QuickBird at 0.6 m) at different scales. They found the QuickBird experiment resulted in lower classification accuracy than the other experiments. OBIA has been frequently used with moderate and coarse resolution data such as the Landsat TM and MODIS (Ranson et al., 2011; Yang et al., 2013; Zhu and Woodcock, 2012).

OBIA studies often adopt a comparison between pixel- and object-based approaches (Chirici et al., 2011; Johansen et al., 2010; Myint et al., 2011; Verbeeck et al., 2012; Watmough et al., 2011). For example, Duro et al. (2012) compared pixel-based methods with OBIA to classify agricultural landscapes using selected machine learning algorithms such as decision trees, random forest, and support vector machines. Multiple scales were tested and the optimum scale was identified using an iterative trial-and-error approach. They found OBIA outperformed the pixel-based methods when the same machine learning approach was used. Quyang et al. (2011) mapped vegetation in a saltmarsh ecosystem using QuickBird imagery based on both pixel- and object-based approaches. They reported that the object-based approach with a two-scale segmentation was superior to the pixel-based one due to membership functions and a hierarchical approach. Whiteside et al. (2011) compared object- and pixel-based classification for mapping savannas from ASTER data using nearest-neighbor and fuzzy classification methods. Multiscale image segmentation was used to generate both coarse and fine objects.

Data from active sensors such as radar and lidar have often been used in OBIA applications (Liu J. et al., 2008; Qi et al., 2012). Arnesen et al. (2013) used multiscale segmentation OBIA to map flood condition by land cover type using the ALOS ScanSAR backscattering signature. Drăgut and Eisank (2011) classified topography from Shuttle Radar Topography Mission (SRTM) data using OBIA. Three scale levels that represent different complexities based on self-adaptive and data-driven techniques were used for topography classification. Im et al. (2008a) introduced an OBIA approach based solely on the analysis of lidar-derived information for land cover classification.

As mentioned in the Section 3.1, object-based change detection is a current focus in OBIA and different change detection approaches have been employed (Conchedda et al., 2008; de Chant and Kelly, 2009; Dronova et al., 2011; Im et al., 2008b; Stow et al., 2008; Wang et al., 2009). Some studies integrated other approaches with OBIA. For example, Bontemps et al. (2008) proposed an automated object-based change detection method based on a probabilistic changed–unchanged threshold procedure from time-series images. The differentiation between inter- and intra-annual dynamics calculated in the method provides efficiency when identifying changes associated with natural variability such as the phenological cycle of vegetation. Doxani et al. (2012) investigated urban changes based on scale-space filtering embedded in OBIA and multivariate alteration detection (MAD) transformation. Selection of optimum scale for object-based change detection is crucial for successful change detection. Dronova et al. (2012) used OBIA to classify the Poyang Lake wetland in China. They determined an optimum segmentation scale using a trial-and-error method and optimized class discrimination based on machine learning approaches.

10.3.2 Algorithm Oriented

Scientists have continued to improve OBIA approaches by fusing different algorithms (Mahmoudi et al., 2013; Sebari and He, 2013). Cai et al. (2009) proposed a segmentation method based on a watershed transformation and multiscale merging. Pan et al. (2009) developed an improved strategy for image segmentation based on the mean-shift segmentation and the fractal net evolution approach (FNEA). Their method reduced the number of objects compared to traditional FNEA, which would benefit the subsequent classification. Tzotsos et al. (2011) described an OBIA approach that integrated nonlinear scale-space filtering into a multiscale segmentation and classification procedure. Their approach does not require testing parameters such as shape, color, and texture that control the segmentation process. OBIA has also been combined with a subpixel mapping approach to map buildings with prior knowledge of building shape (Ling et al., 2012).

Many studies have investigated optimum scale selection in algorithm-oriented OBIA studies (Aguirre-Gutiérrez et al., 2012; Laliberte and Rango, 2009). Tong and Li (2011) investigated the influence of shape parameters on selecting an optimal scale value in multiresolution segmentation using QuickBird imagery. Tong and Li (2011) tested 26 parameter settings for segmentation, with each setting having 10 segmentation scales. They found that different shape parameters had different impacts on the optimum scale and a greater weight on the shape parameter might improve the efficiency of segmentation. Yi et al. (2012) proposed a flexible scale synthesis method to meet different segmentation requirements in image analysis. Their method divides the image into multiple regions where each region contains ground objects at similar scale. They identified the optimal segmentation scale using the method by Crevier (2008). The final segmentation result is achieved through synthesis of the optimum segmentation of each region, which yields more coherent results to ground objects.

Peña-Barragán et al. (2011) investigated 1312 segmentation scenarios with different parameter settings for object-based crop identification and mapping from multiseasonal ASTER data. They applied an empirical discrepancy method (Ortiz and Oliver, 2006; Zhang, 1996) to assess the quality of the segmentation results to identify the optimum parameter settings including the scale. An adaptive spectral matching method to extract thematic objects from Landsat ETM+ imagery was employed in Qiao et al. (2012). Two scales—whole and local—were used for spectral matching through end-member selection. Liu et al. (2010) used a mean-shift algorithm to segment digital aerial imagery. Multiple scales were tested and optimum scale segmentation was selected based on a trial-and-error method. Martha et al. (2011) optimized segmentation based on a plateau objective function derived from spatial autocorrelation and intrasegment variance analysis. Their optimum segmentation results were used in a knowledge-based classification approach to landslide detection. Anders et al. (2011) proposed a stratified OBIA to map alpine geomorphology from airborne lidar data. They used two dimensional frequency distribution matrices of training samples and image objects to identify an optimum scale. Shruthi et al. (2011) extracted gully erosion features using OBIA from IKONOS and GeoEye-1 data. Multiple scales were tested and the estimation of scale parameter (ESP) was used to determine the optimum scale. Drăgut et al. (2010) developed ESP to assist in selecting a suitable range of scales for image segmentation using local variance of object heterogeneity within a scene.

Some studies focused on developing new segmentation algorithms. Wang et al. (2010) proposed a new segmentation algorithm, called the region-based image segmentation algorithm (RISA), based on k-means clustering and region growing and merging. They evaluated RISA through the quantitative assessment of object quality and classification accuracy and found that the RISA performance was similar to the multiscale resolution segmentation algorithm in eCognition. Zhang et al. (2013) proposed a boundary-constrained multiscale segmentation approach and tested it with a set of high-resolution images, including QuickBird, WorldView, and aerial image data. Their approach can produce nested multiscale segmentations and is especially good at delineating boundaries smoothly.

10.4 OPTIMUM SCALES IN OBIA

Many studies pointed out the importance of scale in OBIA as it directly affects the overall size of resultant image objects and consequently the performance of the subsequent analyses, such as classification, feature extraction, and change detection (Addink et al., 2007; Ke et al., 2010). However, the selection of an optimum scale is often challenging. Based on the review of the 76 papers on OBIA published since 2008, there are three basic methods commonly adopted to select optimum scale in OBIA by testing multiple scales with a given segmentation algorithm (Table 10.1). The simplest but still widely adopted method is qualitative visual interpretation of image objects (Aguirre-Gutiérrez et al., 2012; Hernando et al., 2012; Kim et al., 2011a, 2011b; Laliberte et al., 2012; Lamonaca et al., 2008; Vieira et al., 2012). For example, Lamonaca et al. (2008) explored forest structure using multiscale segmentation of very high resolution (VHR) imagery. Three segmentation levels were applied to the image through visual inspection guided by ecological consideration on the size of meaningful objects. Results show that multiscale segmentation was appropriate for identifying scale-dependent forest structural patterns. Accuracy metrics from subsequent analyses, such as classification, are also used to select optimum scales (Ke et al., 2010; Laliberte and Rango, 2009; Stumpf and Kerle, 2011; Wang et al., 2010). In Aksoy and Ercanoglu (2012), an optimum scale in OBIA was determined based on a trial-and-error method to classify a landslide-prone area in Landsat ETM+ images using fuzzy logic. Another method is to quantitatively assess image objects with different scales and select the scale that results in the greatest similarity with real-world ground objects as optimum scales (Ke et al., 2010; Möller et al., 2007; Tong et al., 2012). Zhang and Xie (2012) used a neural network to combine object-based texture metrics for vegetation mapping in the Everglades in South Florida using Airborne Visible/Infrared Imaging Spectrometer (AVIRIS) hyperspectral imagery. Ten different scale parameters were tested and the optimum scale was identified based on an unsupervised image segmentation evaluation approach developed by Johnson and Xie (2011). Some

TABLE 10.1 Methods to Determine Optimum Scales for Image Segmentation Based on 76 Papers Reviewed

Method to Determine Optimum Scale	Number of Papers
Qualitative assessment	24
Visual interpretation of image objects based on knowledge of features within scene	
Consideration of average area (size) of image objects	
Quantitative assessment: accuracy metrics from subsequent analyses such as classification based on multiple scales	12
Quantitative assessment: assessment of image objects using real-world ground objects as reference data based on quantitative metrics	14
Not specifically described how an optimum scale was determined or not applicable	26

studies used a different optimum scale for each type of land cover or land use objects (Huang and Qi, 2010; Mallinis et al., 2008; Wang et al., 2009). Ke et al. (2010) argued that for a particular study there may be no single optimum scale and that a range of optimum scales may exist.

There have been efforts to develop a feasible framework for the selection of optimum scale that can be applied to a wide range of OBIA applications. Lowe and Guo (2011) investigated an optimum scale parameter in OBIA using semivariogram and spatial autocorrelation concepts. Multiple scales were tested and the optimal scale was identified when the average distance between neighboring image object centroids was near the semivariogram lag distance. Estimation of scale parameter by Drăgut et al. (2010) provides useful information in selecting a suitable range of scales for image segmentation using local variance of object heterogeneity within a scene. Crevier (2008) proposed a natural extension to the concept of precision-recall curves and computationally efficient match measures to compare multiple segmentations and identify optimum scale. In particular, the proposed statistical tool can be effectively used to discriminate between segmentations of different images.

In general, there appears to be a consensus on the need of a universally applicable method to identify optimum scale in OBIA and to compare scales among segmentations of different remote sensing data. Image segmentation algorithms have different approaches (i.e., scales) that control the size of image objects. Depending on the segmentation algorithm used, many factors can affect object size, including pixel size, number of input layers, weighting scheme, or radiometric resolution. In order to facilitate the selection of optimum scale or compare segmentation results from different algorithms or data (Figure 10.4), it would be good to provide additional metrics along with the arbitrary scale values from the software tool used. Such metrics include the number of objects, the number of objects divided by total area, average area of objects (e.g., in square meters), and the average area of objects divided by pixel size. Metrics that integrate the information could be used as surrogate variables to arbitrary scale values that directly came from the software tools used. That way, it would be more straightforward and intuitive to deal with the optimum scale concept in OBIA in that the metrics can be compared even when different algorithms or data are used. In fact, some studies provided additional information related to scale such as the average area and number of image objects. For example, Ke et al. (2010) provided the number of segmented objects of interest by scale and Aguirre-Gutiérrez et al. (2012) reported the number of objects, area of biggest object, and average object area.

According to Blaschke (2010), there is no recognizable relationship between the scale parameter used in an image segmentation process and the spatial measures specific to the image objects of interest. In addition, a single scale would not successfully characterize the multitude of different image components; thus there is a strong need for a universally applicable strategy to determine optimum scales for different image components such as relatively homogeneous forests and heterogeneous urban features.

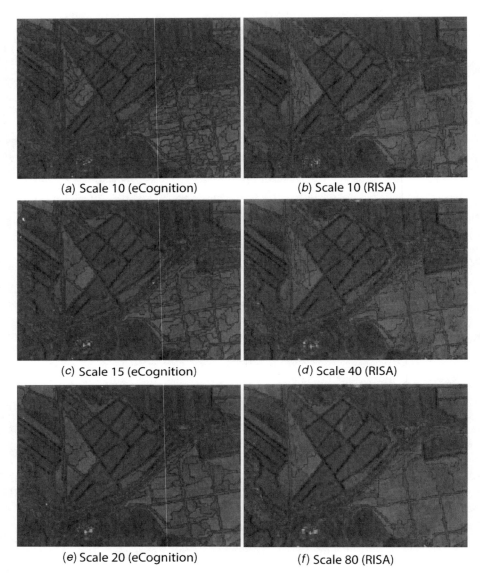

(a) Scale 10 (eCognition)   (b) Scale 10 (RISA)

(c) Scale 15 (eCognition)   (d) Scale 40 (RISA)

(e) Scale 20 (eCognition)   (f) Scale 80 (RISA)

FIGURE 10.4 Multiscale image segmentation comparison using multiresolution image segmentation algorithm in eCognition and RISA from SPOT 5 data. (Adapted from Wang et al., 2010.) (See the color version of this figure in Color Plates section.)

10.5 CONCLUSIONS

This chapter provided a review of recent publications in OBIA and examined the use of scales in those publications. While a simple trial-and-error method and visual interpretation of image objects are still commonly used to identify optimum scale, there is also a need to improve the procedure of optimum scale selection. This may include automated determination of optimum multiscales, development of more sophisticated ways to assess image objects, and development of metrics to describe object size regardless of the segmentation algorithms or data used.

Computational efficiency associated with selecting optimum scale should also be explored. The improvements related to optimum scale selection will likely be a major trend in OBIA within the remote sensing field for the next years.

REFERENCES

Addink, E., de Jong, S., and Pebesma, E. 2007. The importance of scale in object-based mapping of vegetation parameters with hyperspectral imagery. *Photogrammetric Engineering & Remote Sensing* 73(8):905–912.

Aguirre-Gutiérrez, J., Seijmonsbergen, A., and Duivenvoorden, J. 2012. Optimizing land cover classification accuracy for change detection, a combined pixel-based and object-based approach in a mountainous area in Mexico. *Applied Geography* 34:29–37.

Aksoy, B., and Ercanoglu, M. 2012. Landslide identification and classification by object-based image analysis and fuzzy logic: An example from the Azdavay region (Kastamonu, Turkey). *Computers & Geosciences* 38:87–98.

Anders, N., Seijmonsbergen, A., and Bouten, W. 2011. Segmentation optimization and stratified object-based analysis for semi-automated geomorphological mapping. *Remote Sensing of Environment* 115:2976–2985.

Arnesen, A., Silva, T., Hess, L., Novo, E., Rudorff, C., Chapman, B., and McDonald, K. 2013. Monitoring flood extent in the lower Amazon River floodplain using ALOS/PALSAR ScanSAR images. *Remote Sensing of Environment* 130:51–61.

Baatz, M., Benz, U., Dehghani, S., and Heynen, M. 2004. *eCognition User Guide 4.* Munich: Definiens Imagine.

Baatz, M., and Schäpe, A. 2000. *Multiresolution segmentation—an optimization approach for high quality multi-scale image segmentation.* In J. Strobl, T. Blaschke, and G. Griesebner, (Eds.) Heidelberg: Angewandte Geographische Informationsverarbeitung XII, Wichmann, pp. 12–23.

Benz, U., Hofmann, P., Willhauck, G., Lingenfelder, I., and Heynen, M. 2004. Multiresolution, object-oriented fuzzy analysis of remote sensing data for GIS-ready information. *ISPRS Journal of Photogrammetry and Remote Sensing* 58:239–258.

Blaschke, T. 2005. A framework for change detection based on image objects. In S. Erasmi, B. Cyffka, and M. Kappas (Eds.) *Göttinger Geographische Abhandlungen*, 113,Göttingen, pp. 1–9.

Blaschke, T. 2010. Object based image analysis for remote sensing. *ISPRS Journal of Photogrammetry and Remote Sensing* 65:2–16.

Bontemps, S., Bogaert, P., Titeux, N., and Defourny, P. 2008. An object-based change detection method accounting for temporal dependences in time series with medium to coarse spatial resolution. *Remote Sensing of Environment* 112:3181–3191.

Cai, Y., Tong, X., and Shu, R. 2009. Multi-scale segmentation of remote sensing image based on watershed transformation. *Urban Remote Sensing Joint Event*, May 20–22.

Chen, J., Li, J., Pan, D., Zhu, Q., and Mao, Z. 2012. Edge-guided multiscale segmentation of satellite multispectral imagery. *IEEE Transactions on Geoscience and Remote Sensing* 50(11):4513–4520.

Cheng, H., Jiang, X., Sun, Y., and Wang, J. 2001. Color image segmentation: advances and prospects, *Pattern Recognition* 34:2259–2281.

Chirici, G., Giuliarelli, D., Biscontini, D., Tonti, D., Mattioli, W., Marchetti, M., and Corona, P. 2011. Large-scale monitoring of coppice forest clearcuts by multitemporal very high resolution satellite imagery: A case study from central Italy. *Remote Sensing of Environment* 115:1025–1033.

Clinton, N., Holt, A., Scarborough, J., Yan, L., and Gong, P. 2010. Accuracy assessment measures for object-based image segmentation goodness. *Photogrammetric Engineering & Remote Sensing* 76(3):289–299.

Conchedda, G., Durieux, L., and Mayaux, P. 2008. An object-based method for mapping and change analysis in mangrove ecosystems. *ISPRS Journal of Photogrammetry & Remote Sensing* 63:578–589.

Crevier, D. 2008. Image segmentation algorithm development using ground truth image data sets. *Computer Vision and Image Understanding* 112(2):143–159.

de Chant, T., and Kelly, M. 2009. Individual object change detection for monitoring the impact of a forest pathogen on a hardwood forest. *Photogrammetric Engineering & Remote Sensing* 75(8):1005–1013.

Definiens Imaging. 2005. *eCognition User Guide*, Munich: Definiens Imaging.

Doxani, G., Karantzalos, K., and Tsakiri-Strati, M. 2012. Monitoring urban changes based on scale-space filtering and object-oriented classification. *International Journal of Applied Earth Observation and Geoinformation* 15:38–48.

Drăgut, L., and Eisank, C. 2011. Object representations at multiple scales from digital elevation models. *Geomorphology* 129:183–189.

Drăgut, L., Tiede, D., and Levick, R. 2010. ESP: A tool to estimate scale parameter for multiresolution image segmentation of remotely sensed data. *International Journal of Geographical Information Science* 24:859–871.

Dronova, I., Gong, P., Clinton, N., Wang, L., Fu, W., Qi, S., and Liu, Y. 2012. Landscape analysis of wetland plant functional types: The effects of image segmentation scale, vegetation classes and classification methods. *Remote Sensing of Environment* 127:357–369.

Dronova, I., Gong, P., and Wang, L. 2011. Object-based analysis and change detection of major wetland cover types and their classification uncertainty during the low water period at Poyang Lake, China. *Remote Sensing of Environment* 115:3220–3236.

Durieux, L., Lagabrielle, E., and Nelson, A. 2008. A method for monitoring building construction in urban sprawl areas using object-based analysis of Spot 5 images and existing GIS data. *ISPRS Journal of Photogrammetry & Remote Sensing* 63:399–408.

Duro, D., Franklin, S., and Dube, M. 2012. A comparison of pixel-based and object-based image analysis with selected machine learning algorithms for the classification of agricultural landscapes using SPOT-5 HRG imagery. *Remote Sensing of Environment* 118:259–272.

Hall, L., Bensaid, A., Clarke, L., Velthuizen, R., Silbiger, M., and Bezdek, J. 1992. A comparison of neural network and fuzzy clustering techniques in segmenting magnetic resonance images of the brain. *IEEE Transactions on Neural Networks* 3(5):672–681.

Hernando, A., Arroyo, L., Velazquez, J., and Tejera, R. 2012. Object-based image analysis for mapping natural 2000 Habitats to improve forest management. *Photogrammetric Engineering & Remote Sensing* 78(9):991–999.

Hofmann, T., Puzicha, J., and Buhmann, J. 1998. Unsupervised texture segmentation in a deterministic annealing framework. *IEEE Transactions on Pattern Analysis and Machine Intelligence* 20(8):803–818.

Holt, A., Seto, E., Rivard, T., and Gong, P. 2009. Object-based detection and classification of vehicles from high resolution aerial photography. *Photogrammetric Engineering & Remote Sensing* 75(7):871–880.

Huang, F., and Qi, Y. 2010. Object-oriented land cover extraction in Changbai Natural Reserve from IKONOS image. *18th International Conference on Geoinformatics*, June 18–20.

Im, J., Jensen, J., and Hodgson, M. 2008a. Object-based classification using high posting density lidar data. *GIScience and Remote Sensing* 45(2):209–228.

Im, J., Jensen, J., and Tullis, J. 2008b. Object-based change detection using correlation image analysis and image segmentation techniques. *International Journal of Remote Sensing* 29(2):399–423.

Jacquin, A., Misakova, L., and Gay, M. 2008. A hybrid object-based classification approach for mapping urban sprawl in periurban environment. *Landscape and Urban Planning* 84:152–165.

Jensen, J., Garcia-Quijano, M., Hadley, B., Im, J., and Wang, Z. 2006. Remote sensing agricultural crop type for sustainable development in South Africa. *Geocarto International* 21(2):5–18.

Johansen, K., Arroyo, L., Phinn, S., and Witte, C. 2010. Comparison of geo-object based and pixel-based change detection of riparian environments using high spatial resolution multi-spectral imagery. *Photogrammetric Engineering & Remote Sensing* 76(2):123–136.

Johnson, B., and Xie, Z. 2011. Unsupervised image segmentation evaluation and refinement using a multi-scale approach. *ISPRS Journal of Photogrammetry and Remote Sensing* 66:473–483.

Karantzalos, K., and Argialas, D. 2009. A region-based level set segmentation for automatic detection of man-made objects from aerial and satellite images. *Photogrammetric Engineering & Remote Sensing* 75(6):667–677.

Ke, Y., Quackenbush, L., and Im, J. 2010. Synergistic use of QuickBird multispectral imagery and LIDAR data for object-based forest species classification. *Remote Sensing of Environment* 114:1141–1154.

Kim, M., Holt, J., Eisen, R., Padgett, K., Reisen, W., and Croft, J. 2011a. Detection of swimming pools by geographic object-based image analysis to support west nile virus control effects. *Photogrammetric Engineering & Remote Sensing* 77(11):103–113.

Kim, M., Warner, T., Madden, M., and Atkinson, D. 2011b. Multi-scale GEOBIA with very high spatial resolution digital aerial imagery: scale, texture, and image objects. *International Journal of Remote Sensing* 32(10):2825–2850.

Laliberte, A., Browning, D., and Rango, A. 2012. A comparison of three feature selection methods for object-based classification of sub-decimeter resolution UltraCam-L imagery. *International Journal of Applied Earth Observation and Geoinformation* 15:70–78.

Laliberte, A., and Rango, A. 2009. Texture and scale in object-based analysis of subdecimeter resolution unmanned aerial vehicle (UAV) imagery. *IEEE Transactions on Geoscience and Remote Sensing* 47(3):761–770.

Lamonaca, A., Corona, P., and Barbati, A. 2008. Exploring forest structural complexity by multi-scale segmentation of VHR imagery. *Remote Sensing of Environment* 112:2839–2849.

Lein, J. 2012. *Object-Based Analysis, Environmental Sensing: Analytical Techniques for Earth Observation.* London: Springer, pp. 259–278.

Lian, L., and Chen, J. 2011. Research on segmentation scale of multi-resources remote sensing data based on object-oriented. *Procedia Earth and Planetary Science* 2:352–357.

Lillesand, T., Kiefer, R., and Chipman, J. 2008. *Remote Sensing and Image Interpretation.* Hoboken, NJ: Wiley.

Ling, F., Li. X., Xiao. F., Fang. S., and Du, Y. 2012. Object-based sub-pixel mapping of buildings incorporating the prior shape information from remotely sensed imagery. *International Journal of Applied Earth Observation and Geoinformation* 18:283–292.

Liu, J., Pattey, E., and Nolin, M. 2008. Object-based classification of high resolution SAR images for within field homogeneous zone delineation. *Photogrammetric Engineering & Remote Sensing* 74(9):1159–1168.

Liu, Y., Cai, W., Li, M., Hu, W., and Wang, Y. 2010. Multi-scale urban land cover extraction based on object oriented analysis. 18th International Conference on Geoinformatics, June 18–20.

Liu, Y., Guo, Q., and Kelly, M. 2008. A framework of region-based spatial relations for non-overlapping features and its application in object based image analysis. *ISPRS Journal of Photogrammetry & Remote Sensing* 63:461–475.

Lizarazo, I., and Barros, J. 2010. Fuzzy image segmentation for urban land-cover classification. *Photogrammetric Engineering & Remote Sensing* 76(2):151–162.

Lowe, S., and Guo, X. 2011. Detecting an optimal scale parameter in object-oriented classification. *IEEE Journal of Selected Topics in Applied Earth Observations and Remote Sensing* 4(4):890–895.

Mahmoudi, F., Samadzadegan, F., and Reinartz, P. 2013. Object oriented image analysis based on multi-agent recognition system. *Computers & Geosciences* 54:219–230.

Mallinis, G., Koutsias, N., Tsakiri-Strati, M., and Karteris, M. 2008. Object-based classification using Quickbird imagery for delineating forest vegetation polygons in a Mediterranean test site. *ISPRS Journal of Photogrammetry & Remote Sensing* 63:237–250.

Martha, T., Kerle, N., van Westen, C., Jetten, V., and Kumar, K. 2011. Segment optimization and data-driven thresholding for knowledge-based landslide detection by object-based image analysis. *IEEE Transactions on Geoscience and Remote Sensing* 49(12):4928–4943.

Meng, X., Currit, N., Wang, L., and Yang, X. 2012. Detect residential buildings from lidar and aerial photographs through object-oriented land use classification. *Photogrammetric Engineering & Remote Sensing* 78(1):35–44.

Möller, M., Lymburner, L., and Volk, M. 2007. The comparison index: A tool for assessing the accuracy of image segmentation. *International Journal of Applied Earth Observation and Geoinformation* 9(3):311–321.

Myint, S., Gober, P., Brazel, A., Grossman-Clarke, S., and Weng, Q. 2011. Per-pixel vs. object-based classification of urban land cover extraction using high spatial resolution imagery. *Remote Sensing of Environment* 115:1145–1161.

Ortiz, A., and Oliver, G. 2006. On the use of the overlapping area matrix for image segmentation evaluation: A survey and new performance measures. *Pattern Recognition Letters* 27(16):1916–1926.

Ouyang, Z., Zhang, M., Xie, X., Shen, Q., Guo, H., and Zhao, B. 2011. A comparison of pixel-based and object-oriented approaches to VHR imagery for mapping saltmarsh plants. *Ecological Informatics* 6:136–146.

Pan, P., Xiuguo, L., Wei, G., and Qihao, C. 2009. An improved strategy for object-oriented multi-scale remote sensing image segmentation. *The 1st International Conference on Information Science and Engineering*, Nanjing, China, December 18-20 2009, pp. 1149–1152.

Peña-Barragán, J., Ngugi, M., Plant, R., and Six, J. 2011. Object-based crop identification using multiple vegetation indices, textural features and crop phenology. *Remote Sensing of Environment* 115:1301–1316.

Qi, Z., Yeh, A., Li, X., and Lin, Z. 2012. Integration of polarimetric decomposition, object-oriented image analysis, and decision tree algorithms for land-use and land-cover classification using RADARSAT-2 polarimetric SAR data. *Photogrammetric Engineering & Remote Sensing* 78(2):169–181.

Qiao, C., Luo, J., Shen, Z., Zhu, Z., and Ming, D. 2012. Adaptive thematic object extraction from remote sensing image based on spectral matching. *International Journal of Applied Earth Observation and Geoinformation* 19:248–251.

Quyang, Z., Zhang, M., Xie, X., Shen, Q., Guo, H., and Zhao, B. 2011. A comparison of pixel-based and object-oriented approaches to VHR imagery for mapping saltmarsh plats. *Ecological Informatics* 6:136–146.

Radoux, J., Bogaert, P., and Defourny, P. 2010. Overall accuracy estimation for geographic object-based image classification. In N. J. Tate and P. F. Fisher (Eds.), *Ninth International Symposium on Spatial Accuracy Assessment in Natural Resources and Environmental Sciences International Spatial Accuracy Research Association*, Leicester, United Kingdom.

Ranson, K., Montesano, P., and Nelson, R. 2011. Object-based mapping of circumpolar taiga-tundra ecotone with MODIS tree cover. *Remote Sensing of Environment* 115:3670–3680.

Schöpfer, E., Lang, S., and Albrecht, F. 2008. Object-fate analysis: Spatial relationships for the assessment of object transition and correspondence. In T. Blaschke, S. Lang, and G. J. Hay (Eds.), *Object-Based Image Analysis: Spatial Concepts for Knowledge-Driven Remote Sensing Applications*. Berlin: Springer, pp. 785–801.

Sebari, I., and He, D. 2013. Automatic fuzzy object-based analysis of VHSR images for urban objects extraction. *ISPRS Journal of Photogrammetry and Remote Sensing* 79:171–184.

Shih, F., and Cheng, S. 2005. Automatic seeded region growing for color image segmentation. *Image and Vision Computing* 23:877–886.

Shruthi, R., Kerle, N., and Jetten, V. 2011. Object-based gully feature extraction using high spatial resolution imagery. *Geomorphology* 134:260–268.

Stow, D. 2010. In M. M. Fischer and A. Getis (Eds.), *Geographic Object-based Image Change Analysis Handbook of Applied Spatial Analysis*, Springer: Berlin, pp. 565–582.

Stow, D., Hamada, Y., Coulter, L., and Anguelova, Z. (2008). Monitoring shrubland habitat changes through object-based change identification with airborne multispectral imagery. *Remote Sensing of Environment* 112:1051–1061.

Stow, D., Lippitt, C., and Weeks, J. 2010. Geographic object-based delineation of neighborhoods of Accra, *Ghana using QuickBird satellite imagery. Photogrammetric Engineering & Remote Sensing* 76(8):907–914.

Stumpf, A., and Kerle, N. 2011. Object-oriented mapping of landslides using random forests. *Remote Sensing of Environment* 115:2564–2577.

Tiede, D., Lang, S., Albrecht, F., and Holbling, D. 2010. Object-based class modeling for cadaster-constrained delineation of geo-objects. *Photogrammetric Engineering & Remote Sensing* 76(2):193–202.

Tong, B., and Li, J. 2011. Influence of shape parameters on optimal scale selection in multi-resolution segmentation. *International Conference on Image Analysis and Signal Processing*, Hubei, China Oct. 21–23, pp. 235–239.

Tong, H., Maxwell, T., Zhang, Y., and Dey, V. 2012. A supervised and fuzzy-based approach to determine optimal multi-resolution image segmentation parameters. *Photogrammetric Engineering & Remote Sensing* 78(10):1029–1044.

Tzotsos, A., Karantzalos, K., and Argialas, D. 2011. Object-based image analysis through nonlinear scale-space filtering. *ISPRS Journal of Photogrammetry and Remote Sensing* 66:2–16.

Verbeeck, K., Hermy, M., and Orshoven, J. 2012. External geo-information in the segmentation of VHR imagery improves the detection of imperviousness in urban neighborhoods. *International Journal of Applied Earth Observation and Geoinformation* 18:428–435.

Vieira, M., Formaggio, A., Renno, C., Atzberger, C., Aguiar, D., and Mello, M. 2012. Object based image analysis and data mining applied to a remotely sensed Landsat time-series to map sugarcane over large areas. *Remote Sensing of Environment* 123:553–562.

Wang, W., Zhao, Z., and Zhu, H. 2009. Object-oriented change detection method based on multi-scale and multi-feature fusion. *2009 Urban Remote Sensing Joint Event*. Shanghai, China May 20–22.

Wang, Z., Jensen, J., and Im, J. 2010. An automatic region-based image segmentation algorithm for remote sensing applications. *Environmental Modelling and Software* 25:1149–1165.

Watmough, G., Atkinson, P., Hutton, C. 2011. A combined spectral and object-based approach to transparent cloud removal in an operational setting for Landsat ETM+. *International Journal of Applied Earth Observation and Geoinformation* 13:220–227.

Whiteside, T., and Ahmad, W. 2005. A comparison of object-oriented and pixel-based classification methods for mapping land cover in Northern Australia. *Proceedings of SSC 2005 Spatial Intelligence, Innovation and Praxis, Melbourne*. September 2005.

Whiteside, T., Boggs, G., and Maier, S. 2011. Comparing object-based and pixel-based classifications for mapping savannas. *International Journal of Applied Earth Observation and Geoinformation* 13:884–893.

Wilschut, L., Addink, E., Heesterbeek, J., Dubyanskiy, V., Davis, S., Laudisoit, A., Begon, M., Burdelov, L., Atshabar, B., and de Jong, S. 2013. Mapping the distribution of the main host for plague in a complex landscape in Kazakhstan: An object-based approach using SPOT-5 XS, Landsat 7 ETM+, SRTM and multiple random forests. *International Journal of Applied Earth Observation and Geoinformation* 23:81–94.

Yang, G., Pu, R., Zhang, J., Zhao, C., Feng, H., and Wang, J. 2013. Remote sensing of seasonal variability of fractional vegetation cover and its object-based spatial pattern analysis over mountain areas. *ISPRS Journal of Photogrammetry and Remote Sensing* 77:79–93.

Yi, L., Zhang, G., and Wu, Z. 2012. A scale-synthesis method for high spatial resolution remote sensing image segmentation. *IEEE Transactions on Geoscience and Remote Sensing* 50(10):4062–4070.

Zhang, C., and Xie, Z. 2012. Combining object-based texture measures with a neural network for vegetation mapping in the Everglades from hyperspectral imagery. *Remote Sensing of Environment* 124:310–320.

Zhang, X., Xiao, P., Song, X., and She, J. 2013. Boundary-constrained multi-scale segmentation method for remote sensing images. *ISPRS Journal of Photogrammetry and Remote Sensing* 78:15–25.

Zhang, Y. 1996. A survey on evaluation methods for image segmentation. *Pattern Recognition* 29(8):1335–1346.

Zhou, W., Huang, G., Troy, A., and Cadenasso, M. 2009. Object-based land cover classification of shaded areas in high spatial resolution imagery of urban areas: A comparison study. *Remote Sensing of Environment* 113:1769–1777.

Zhu, Z., and Woodcock, C. 2012. Object-based cloud and cloud shadow detection in Landsat imagery. *Remote Sensing of Environment* 118:83–94.

PART IV

SCALE AND LAND SURFACE PATTERNS

11

SCALING ISSUES IN STUDYING THE RELATIONSHIP BETWEEN LANDSCAPE PATTERN AND LAND SURFACE TEMPERATURE

Hua Liu and Qihao Weng

11.1 INTRODUCTION

Landscape pattern is an important concept in landscape ecology (Forman and Godron, 1986). As an important determinant of ecosystem function, land use and land cover (LULC) patterns can be considered as the representative of landscape patterns (Bain and Brush, 2004). LULC patches in a landscape may have different characteristics (e.g., sizes, shapes, and spatial arrangements). LULC categories are associated with various surface temperatures in urban environment (Voogt and Oke, 1997). Land surface temperatures (LST) reflect surface–atmosphere interactions and energy fluxes between the ground and the atmosphere (Wan and Dozier, 1996). It has been used to measure the urban heat island parameters (e.g., magnitudes, spatial extents, and directions of heat movements) (Streutker, 2002, 2003). Characteristic natural and human-involved patches have ecological implications at various spatial levels and influence the distribution of habitats and material flows (Peterjohn and Correll, 1984; Turner, 1990). Relationship between landscape pattern and LST is an important topic in the study of urban heat islands (Weng et al., 2004). Research is needed to examine the causes, processes, and ecological implications of LULC changes at various spatial scales (Weng et al., 2004; Alberti, 2005), especially the relationship with LST.

It is needed to quantify the spatial pattern and temporal change in order to understand patterns and processes in a landscape (Wu et al., 2000). A great number of landscape metrics have been developed to characterize the spatial patterns of

Scale Issues in Remote Sensing, First Edition. Edited by Qihao Weng.
© 2014 John Wiley & Sons, Inc. Published 2014 by John Wiley & Sons, Inc.

217

landscapes and to compare ecological quality across the landscapes (O'Neill et al., 1988; Ritters et al., 1995; McGarigal and Marks, 1995; Gustafson, 1998). Some software has integrated landscape metrics to facilitate metrics calculation, for example, FRAGSTATS (McGarigal et al., 2002).

Scale is an important issue in remote sensing and GIS studies. Scale influences the examination of the landscape patterns in a region. Spatial scale is thought to affect the quantification of landscape metrics (Turner, 1990; Wang et al., 1999). Choosing different remote sensing sensors may cause various research results in the regional analysis with diverse spatial resolutions. Because the spatial characteristics of large regions on the level of the landscape have attracted researchers' attentions (Franklin and Forman, 1987; Turner, 1990), methods and technologies to examine the spatial arrangements at broad spatial scales have become more and more important. Turner (1990) developed a neutral model to examine the relationship between landscape patterns and ecological processes across spatial scales. Krönert et al. (2001) believed that up/downscaling was a fundamental operation of transmission. They summarized the classification of up/downscaling methods: Upscaling includes (i) averaging of observations or output variables, (ii) finding representative parameters, (iii) averaging of model equations, and finally (iv) model simplification. Downscaling includes (i) empirical functions, (ii) mechanisitc models, and (iii) fine scale auxiliary information. Scaling issue has been considered an important topic in landscape mapping. Imagery with finer resolution includes greater spatial information, which, in turn, enables the description of smaller features than imagery with lower spatial resolution. This scale problem challenges the research of the relationship between spatial patterns and processes (Meentemeyer, 1989).

Despite much research focused on human-related environment and related issues at global and regional scales, some questions remain to be answered: What are the methods to scale up or scale down across the landscape based on different data resources? What are the scaling effects on the processes? We conducted a case study to examine the changes in spatial configurations on landscape patterns and land surface temperatures at different spatial resolutions. A logical approach was used to examine the scaling-up effect on the measurements of LULC and LSTs by the use of landscape metrics. The results helped to further understand the relationships between landscape patterns and LSTs in urban environments.

11.2 CASE STUDY

11.2.1 Study Area

The city of Indianapolis, Indiana (Marion County, 39°47'N, 86°09'W) was selected as study area in the case study (Figure 11.1). It is the nation's 12th largest city and the capital city of Indiana with a population of 0.8 million according to the U.S. Census 2010 (about 1.8 million in the metropolitan area). The city has clear seasonal changes, but no pronounced wet or dry seasons. Its annual average temperature is 11.3 °C, and the average monthly temperature is -3.3 °C in January and 23.9 °C in July. The

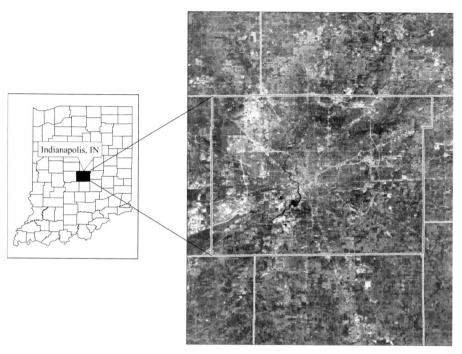

FIGURE 11.1   Geographical location of the case study area, the city of Indianapolis, Indiana.

average annual precipitation is 39.9 in., with about 2.3 in. in January and about 4.6 in. in July. It is located on a flat plain, and is relatively symmetrical, having possibilities of expansion in all directions.

11.2.2   Data Processing

Two Advanced Spaceborne Thermal Emission and Reflection Radiometer (ASTER) imagery were used to derive LULC types and LSTs for the city on two image dates, June 16, 2001 and October 3, 2000 (Table 11.1). Each ASTER images possesses 14 bands from visible to thermal infrared regions (15–90 m spatial resolution) (ASTER online products description, 2005). Cloud cover of each image was below 10%. Both images were geocorrected to Universal Transverse Mercator projection (zone 16).

TABLE 11.1   Two ASTER Images Covering the City of Indianapolis

Acquisition Dates (mm/dd/yyyy) and time (GMT)	Mean LST ($^\circ$C)
06/16/2001, 16:55:29	33.95
10/03/2000, 17:00:51	28.50

Six LULC types were derived from each ASTER image: urban, forest, grassland, agriculture, water, and barren land based on unsupervised classification. Postclassification smoothing process was executed to improve the accuracy of image classification. Image refinement was possessed to manually correct certain confusing pixels to further improve the classification accuracy. The overall accuracy of each classified image was above 85%. The spatial resolution of each LULC map is 15 m.

Existing ASTER land surface kinetic temperature images (90 m spatial resolution) were purchased in order to investigate the scaling effects in studying the relationship between landscape pattern and land surface temperature. Each LST map was divided into six temperature zones by using a standard deviation method of data classification (Smith, 1986). Please refer to Liu and Weng (2008) for statistic details of LST classification. Figure 11.2 shows the LULC and LST maps for each image date.

In order to examine the scaling-up effects in studying the relationship between LULC and LSTs in the landscape, we resampled each image (both LULC and LST) to different spatial resolutions: 30×30, 60×60, 120×120, 250×250, 500×500, and 1000×1000 m. Each resampled LST map was divided into six temperature zones by using the same method as the one used for classifying the LST image with original 90 m resolution (standard deviation method). We intended to better understand their possible relationships at different scales by examining the LULC and LST at various spatial resolutions. As a result, 32 LULC and LST images in all were created for further analysis.

Four class-level landscape metrics (patch percentage, patch density, landscape shape index, and perimeter–area fractal dimension) and four landscape-level metrics (patch density, landscape shape index, perimeter–area fractal dimension, and mean perimeter–area ratio) were derived from all 32 maps (see Table 11.2 for metrics details). According to McGarigal et al. (2002), patch percentage index is the proportion of each patch type within the study area; patch density index implies the number of patches per 100 hectares; landscape shape index measures the aggregation of landscape and its value increases as the patch type becomes more disaggregated; perimeter–area fractal dimension index reflects the shape complexity of patch, class, or the whole landscape. Mean perimeter–area ratio is a simplified method for evaluating the shape complexity. All the indices were computed by the use of FRAGSTATS (McGarigal et al., 2002) that were next used to analyze the scaling-up effect on the analysis of landscape and LST patterns.

11.2.3   Results

11.2.3.1 Scaling-Up Effect on Class-Level Landscape Metrics   Figure 11.3 shows the area percentages of LULC and LST patches across eight spatial scales for two image dates, June 16, 2001 and October 3, 2000. According to the calculations, LULC patch percentages did not possess obvious variations across the spatial scales, although slight changes were captured when the pixel size changed from 500 to 1000 m for urban, forest, and grassland for both image dates. When the scale changed from 500 to 1000 m, an increase was observed in urban patch percentage; however; decreases were identified in both forest and grassland patch

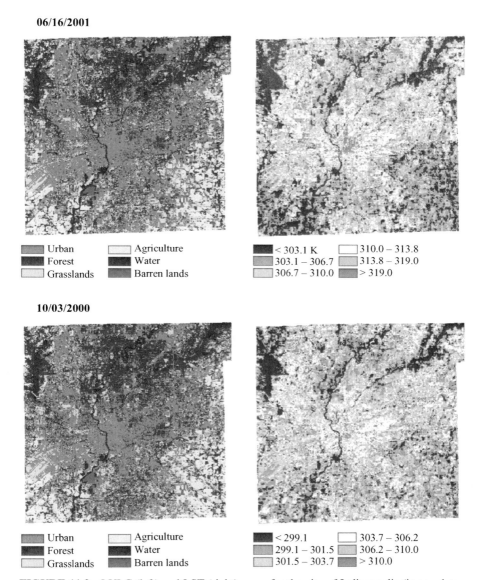

FIGURE 11.2 LULC (left) and LST (right) maps for the city of Indianapolis (image dates: June 16, 2001 and October 3, 2000). (See the color version of this figure in Color Plates section.)

percentages. Within the same window of scale changing, the measurements of patch percentages for water, agriculture, and barren land were more stable. It implies that the assessment of patch percentages for domain LULC types seemed to be more sensitive to the scaling-up process. A possible explanation was that more forest and grass patches were aggregated into urban patches during image resampling than those of

TABLE 11.2 Descriptions of Landscape Metrics Used in the Study (McGarigal et al., 2002)

Metrics	Abbreviations	Definitions	Equations
Patch percentage	PLAND	Proportional abundance of a class	$\text{PLAND} = P_i = \left(1000 \times \sum_{j=1}^{n} a_{ij}\right) / A$ P_i: proportion of the landscape occupied by patch type i. a_{ij}: area (m²) of patch ij A: total landscape area (m²)
Patch density (class- and landscape-level metrics)	PD	Densities of patches	$\text{PD} = n_i \times 10{,}000 \times 100 \div A$ n_i: number of patches in the landscape of patch type i (class level) or total number of patches in the landscape (landscape level) A: total landscape area (m²)
Landscape shape index (class- and landscape-level metrics)	LSI	A measure of class clumpiness	$\text{LSI} = e \div \min e$ e: total length of edge (of particular class) in terms of number of cell surfaces $\min e$: minimum total length of edge of class i in terms of number of cell surfaces
Perimeter–area fractal dimension (class- and landscape-level metrics)	PAFRAC	Shape index based on perimeter and area	$\text{PAFRAC} = 2 \Big/ \Bigg\{ \bigg\{ \left[n_i \times \sum_{j=1}^{n} (\ln P_{ij} \times \ln a_{ij}) \right] - \left[\left(\sum_{j=1}^{n} \ln P_{ij}\right)\left(\sum_{j=1}^{n} \ln a_{ij}\right) \right] \bigg\} \Big/ \bigg\{ \left[n_i \sum_{j=1}^{n} \ln (P_{ij})_2 \right] - \left(\sum_{j=1}^{n} \ln P_{ij}\right)_2 \bigg\} \Bigg\}$ a_{ij}: area (m²) of patch ij P_{ij}: perimeter (m) of patch ij n_i: number of patches in the landscape of patch type i
Mean perimeter–area ratio (landscape level)	PARA_MN	Mean value of a shape index	$\text{PARA}_{MN} = \left(\sum_{i=1}^{m}\sum_{j=1}^{n} (P_{ij}/a_{ij}) \right) / N$ a_{ij}: area (m²) of patch ij P_{ij}: perimeter (m) of patch ij N: total number of patches

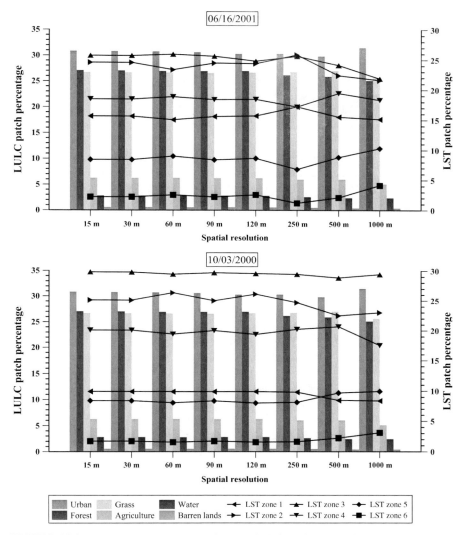

FIGURE 11.3 Patch percentage index (class-level) derived from both LULC and LST maps for two image dates: June 16, 2001 and October 3, 2000. (See the color version of this figure in Color Plates section.)

water, agriculture, and barren land patches. It was suggested that patch percentage is not a scale-sensitive variable in landscape analysis.

More changes were observed in LST patch percentages across the spatial scales for both image dates. For the image date June 16, 2001, patch percentages of zones 1 and 4 were very similar at 250 m spatial scale, and those of zones 2 and 3 had close values at some scales (e.g., 120 and 250 m). The area percentage of zone 1 had a temporary small increase at 250 m, but kept decreasing afterward. Zones 2 and 3 had decreasing patch percentages from 250 m resolution. Zone 4 had a temporary small decrease at

250 m resolution. No obvious changes were observed in zones 5 and 6 across the scales, except a temporary decrease at 250 m indicating that area percentages of both zones seemed to be more affected at the 250 m resolution during the aggregation process. For the image date October 3, 2000, zone 1 had a slight decrease in patch percentage from 250 to 500 m resolution. The percentages of zones 2 and 4 showed much more obvious but still mild changes across the scales. Zone 3 as the largest zone in area did not show significant variations across the scales. There was a slight increase in patch percentage in zone 5 at 250 m. The patch percentage of zone 6 remained consistent with some minor increases after 250 m resolution.

The patch densities of LULC types became lower and lower from 15 to 1000 m resolution for each image date. For each season/image date, urban, forest, and grassland experienced notable decreases before the scale reached 60 m, and then became much gentler toward 250 m resolution. No clear variations could be observed after 250 m. Meanwhile, water, agriculture, and barren lands experienced decreases in patch density across the scales. Patch densities of LST types did not undergo obvious changes as the scale changed from 15 to 90 m followed by notable decreases. It is clear that the higher the patch density, the more variations in the temperature zone across the scales. All LST patch densities seemed to be equal at 1000 m resolution, indicating a maximum patch aggregation at the scale of 1000 m.

No clear seasonal changes could be observed in landscape aggregation based on the measurements of landscape shape index for both LULC and LST maps. The aggregation level of all LULC types kept decreasing across the scales for both image dates. It appears that highly disaggregated LULC types (urban, forest, and grassland) experienced more changes during image resampling compared to those less aggregated LULC types (water, agriculture, and barren lands). A decrease in values was observed in each LST zone since 90 m resolution. Zones with greater aggregation levels (zones 2–4) appeared to be more affected during aggregation process than zones with lower aggregation level (zones 1, 5, and 6).

Unexpected measurements of perimeter–area fractal dimension index were obtained for both LULC and LST maps for each image date. As for the LULC maps, the fractal dimension of each LULC type seemed to increase overall. However, the aggregation process during image resampling intended to decrease but not increase the overall level of fractal dimension. The results of the measurements of perimeter–area fractal dimension indices on LST maps seemed to be more complicated than those on LULC maps. No distinct changes were shown at each temperature zone for each image date as the resolution changed from 15 to 30 m. The values slightly dropped at 60 m resolution and kept increasing since then until 250 m resolution. The changes became inconsistent after 250 m resolution for both image dates.

11.2.3.2 *Scaling-Up Effect on Landscape-Level Landscape Metrics*   According to Figure 11.4, the landscape patch density continuously decreased across the scales for both image dates. It could be explained by the fact that the LULC patches became more and more aggregated and the overall landscape contained fewer and fewer interspersed patches during the scaling-up process. The October image had a

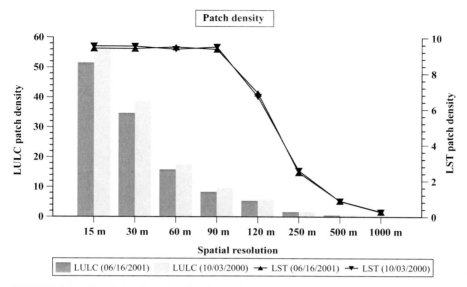

FIGURE 11.4 Patch density index (landscape-level) derived from both LULC and LST maps for two image dates: June 16, 2001 and October 3, 2000. (See the color version of this figure in Color Plates section.)

somewhat higher patch density than the June image at 15 m resolution; however, this difference narrowed with the increase of pixel sizes. It indicates that the possible difference of patch density between two dates was diminished during the scaling process. The patch densities of LSTs for both dates were quite consistent when the scale changed from 15 to 90 m, but the values started to decrease since then and were close to zero at 1000 m resolution. It implies that the number of different LST patches dropped with the increase of pixel sizes.

Similar to patch density, the values of landscape shape index also decreased from 15 to 1000 m for both image dates (Figure 11.5). It implies that the shape of landscape became more and more regular with the increase of pixel aggregation during the scaling-up process. The October image once again had more complicated overall shape than June image at 15 m resolution, but the difference became less and less close to 1000 m resolution. For two LST maps, no obvious changes were observed in shape complexity from 15 to 90 m, but the values significantly dropped after 90 m resolution. It is an indication of shape simplification among LST patches when the pixel sizes became larger.

The measurement of perimeter–area fractal dimension index showed unexpected results (Figure 11.6). The fractal dimension of landscape kept increasing from 15 to 1000 m for both image dates, which seems to associate with the fact that the landscape became more and more complex with the increase of pixel sizes. The finding appears to be inconsistent with the calculations of patch density and landscape shape index that provide more reasonable results. The same increase was observed in LST fractal dimension. These apparent contradictions indicate that landscape-level fractal dimension index can be unreliable during scaling-up process.

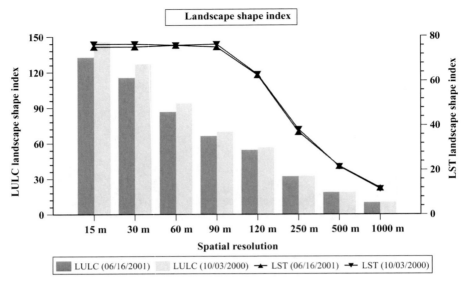

FIGURE 11.5 Landscape shape index (landscape-level) derived from both LULC and LST maps for two image dates: June 16, 2001 and October 3, 2000. (See the color version of this figure in Color Plates section.)

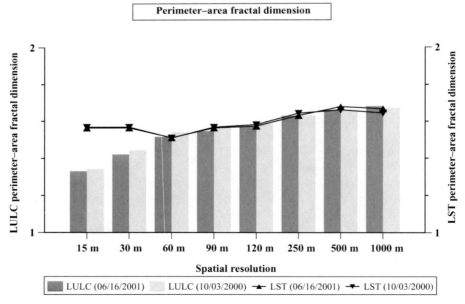

FIGURE 11.6 Perimeter–area fractal dimension index (landscape-level) derived from both LULC and LST maps for two image dates: June 16, 2001 and October 3, 2000. (See the color version of this figure in Color Plates section.)

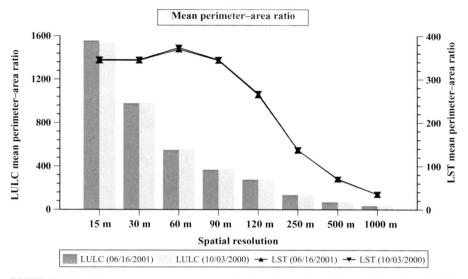

FIGURE 11.7   Mean perimeter–area ratio index (landscape-level) derived from both LULC and LST maps for two image dates: June 16, 2001 and October 3, 2000.

As shown in Figure 11.7, the landscape mean perimeter–area ratio kept decreasing from 15 to 1000 m on both image dates. It indicates that the landscape complexity became more and more regular during scaling-up process. The lower shape complexity can be associated with the increase of landscape aggregation across the scales. Two image dates seem to have very similar mean perimeter–area ratio under all scale levels. The LST mean perimeter–area ratios remained consistent from 15 to 90 m and then kept decreasing to 1000 m, with an exception that there was a slight increase at 60 m resolution. It indicates that there was no obvious difference between two image dates across the scales with regard to LST mean perimeter–area ratios.

11.3   SUMMARY

This chapter explored the scaling issues in studying the relationship between landscape pattern and LST. The case study examined the scaling-up effect on the relationship in the city of Indianapolis, Indiana. The finding would contribute to the study of urban energy budget and natural resource management by the use of remote sensing and other geospatial technologies, for example, GIS. This result is of benefit to the study of regional landscape patterns and urban heat islands by the use of landscape metrics in diverse scales.

Urban, forest, and grassland were more easily affected by the scaling-up process in patch density and landscape shape index compared to water, agriculture, and barren land. The area percentages of LST patches changed across the scales. The overall complexities and aggregations of LULC types and LST zones decreased across the scales based on the measurements of patch density, landscape shape index, and mean perimeter–area ratio at landscape level. However, the calculations of perimeter–area fractal dimension (both class and landscape levels) showed unexpected results for both LULC and LST maps across the scales. Improved fractal dimension indices could be used to improve the result (Frohn, 1998).

Additional metrics, like edge contrast index can be included in determining an optimal spatial scale for the examination of relationships between landscape patterns and LSTs. For example, edge can affect the relationship of one patch with its neighbor patches, making it especially significant in the study of energy transfer in the urban area. However, how different temperature zones impact each other at edges is little understood, and how to determine the contrast weights for the focal LST patch remains unanswered. Future study may include the identification of edge effects between LST zones.

REFERENCES

Alberti, M. 2005. The effects of urban patterns on ecosystem function. *International Regional Science Review* 28(2):168–192.

ASTER online products description. 2005. Available at http://asterweb.jpl.nasa.gov/content/ 03_data/01_Data_Products/SurfaceTemperature.pdf

Bain, D.J., and Brush, G.S. 2004. Placing the pieces: reconstructing the original property mosaic in a warrant and patent watershed. *Landscape Ecology* 19(8):843–856.

Forman, R.T.T., and Godron, M. 1986. *Landscape Ecology*. New York, NY: John Wiley & Sons, Inc.

Franklin, J.F., and Forman, R.T.T. 1987. Creating landscape patterns by forest cutting: ecological consequences and principles. *Landscape Ecology* 1:5–18.

Frohn, R.C. 1998. *Remote Sensing for Landscape Ecology: New Metric Indicators for Monitoring, Modelling, and Assessment of Ecosystems*. Boca Raton, FL: Lewis Publishers.

Gustafson, E.J. 1998. Quantifying landscape spatial pattern: What is the state of the art? *Ecosystems* 1:143–156.

Liu, H., and Weng, Q. 2008. Seasonal variations in the relationship between landscape pattern and land surface temperature in Indianapolis, USA. *Environmental Monitoring and Assessment* 144(1–3):199–219.

Krönert, R., Steinhardt, U., and Volk, M. 2001. *Landscape Balance and Landscape Assessment*. Berlin: Springer.

McGarigal, K., and Marks, B.J. 1995. *FRAGSTATS: spatial pattern analysis program for quantifying landscape structure*. General Technical Report PNW-GTR-351. USDA Forest Service. Pacific Northwest Research Station. Portland, OR.

McGarigal, K., Cushman, S.A., Neel, M.C., and Ene, E. 2002. *FRAGSTATS: spatial pattern analysis program for categorical maps*. Computer software program produced by the authors at the University of Massachusetts, Amherst. Available at www. umass.Edu/ landeco/research/fragstats/fragstats.

Meentemeyer, V. 1989. Geographical perspectives of space, time, and scale. *Landscape Ecology* 3(4):163–173.

O'Neill, R.V., Krummel, J.R., Gardner, R.H., Sugihara, G., Jackson, B., DeAngelis, D.L., Milne, B.T., Turner, M.G., Zygmunt, B., Christensen, S., Dale, V.H., and Graham, R.L. 1988. Indices of landscape pattern. *Landscape Ecology* 1:153–162.

Peterjohn, W.T., and Correll, D.L. 1984. Nutrient dynamics in an agricultural watershed: observations on the role of a riparian forest. *Ecology* 65:1466–1475.

Ritters, K.H., O'Neill, R.V., Hunsaker, C.T., Wickham, J.D., Yankee, D.H., and Timmins, S.P. 1995. A factor analysis of landscape pattern and structure metrics. *Landscape Ecology* 10(1):23–39.

Smith, R.M. 1986. Comparing traditional methods for selecting class intervals on choropleth maps. *Professional Geographer* 38(1):62–67.

Streutker, D. 2002. A remote-sensing study of the urban heat island of Houston, Texas. *International Journal of Remote Sensing* 23:2595–2608.

Streutker, D. 2003. Satellite-measured growth of the urban heat island of Houston, Texas. *Remote Sensing of Environment* 85:282–289.

Turner, M.G. 1990. Spatial and temporal analysis of landscape patterns. *Landscape Ecology* 4(1):21–30.

Voogt, J.A., and Oke, T.R. 1997. Complete urban surface temperatures. *Journal of Applied Meteorology* 36:1117–1132.

Wan, Z., and Dozier, J. 1996. A generalized split-window algorithm for retrieving land-surface temperature from space. *IEEE Transactions on Geoscience and Remote Sensing* 34(2):892–905.

Wang, Y., Zhang, X., Liu, H., and Ruthie, H.K. 1999. Landscape characterization of metropolitan Chicago region by Landsat TM. *Proceedings of the ASPRS Annual Conference, May 17–21, 1999, Portland, OR, USA, pp.* 238–247.

Weng, Q., Lu, D., and Schubring, J. 2004. Estimation of land surface temperature–vegetation abundance relationship for urban heat island studies. *Remote Sensing of Environment* 89:467–483.

Wu, J., Jelinski, D.E., Luck, M., and Tueller, P.T. 2000. Multiscale analysis of landscape heterogeneity: scale variance and pattern metrics. *Geographic Information Sciences* 6(1):6–19.

12

MULTISCALE FRACTAL CHARACTERISTICS OF URBAN LANDSCAPE IN INDIANAPOLIS, USA

Bingqing Liang and Qihao Weng

12.1 INTRODUCTION

A major problem in urban remote sensing is the heterogeneity of the urban environment. For example, low-density residential areas may be composed of tree crowns, rooftops, lawns, paved streets, driveways, and parking lots. Such a heterogeneous landscape pattern is usually displayed as a spatial variation of spectral responses and structural features like texture in remotely sensed images, presenting a challenge for image analysis and interpretation. In this situation, it is necessary to focus on the overall spatial pattern of variation that characterizes each urban category. However, remote sensing analysis based on spatial information in general has not been well understood. One of the challenges is how to characterize and derive the overall spatial complexity of different land surface features either individually or as a whole from images. Ideally, the extracted spatial pattern should represent those that may present at their operational scale, or the scale at which the features operate in the real world. The fast-growing remote sensing community has been able to provide numerous images with varied spatial, spectral, radiometric, and temporal resolutions. These resolutions can be interpreted as the scale of observation or measurement and they may or may not match up with the operational scale for a given spatial feature. Therefore, depending on the scene's resolutions, an individual object's spatial pattern may appear heterogeneous at one level but homogeneous at another and is not always comparable to that in nature (Turner and Garnder, 1991). For a given spatial object, a clear knowledge

Scale Issues in Remote Sensing, First Edition. Edited by Qihao Weng.

about how to sufficiently measure its textural property in images at a single scale/resolution or across multiple scales/resolutions is not yet available. Characterized by irregularity and scale independence, fractal measurement can offer great potential in describing the spatial complexity of landscape features in images and in providing insights into the issue of scale and resolution in remote sensing.

Spatial objects in an image scene are often characterized by their dimensionality. In classical or Euclidean geometry, integers are used to define this characteristic since the shapes of spatial objects are assumed to be composed of straight lines, for instance, one dimension for a road, two dimensions for a lake, and three dimensions for a building. However, the shapes of many natural and man-made objects tend to be irregular and rarely are composed of straight lines. To avoid the gross simplification embedded in Euclidean geometry, fractal measurement makes use of the fractional value reported by the fractal dimension (FD) to define the dimensionality at a more detailed degree that is otherwise impossible with classic geometry. Typically, FD values range from 1.0 for a simple curve to 2.0 for a tortuous curve that ultimately fills two spaces and appears two dimensional. The values can also change from 2.0 for a simple field to 3.0 for a field that is so complex that it finally fills three spaces and looks three dimensional. In remote sensing, images can be viewed as a "hilly terrain surface" whose elevations are represented by the digital numbers (DNs). So the FD value varies from 2.0 for a perfectly smooth surface to 3.0 for a very rugged surface that eventually fills a volume. Overall, the more spatially complex an object or image surface, the higher its FD. FD values can therefore be used as an indicator of the overall spatial pattern in terms of the degree of irregularity or complexity of an object or image surface.

To calculate FD, the characteristic of self-similarity or scale independence, meaning invariance with respect to scale, must be considered. For an ideal self-similar object, or a fractal, it is made up of copies of an infinite number of copies of itself at reduced scale(s). Research has indicated that many natural and man-made objects such as a mountain, a forest, drainage, agriculture fields, and an urban landscape exhibit this self-similarity property (Mandelbrot, 1983). The underlying principle of fractal measurement is to capture the self-similarity of an object/surface, using statistics, to determine its FD. However, most spatial objects, including remote sensing images, are not pure fractals at all scales. Rather, they only exhibit a certain degree of self-similarity over limited ranges of scale, which can have important implication for the understanding of the operational scale of those spatial objects. With this in mind, the spatial complexity suggested by a FD value can be assumed to result directly from the spatial processes of a given object operated at a certain scale. Therefore, the higher FD may imply a scale where more processes operate or at a level more approximating to the operational scale of a single process. The significance of FD for studies concerning not only spatial analysis but also the scale resolution issue mentioned above is thus apparent. Therefore, it is not surprising to see a growing number of researches utilize fractal measurement in remote sensing analysis (De Jong and Burrough, 1995; Emerson et al., 1999, 2002, 2005; Lam, 1990; Lam and De Cola, 1993; Myint, 2003, 2007; Qiu et al., 1999; Read and Lam, 2002; Weng, 2003). A review of the literature indicates that FDs have been frequently employed to characterize the spatial complexity recorded in raw satellite images (Lam, 1990; Read and Lam, 2002; Weng, 2003; Qiu et al., 1999), to examine the spatial patterns involved in land use/land cover (LULC) classified images (De Cola, 1989), to assist in image classification (Myint, 2003; Emerson et al., 2005), and to study the scaling effect of remotely sensed images (Emerson et al., 1999; Myint, 2003).

To successfully apply FD in remote sensing, the most critical step is the selection of many algorithms to calculate the FD of imagery. All algorithms were developed based on the following equation used to define FD for strictly self-similar objects (Mandelbrot, 1977):

$$D = \frac{\log(N_r)}{\log(1/r)} \tag{12.1}$$

where N_r refers to the parts of an object scaled down by a ratio r and D is the shape similarity dimension. Overall, the definition of FDs relates to the size of the measured units and the number of units to traverse the spatial data set (Brown, 1995). For remote sensing data, the calculation of FD relies on the fractional Brownian motion (fBm) model (Zhao, 2001). Generally speaking, three steps are involved to compute a FD for remotely sensed images: (1) determine the quantities of the study target using different step sizes; (2) generate a least-squares regression model based on the log transformations of two variables, measured quantities and step sizes; and (3) derive a FD based on the slope of the regression line. Most FD algorithms currently available are derived from empirical studies and can be grouped into two categories (Zhao, 2001): those based on recursive length/area units to match a curve or surface at different scales and those based on approximation of a curve/surface to a known fractal function or statistical property like the fBm model.

It should be noted the fractal nature of an image is displayed in a variety of aspects, for example, size, shape, area, distance, correlation, and power spectra. Individual FD methods often can only address certain aspects of the spatial structure of a fractal object. Therefore, it is possible that different FD algorithms would produce different results for the same object. As a result, to validate the reliability of individual methods before they can be further applied for other applications, comparative analyses with various methods using a single type of data set seem to be unavoidable in many studies using FD (Emerson et al., 1999, 2002, 2005; Jaggi et al., 1993; Lam and De Cola, 1993; Lam et al., 1997, 2002; Liang and Weng, 2013; Myint, 2003; Qiu et al., 1999; Read and Lam, 2002). Also, even for the same method, its results can be affected by a number of factors, including the type of input data, the specification of the algorithm's various parameters, the spatial pattern recorded in the scene, and the intended application. Consequently, the robustness of a selected algorithm must be examined with more extensive multiscale (both spatial and temporal) images acquired by a variety of sensors.

This chapter presents an evaluation of the effectiveness of the selected triangular prism (TP) FD algorithm for characterizing urban landscape in Indianapolis, Indiana, at multiple spatial and temporal scales. Specifically, the objectives of the study are (1) to analyze the performance of the TP approach in capturing different fractal properties recorded in remotely sensed data acquired by various sensors, that is, Landsat Multispectral Scanner (MSS), Landsat Thematic Mapper (TM)/Enhanced Thematic Mapper Plus (ETM+), Advanced Spaceborne Thermal Emission and Reflection (ASTER), and IKONOS; (2) to examine fractal properties of the urban landscape derived by using raw and classified images; (3) to study the effect of scale change on image properties with resampled images; and (4) to investigate the potential of FD in revealing landscape changes between 1985 and 2000 by four Landsat TM/ETM+ images.

12.2 STUDY AREA AND DATA SETS

The study area is Indianapolis, located in Marion County, Indiana, in the United States (Figure 12.1). With a consolidated city-county structure (Nunn, 1995), the city of Indianapolis and Marion County are often loosely considered the same. The city has a total area of 963.5 km^2 (98.14% of land and 1.86% of water). With a total population of 829,718 (Census, 2010), Indianapolis is the 12th largest city in the nation, the geographical center of Indiana, and the capital of the state. With roads leading out of the city center at all directions, Indianapolis has been famous as "The Crossroads of America." Since the city is located on a flat plain (the Central Till Plains) and is

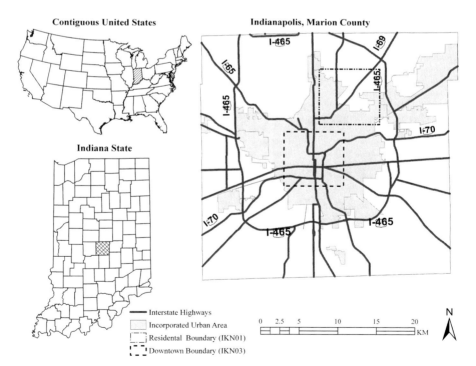

FIGURE 12.1 Study area, Indianapolis, Marion County, Indiana.

relatively symmetrical, it has the possibility to expand in all directions. Like many other cities in the United States, its urban area enlarges at the expense of the loss of adjacent agricultural areas and forestlands. Indianapolis is now the most populous city of the state and the second most populous capital in the United States.

Data sets used in this chapter consisted of eight satellite images that were acquired by five types of sensors (Table 12.1). For these data, only MSS, TM, and ASTER image scenes cover the whole study area. The two IKONOS scenes roughly cover the downtown and residential areas of the city. Since the two images partially overlap each other, to avoid confusion that may result from this overlap on the fractal analysis, the two IKONOS data were used to create two subscenes (7×7 km^2) to include only the downtown and residential land use of the city (Figure 12.2). All satellite images were first georectified to a common Universal Transverse Mercator (UTM) coordinate system using the $1:24{,}000$ scale topographic maps as reference. For each image, 25 ground control points were selected to generate coefficients for a first-order polynomial, and a nearest-neighbor method was applied to resample the image according to their nominal spatial resolution (Table 12.1). The resultant values of the root mean-square error (RMSE) were all found to be less than 0.4 pixel.

Eight classified images were derived by performing the ISODATA unsupervised classification on the spectral bands that were available from all sensors [green, red, and near-infrared (NIR) bands]. Five LULC types were then identified: cropland and pasture (CroPas), water, urban built-up lands (Build-up), forest, and grass. Accuracy assessment was conducted for the resultant classified images. Aerial photographs ranging from $1:80{,}000$ (for early photos) to $1:12{,}000$ (for recent photos) scales were used as the reference data. For each class, 50 points were randomly selected and compared with the reference points to generate the producer's accuracy and user's accuracy. The overall accuracy for each classified images was also computed.

The effect of changes of spatial scale and resolution on detecting landscape patterns and changes are central to geography and mapping science such as remote sensing (Dell'Acqua and Gamba, 2006; Frohn, 1998; Lam and Quattrochi, 1992). In order to provide some insights into the scaling effect on the analytical results, two image groups were used to perform the fractal analysis. The first image group consisted of resampled images by smoothing the image spatial resolution. All the selected reflective bands were resampled to different levels of pixel sizes. Specifically, starting from their own nominal spatial resolutions, all MSS, ETM+, and ASTER images were resampled up to 960 m and the two IKONOS images were resampled up to 240 m. A total of 33 resampled images were thus generated. Besides, the underlying dominant landscape pattern (e.g., downtown vs. residential) can potentially change the analytical results dealing with scale and resolution. To reduce this effect, the second group of data sets comprised only images covering the downtown and residential landscapes but collected by sensors that are inherently different in their spatial resolutions. Only images acquired by ETM+, ASTER, and IKONOS sensors were considered as they were collected about the same time, which could reduce any confusion that may result from the time factor. An additional 30 subscenes from the ETM+ and ASTER images were created to cover the same areas as the 2 IKONOS image subscenes. This included 12 subscenes from the three reflective bands, 4 from

TABLE 12.1 Information of Satellite Images Used in This Chapter

Images	Sensors	Acquisition Dates	Julian Days	θ_{sun}	Φ_{sun}	Earth-sun Distance (astronomic units)	Spatial resolution (m)	Radiometric Resolution	Spectral Bands (nm)[a]	# of Rows[b]	# of Columns[c]
MSS75	MSS	Sep 14, 1975	160	30	120	1.0150	79	7 bits	500–600 (G) 600–700 (R) 700–800 (NIR) 800–1100 (SIR1)	421	425
TM85	TM	Jul 23, 1985	204	33	120	1.0160	30	8 bits	450–520 (B) 520–600 (G) 630–690 (R) 760–900 (NIR)	1105	1115
TM91		Jun 6, 1991	157	30.02	116.92	1.0146					
TM95		Jul 3, 1995	184	33.9	110.24	1.0167			1550–1750 (SIR1) 2080–2350 (SIR2)		
ETM+00	ETM+	Jun 22, 2000	174	25.3	124.55	1.0163	30	8 bits	520–600 (G) 630–690 (R)		
AST01	ASTER	Jun 16, 2001	167	19.57	142.95	1.0158	15	8 bits	780–860 (NIR) 1600–1700 (SIR1) 2145–2185 (SIR2) 2185–2225 (SIR3) 2235–2285 (SIR4) 2295–2365 (SIR5) 2360–2430 (SIR6)		
							30				
IKN03[d]	IKONOS	Oct 6, 2003	279	45.42	168.51	0.9997	4	11 bits	450–520 (B) 520–600 (G) 630–690 (R) 760–900 (NIR)	3702	3154
IKN01[e]		Sep 30, 2001	273	44.34	161	1.0014				3203	2986

Note: Including image acquisition dates, Julia days, sun view and sun geometries (θ = zenithal angle; Φ = azimuthal angle), spatial, spectral (reflected bands only), and radiometric sensor characteristics, along with numbers of rows and columns that study area covers for different images.

[a] G—green band, R—red band, NIR—near-infrared band, SIR—short-wave infrared band.
[b] Number of rows that study area covers for reflective band.
[c] Number of columns that study area covers for reflective band.
[d] IKONOS image covering downtown area of city of Indianapolis.
[e] IKONOS image covering a rural area of city of Indianapolis.

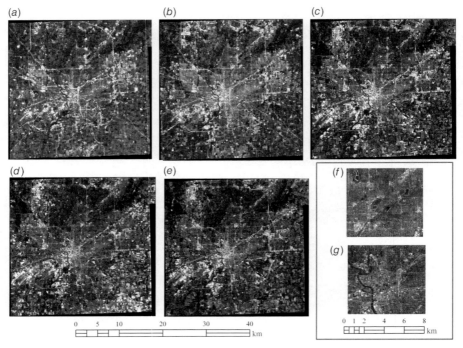

FIGURE 12.2 Red bands of Landsat MSS75 (*a*), Landsat TM85 (*b*), Landsat TM95 (*c*), Landsat ETM + 00 (*d*), AST01 (*e*), IKN01 subscene for residential area (*f*), and IKN03 subscene for downtown area (*g*).

the classified images, and 14 from the resampled images (at the aggregation levels between 30 and 240 m) of the ETM+ and ASTER images. In doing so, the efficiency of various spatial techniques in measuring and characterizing land covers in response to scale and resolution change could be more thoroughly examined.

12.3 LANDSCAPE CHARACTERIZATION AT MULTIPLE SCALES BY FRACTAL MEASUREMENT

Among the many FD algorithms, the most popular one seems to be the TP algorithm due to its robustness (Emerson et al., 2005; Liang and Weng, 2013; Soille and Rivest, 1996). Developed by Clarke (1986), this method employs the idea of interpreting images from a three-dimensional perspective by viewing all the pixel values as "elevation." Taking the four adjacent pixels that make up the four corners of a square and their mean value to build a vertical line which locates centrally, a prism with four triangular facets is thus constructed. The method is implemented by using these triangular facets as the basic tool to fill an image surface until the total surface area is calculated. The same procedure is repeated for each step size that equals the number of pixels on a side of the square. The regression model for this method is given below:

$$\log A = C + \beta \log r \tag{12.2}$$

$$FD = 2 - \beta \tag{12.3}$$

where β is the slope of the regression and r is the step size. All the original, resampled, and classified images and image subscenes were examined by fractal measurement using the TP method given in the Image Characterization and Modeling System (ICAMS) software due to its popularity and robustness (Emerson et al., 2005; Liang and Weng, 2013). Each image would produce a single FD value. The results would then be compared with each other.

12.4 RESULTS

12.4.1 Fractal Analysis Using Raw Images

Since the TM data share the same spectral, spatial, and radiometric resolutions as the ETM+ images, only MSS75, ETM + 00, AST01, IKN01, and IKN03 images were used in this part of the analysis. Before performing fractal analysis in ICAMS, image values were normalized to a range of 0–255. This process facilitated the subsequent comparison analysis not only between image textures but also between computed FDs. For the TP method, only one parameter—the number of steps—is required to implement the algorithm. Liang and Weng (2013) indicated that, with the same tested images, the largest FDs were always produced by the number of steps equaling either 4 or 6, which were very close to the default value of 5. So this default parameter value was used to calculate FD.

Figure 12.3 shows the FDs of the tested images by bands for four image types. Generally speaking, for the MSS and ETM+ data, the highest FDs all went to red bands while the lowest FDs were associated with the green band for the former but the NIR band for the latter. Yet for the ASTER image, its FDs for the green and NIR bands were slightly higher than that from the red band and the lowest FD was tied to the short-wave infrared (SIR) bands. For the two IKONOS image subscenes, however, FDs increased as the wavelength became longer. Overall, the bands containing the most information content were the reflective bands, especially the red bands, in Landsat and ASTER images but the NIR bands in IKONOS images. This suggests the red band for Landsat and ASTER sensors and the NIR band for IKONOS sensor may be better than their other available bands to characterize the overall spatial complexity for the study area.

In order to investigate the influence of the underlying landscape and the scale effect based on the sensor's nominal spatial resolution on fractal measurement, FDs from the three common bands (green, red, NIR) for the subscenes of the ETM+, ASTER, and IKONOS images were also computed (Figure 12.4). The comparison between FD values reported by the images covering different landscapes showed that the results

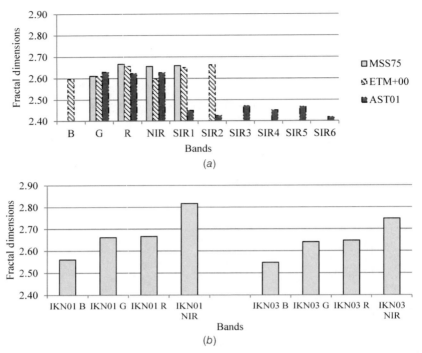

FIGURE 12.3 Fractal dimensions by bands for MSS75, ETM + 00 AST01 (*a*), and IKN01 and IKN03 subscenes (*b*) using triangular prism method. (Refer to Table 12.1 for band abbreviations for each sensor).

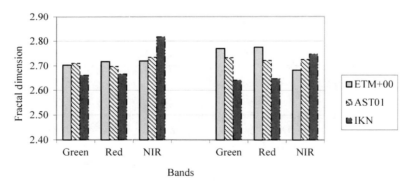

FIGURE 12.4 Fractal dimensions by green, red, and near-infrared bands for ETM+, ASTER, and IKONOS image subscenes for residential (left) and downtown (right) areas using triangular prism method.

varied by sensors as well as the spectral bands being considered. For the three sensors, only the IKONOS subscenes could generate results consistent for all three bands: The computed FD values for the residential area were all higher than those for the downtown area. This indicates the residential landscape is with more textural complexity than the downtown landscape. However, when the ETM+ and ASTER images were taken into account, the same result was observed to be true only with their NIR bands. Therefore, for ETM+ and ASTER images, their NIR bands may be better than the G and R bands to characterize the residential landscape in the scene, while their G and R bands will be a better option than the NIR bands to capture the downtown landscape in the scene.

Obviously, when the spatial resolution increased from 30 to 15 m and then to 4 m, all resultant FD values were different, illustrating the scale effect on fractal measurement due to the sensor's spatial resolution. Besides, the three reflective bands did not behave the same to the fractal measurement. With finer measurement scales, FDs from the NIR bands increased consistently while those from the two visible bands tended to decline instead. Ideally, more spatial details in terms of the level of variability within a single geographic area should be recorded by images with smaller pixel sizes. Potentially, this will allow the creation of images with rougher texture. Yet this only seemed to be true for the NIR data, indicating this may be more applicable than other bands for studying the scale effect on the overall spatial complexity resulting directly from the sensor's spatial resolution. For the two visible bands (G and R bands), however, the recorded shapes of the image texture did not grow more complex with increasing spatial resolution. They thus fail to characterize the spatial complexity captured by better spatial resolutions.

12.4.2 Fractals Analysis Using LULC Maps

Six LULC maps were created from Landsat (MSS, TM, and ETM+) and ASTER images acquired in 1975, 1985, 1991, 1995, 2000, and 2001 and their overall accuracy of classification was 83.20, 90.00, 88.80, 87.30, 89.00, and 89.87%, respectively. The two LULC maps were also derived from the two IKONOS images dated in 2001 and 2003 and they were found to be highly accurate with the overall accuracy of 93.33% for IKN01 and 94% for IKN03. Figure 12.5 illustrates LULC maps derived from the ETM+ 2000 image (*a*), the ASTER 2001 image (*b*), subscenes from the IKONO 2001 (*c*), and the 2003 image (*d*). The LULC change matrix from 1975 to 2000 in Table 12.2 clearly shows the study area's landscape had undergone a significant change during the 25-year period. Urban built-up lands and grassland had each increased in sizes by 42.68 and 67.28%, while the areas of cropland and pasture had decreased by 76.70%. Table 12.2 also summarizes the city population change over the time. Although the Indianapolis population reduced by .91% from 1975 to 1985, its residents increased consistently after 1985, with the greatest change occurring between 1990 and 1995 (3.46%).

12.4.2.1 *Fractals Analysis Using Raw Red Bands Classified by LULC Classes*   In Section 12.4.1 the red band was found to contain more spatial information content than others for most images used in this research, and they were used frequently for the analysis presented in this section and Sections 12.4.3 and 12.4.4. The five Landsat LULC maps were first employed as filters to mask out specific areas covered by various LULC types using their corresponding raw red bands. Table 12.3 presents

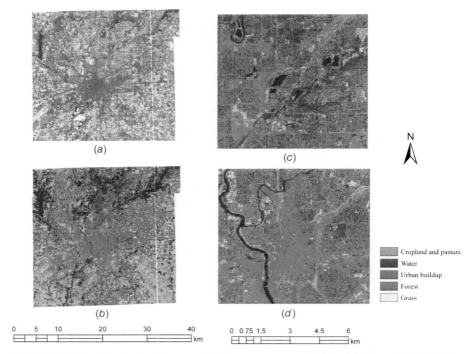

FIGURE 12.5 Sample land use and land cover maps: ETM + 00 (*a*), ASTER01 (*b*), IKONO01 subscene (*c*), and IKONO03 subscene (*d*). (See the color version of this figure in Color Plates section.)

TABLE 12.2   Land Use and Land Cover and Population Changes for 1975, 1985, 1991, 1995, and 2000

	Cropland and Pasture[a]	Water[a]	Urban Build-Up[a]	Forest[a]	Grass[a]	Total LULC Changes[b]	Population[c]
MSS75	368.37	25.63	243.56	168.73	242.22		722715
TM85	238.37	25.58	339.82	122.39	321.52	134.91	716067 (−0.91)
TM91	102.65	30.53	265.89	148.88	496.45	174.08	731327 (+2.13)
TM95	130.84	21.58	337.81	243.31	314.17	183.95	756599 (+3.46)
ETM+00	85.83	25.15	347.53	181.44	405.19	108.19	781870 (+3.34)

[a]Land use and land cover changes in km^2
[b]Total LULC changes in km^2 and include changes from all LULC types for each year (beginning in 1985) as compared to previous year.
[c]1975, 1985, and 1995 population was estimated using the mean values of census 1970 and 1980, census 1980 and 1990, and census 1990 and 2000 data, respectively.

TABLE 12.3 Descriptive Statistics of Raw Red Bands of Five Landsat Data Classified by Land Use and Land Cover Classes

Layers	Min	Max	Mean	SD	CV
MSS 75					
CroPas	10	46	20.25	3.09	0.15
Water	7	31	15.63	4.2	0.27
Build-up	10	84	28.78	6.77	0.24
Forest	10	24	14.72	2.38	0.16
Grass	9	55	19.95	3.78	0.19
TM 91					
CroPas	12	167	78.87	22.06	0.28
Water	14	77	34.77	11.24	0.32
Build-up	20	255	73.16	22.67	0.31
Forest	13	37	26.4	3.73	0.14
Grass	16	90	45.77	10.41	0.23
ETM+00					
CroPas	29	255	70.57	21.34	0.30
Water	24	196	48.41	12.59	0.26
Build-up	24	255	94.54	31.49	0.33
Forest	24	255	47.2	13.56	0.29
Grass	24	255	61.68	16.94	0.27
TM 85					
CroPas	1	247	36.32	13.67	0.38
Water	11	129	26.38	9.53	0.36
Build-up	2	255	60.23	20.25	0.34
Forest	3	194	28.2	9.01	0.32
Grass	1	248	40.33	12.93	0.32
TM 95					
CroPas	19	187	55.97	17.19	0.31
Water	18	117	30.22	6.87	0.23
Build-up	15	255	58.62	17.71	0.30
Forest	18	130	32.92	7.46	0.23
Grass	19	167	42.3	9.64	0.23

Note: SD, standard deviation; CV, coefficient of variation.

descriptive statistics for the resultant unclassified raw images by LULC types. These data were then applied to perform FD analysis using the TP method. From Figure 12.6, it is observed that the LULC types of urban build-up lands, forest, and grassland would result in higher FDs (greater than 2.5) than cropland and pasture and water. This implies that the first three LULCs contained more spatial complexity and higher texture information than the other two. This finding was consistent with those reported by De Cola (1989), Lam (1990), and Read and Lam (2002). Among all tested LULC categories, most of their FDs remained constant throughout the years except those given by the cropland and pasture and water classes. The sharp change of the two classes appeared mostly similar except those that happened in 1991. For the cropland and

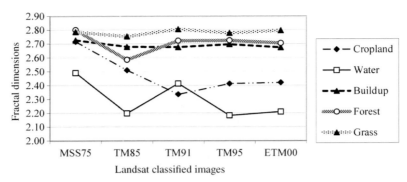

FIGURE 12.6 Fractal dimensions by raw red bands of MSS75, TM85, TM91, TM95, and ETM + 00 images for cropland and pasture, water, urban build-up, forest, and grass using triangular prism method.

pasture, it had one trough in this year; for the water, it had another peak in this year instead. When compared with Table 12.2, it is found that as the two land covers shrunk in size in the classified images, the corresponding spatial complexity of the raw red bands decreased. However, when the two land covers expanded in space in the LULC images, their spatial complexity in the associated raw red bands increased as well. It thus implies the potential of the TP method in examining the temporal changes of individual land covers with unclassified raw images. It should be noted that the images used to perform this part of the fractal analysis were fragmented as they all contained various numbers of "zero" values. It is not clear how those zero values would impact the computational procedure and contribute to the results summarized above. Nevertheless, a comparison with previous studies shows that these fragmented data were useful to reveal the spatial complexity of different urban land covers.

12.4.2.2 Fractal Analysis Using LULC Classes   To compute FDs from classified image, one of the critical issues is how to assign values to individual classes so that the categorical values can be converted to numerical data. Preliminary studies had tried several sets of pixel values for these LULC classes. The results have showed that, although the resultant mean FDs always varied according to the assigned pixel values, the general trends tended to be similar. The current chapter only reported those produced using the values of 1–5. Generally speaking, the values of 1–5 were individually assigned to each LULC class. For each LULC image, a total of 120 versions were generated and were separately applied to perform the TP-based fractal analysis. Results are graphed in Figure 12.7. Evidently, for both landscapes, FDs tended to decline with LULC maps derived from the ETM+, ASTER, and IKONOS images. The only exception was that the FD was higher in the IKONOS residential LUCL subscene than that from the ASTER subscene. Consequently, when the per-pixel unsupervised technique was used, the resultant LULC classified images derived from data sets with higher spatial resolution did not always present surfaces with more irregular shape complexity. To better understand these results, Moran's I index was also calculated from these images. From Figure 12.7, it is clear that there is a reverse

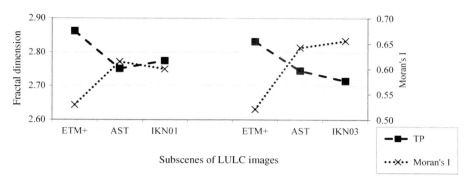

FIGURE 12.7 Moran's I index and fractal dimensions by LULC maps derived from subscenes of ETM+, ASTER, and IKONOS images for residential (left) and downtown (right) areas using TP method.

relationship between spatial patterns suggested by the FD and Moran's I index. For LULC maps derived from images with increasing pixel size, their individual LULC classes tended to have stronger spatial autocorrelation with each other, as suggested by the higher Moran's I, while the shape of their LULC landscape as a whole was inclined to grow smoother as indicated by lower FDs. Yet this is completely opposite when LULC maps were derived from coarser images. This scale change is more obvious with the downtown landscape than the residential landscape whose spatial complexity can be subject to the influence of other factors that are not accounted by FD and Moran's I index (e.g., the form, size, and distribution of individual LULC classes).

12.4.3 Fractal Analysis Using Resampled Raw Red Images

The DN values of red bands from MSS, ETM+, ASTER, and IKONOS images were resampled to four to six different levels of their nominal pixel sizes before their FD values were calculated using the TP approach (Figure 12.8). Clearly, different FDs were generated across varying pixel sizes for all tested images. This scale effect suggests remote sensing images are seldom self-similar and are therefore not true fractals. While the image's spatial details were smoothed out by larger and larger pixel sizes, individual objects found within the urban landscape in the scene were instead more irregularly shaped, as suggested by consistent higher FD values, and therefore generated a surface very much like a cube as a whole. As the spatial resolution decreased, the TP algorithm also became unstable and produced results higher than the theoretical limit of 3.0. This all happened at the resampled level of 240 m, which was the largest scale for analysis considering all six subscenes extracted from the EMT+, ASTER, and IKONOS images (Figures 12.8b,c). This may imply that the shape of underlying landscape features created by these data sets could exceed their most complex form at 240 m. However, it is not clear whether this is a function of the dominant landscape pattern being covered (e.g., downtown vs. residential) or/and just a matter of the size of the geographic extent or/and simply a technical failure. A close

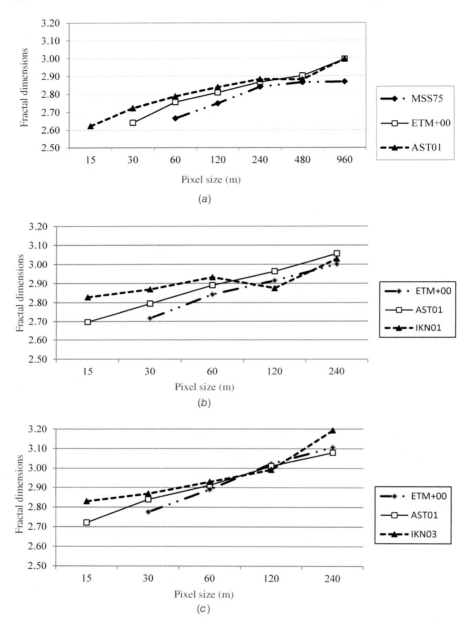

FIGURE 12.8 Fractal dimensions by original and resampled red bands for MSS75, ETM + 00, AST01, and IKONOS images using triangular prism method with whole image scenes (*a*) and subscenes (*b*, *c*) (the fractal dimension at 60 m for MSS75 was calculated using the 79-m image).

look at Figure 12.8a indicates that some FD values appeared to be more comparable at certain aggregation levels between images acquired by varying sensors, for instance, those at 480 m for the MSS, ETM+, and ASTER data sets and those at 30–120 m for all image subscenes.

To facilitate the interpretation, a distance analysis was performed with a chi-square statistic using SPSS. This method allowed the creation of a dissimilarity matrix that could be used to determine how FDs calculated from the original and resampled images for a given data set were different from each other. The analytical results were reported in Tables 12.4 and 12.5. Overall, the results calculated from the whole image

TABLE 12.4 Dissimilarity Matrix for MSS, ETM+, and ASTER Original and Resampled Image Scenes

Pixel size (m)	15	30	60[a]	120	240	480	960
MSS75							
15							
30							
60a			0.000	0.142	0.000	0.000	0.000
120			0.142	0.000	0.000	0.628	0.337
240			0.000	0.000	0.000	1.000	0.534
480			0.000	0.628	1.000	0.000	0.416
960			0.000	0.337	0.534	0.416	0.000
Total difference 1			0.142	1.107	1.534	2.043	1.287
Total difference 2				0.965	1.534	2.043	1.287
ETM+00							
15	0.000	0.903	0.410	0.406	0.000	0.770	0.000
30	0.903	0.000	0.608	0.843	0.000	0.000	0.903
60	0.410	0.608	0.000	0.265	0.871	0.000	0.410
120	0.406	0.843	0.265	0.000	0.720	0.000	0.406
240	0.000	0.000	0.871	0.720	0.000	1.000	0.000
480	0.770	0.000	0.000	0.000	1.000	0.000	0.770
960	0.000	0.903	0.410	0.406	0.000	0.770	0.000
Total difference 1		2.489	2.354	2.154	2.234	2.591	1.770
Total difference 2			1.451	1.744	1.828	2.591	1.000
ASTER01							
15	0.000	0.000	0.000	0.870	1.000	0.567	0.000
30	0.000	0.000	0.414	0.936	0.306	0.401	0.598
60	0.000	0.414	0.000	0.462	0.940	0.660	0.872
120	0.870	0.936	0.462	0.000	0.000	0.000	0.000
240	1.000	0.306	0.940	0.000	0.000	0.874	0.000
480	0.567	0.401	0.660	0.000	0.874	0.000	0.827
960	0.000	0.598	0.872	0.000	0.000	0.827	0.000
Total difference 1	2.437	2.655	3.349	2.268	3.120	3.329	2.297
Total difference 2		2.655	3.349	1.398	2.120	2.763	2.297

Note: Total difference 1 = Sum of results from all images including original image scene. Total difference 2 = sum of results only from resampled image scenes.
[a]This pixel size represents 79 m for the MSS75 image scenes.

TABLE 12.5 Dissimilarity Matrix for ETM+, ASTER, and IKONOS Original and Resampled Image Subscenes

Pixel size (m)	Residential Area					Downtown Area				
	15[a]	30	60	120	240	15[a]	30	60	120	240
ETM+										
15	0.000	0.000	0.000	0.000	0.000	0.000	0.000	0.000	0.000	0.745
30	0.000	0.000	0.000	0.000	0.000	0.000	0.000	0.000	0.000	0.000
60	0.000	0.000	0.000	0.000	0.604	0.000	0.000	0.000	0.000	0.000
120	0.000	0.000	0.000	0.000	1.000	0.000	0.000	0.000	0.000	1.000
240	0.000	0.000	0.604	1.000	0.000	0.745	0.000	0.000	1.000	0.000
Total difference 1	0.000	0.000	0.604	1.000	1.604	0.745	0.000	0.000	1.000	1.745
Total difference 2		0.000	0.604	1.000	1.604		0.000	0.000	1.000	1.000
AST										
15	0.000	0.634	0.994	0.000	0.000	0.000	0.000	0.000	1.000	0.886
30	0.634	0.000	0.989	0.000	0.000	0.000	0.000	0.000	0.930	0.772
60	0.994	0.989	0.000	1.000	0.000	0.000	0.000	0.000	0.000	0.868
120	0.000	0.000	1.000	0.000	0.000	1.000	0.930	0.000	0.000	0.000
240	0.000	0.634	0.994	0.000	0.000	0.886	0.772	0.868	0.000	0.000
Total difference 1	1.628	1.623	2.983	1.000	0.000	1.886	1.702	0.868	1.930	2.525
Total difference 2		0.989	1.989	1.000	0.000		1.702	0.868	0.930	1.640
IKN										
4	0.000	0.868	0.000	0.000	0.000	0.000	0.966	1.000	0.000	0.000
30	0.868	0.000	0.000	0.000	0.000	0.966	0.000	0.997	0.255	0.000
60	0.000	0.000	0.000	1.000	0.182	1.000	0.997	0.000	0.000	0.000
120	0.000	0.000	1.000	0.000	0.000	0.000	0.255	0.000	0.000	0.000
240	0.000	0.000	0.182	0.000	0.000	0.000	0.000	0.000	0.000	0.000
Total difference 1	0.868	0.868	1.182	1.000	0.182	1.966	2.218	1.997	0.255	0.000
Total difference 2		0.000	1.182	1.000	0.182		1.252	0.997	0.255	0.000

Note: Total difference 1 = sum of results from all image subscenes, including original image. Total difference 2 = sum of results only from resampled image subscenes.

[a]This pixel size represents 4 m for IKONOS subscenes.

scenes of Landsat MSS, ETM+, and ASTER data tended to be consistent (Table 12.4). With the highest dissimilarity values of 0.142, 0.903, and 1.0, the original image scenes were mostly different from its resample images at the first aggregation level for both Landsat MSS (120 m) and ETM+ (60 m) data but at the fourth aggregation level for the ASTER image (240 m). The zero dissimilarity values reported along the original image row/column in Table 12.4 may suggest that at certain aggregation levels the fractal complexity of the original scenes could be less dissimilar with those found within their resampled images. These scales were found to be 240–960 m for the MSS, 480 m for the ETM+, and 30–60 m and 960 m for the ASTER images. When all three image types were considered, it should be noted that at the 480 m aggregation level, the dissimilarity values were either the lowest (for the MSS and ETM+ images) or the second lowest (for the ASTER image). This implies that the textural complexity of all three image types was mostly alike after being resampled to the 480 m pixel size. This corresponds to the finding from Figure 12.8a. The total dissimilarity values for all original and all resampled images were also calculated for each image type. When only the resampled images were taken into account, the highest total dissimilarity values were found to be 2.043, 2.591, and 2.763 for the MSS, ETM+, and ASTER data, respectively. Interestingly, these all corresponded to the scale of 480 m, although the ones from the 60 m (3.349) were higher for the ASTER image. This suggests that for all three image types this aggregation level could generate images that were more distinct from those from other aggregation levels (Figure 12.8a).

More discrepancies were observed from image subscenes derived from the ETM+, ASTER, and IKONOS images (Table 12.5). Since FDs calculated from the 240 m level were all higher than the theoretical limit of 3.0, only dissimilarity values reported for 15–30 m were analyzed. In general, the resampled images mostly/least dissimilar from the original image subscenes tended to vary by image types as well as the dominant landscape (downtown and residential). With zero dissimilarity values between the original and resampled images for both landscapes, the resampling process did not seem to significantly change the fractal properties recorded in the ETM+ images. For the ASTER subscenes, with the highest dissimilarity values of 0.994, and 1.0, the original images were mostly different from its resample images at the second aggregation level (60 m) for the residential landscape but the third level (120 m) for the downtown landscape. When the lowest dissimilarity values were examined, it was found that the two ASTER subscenes were not similar across all aggregation scales. The results from the IKONOS subscenes tended to be more consistent. With the higher dissimilarity values of 0.868 and 0.966, the original scenes were both observed to be greatly different from their resampled images at the third aggregation level (30 m), although those from the four aggregation level (60 m) was even higher for the downtown subscene (1.0). Besides, both original scenes were least dissimilar to their resampled images at the 15 and 120 m levels, suggesting the difference in landscape may have the least impact on the scaling change on the textural complexity of the IKONOS image at the two scales. When all three image types were considered, the dissimilarity values between the original and resampled subscenes tended to be the lowest at the 120 m aggregation level for the two landscapes. The only exception was the ASTER subscene for the downtown landscape. It thus implies that the residential and downtown textural complexity recorded by the three image types was mostly similar when resampled to the 120 m pixel size. This corresponds to the finding from Figures 12.8*b,c*. When only the resampled images were taken into account, those more distinct from others were both observed to be the 120 m level for the ETM+ data regardless of their dominant landscape as each displayed a total difference of 1.0. However, for image subscenes from the ASTER and IKONOS data, the aggregation scales that were mostly different from others were the 60 m for the residential landscape (total difference values of 1.989 and 1.182) but 30 m for the downtown landscape (total difference values of 1.702 and 1.252). As a result, for all three image types used to represent the two urban landscapes, aggregation levels of 120, 60, and 30 m could generate image surfaces that were more distinct from those from other aggregation levels (Figure 12.8*a*).

12.4.4   Fractal Analysis for Temporal Change Characterization

In order to assess the utility of FD analysis for characterizing the temporal change of the landscape in the study area, all four Landsat images, including three TM and one ETM+, were applied since together they represented a complete time series of 25 years. The data analyzed included their raw red bands and LULC maps (Figure 12.9). Ideally, if fractal analysis is capable of denoting the temporal change of the landscape, the computed FD values should change with time consistently.

The raw red bands were first employed because they displayed the most spatial complexity among all reflected bands. For all five LULC maps, their FDs were calculated using the method described in Section 12.4.2.2. The mean FD value for each data set is plotted in Figure 12.9. Clearly, the two sets of FD values did not change significantly over time yet overall the two FD lines did follow a similar trend. That is, the FD values decreased steadily before 1991 but increased gradually after 1991. Obviously, although significant landscape changes were revealed by examining the LULC maps at the same period (Table 12.2), increased urban development does

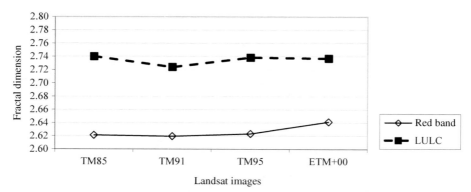

FIGURE 12.9   Fractal dimensions by red bands and LULC maps of TM85, TM91, TM95, and ETM + 00 images using triangular prism method.

not necessarily lead to more fragmented fractal patterns as suggested by higher FDs. It is still uncertain how effectively FD values can reveal the true change over time, since the discrepancy in FD values may also be caused by other factors, such as the changes in spectral reflectance of an individual band in case of raw images and the spatial composition of LULC classes in LULC images. More future research efforts are needed to examine this uncertainty.

12.5 DISCUSSION AND CONCLUSION

This chapter sought to explore the potential of fractal measurement in characterizing the dynamic changes of urban landscape of Indianapolis, Indiana, at multiple spatial and temporal scales. Eight satellite images collected by various sensors (MSS, TM, ETM+, ASTER, and IKONOS) in multiple years (from 1975 to 2003) and presented in different formats (raw, classified, and resampled images) were analyzed extensively using the TP algorithm in ICAMS. Overall, the fractal analysis based on the TP method was found to be useful in discriminating the geometric complexity of land surface features in both the unclassified and classified images collected by the five sensors. The examination of the original raw images shows that their overall spatial complexity will vary by the spectral bands, the sensor/image types, as well as the application being considered. For images collected by MSS, ETM+, and ASTER covering the whole city, the reflective bands, the red and NIR bands in particular, tended to create more spatially complex surfaces than other spectral bands for all image types being examined. Hence, these two bands may be better than others to characterize the texture of the overall landscape for the study area. However, for images dominated by a given LULC type such as downtown and residential land use, their textural complexity will depend on both sensor/image types and individual spectral bands. Among the three sensors—ETM+, ASTER, and IKONOS—the G, R, and NIR images collected by the IKONOS sensor are more suitable than those by the ETM+ and ASTER sensors to reveal the rougher textural pattern in the residential landscape as compared to the downtown landscape. Nevertheless, when only the ETM+ and ASTER images are taken into account, their G and R bands may be better than the NIR bands to capture the geometric complexity in the downtown land use, while their NIR bands will be more useful than the G and R bands to characterize the residential land use. On the other hand, the results based on the LULC maps indicated that fractal analysis was useful in identifying the spatial complexity of individual LULC classes recorded in the original image. The use of fragmented raw red bands filtered by LULC types showed that grassland, forest, and urban built-up land were more fragmented than other LULCs. Additionally, with only the LULC maps, the calculated FD results could illustrate that the region's urban landscape had experienced changes over time. Nevertheless, the fractal analysis alone was somewhat ineffective to detect the actual temporal change. Detailed quantitative change information is needed to assist in the interpretation of resultant FD values.

The potential of fractal measurement for illustrating the scale effect due to a sensor's nominal spatial resolution was examined. Based on image subscenes collected by the ETM+, ASTER, and IKONOS sensors, it was shown that, when the spatial resolution increased from 30 to 15 m and then to 4 m, all resultant FD values were different, illustrating the scale effect on fractal measurement resulted from the sensor's spatial resolution. Overall, it was found that the shapes of the image texture of individual spectral bands do not always grow more complex with data acquired by sensors having finer spatial resolution. Only the NIR bands of the three sensors can characterize rougher texture in images collected with smaller pixel size and thus are superior to others (G and R in particular) for this purpose. A similar trend was also noticed when the LULC images derived from the same image types were employed. Using the traditional ISODATA unsupervised algorithm, the classified images derived from the data sets with higher spatial resolution (e.g., IKONOS) grew smoother instead rougher as compared to those from images with coarser spatial resolution, such as ETM+. Obviously, there exists a discrepancy between the more detailed spatial information recorded in images with better spatial resolution and the less fragmented image surface as suggested by lower FD values. This can be attributed to a major drawback of fractal measurement in describing the spatial complexity. It only considers the *shape* of a feature, which is just one of the many properties of image texture. Other factors such as the form, size, and distribution of a given spatial object can all potentially help or hinder the separation of one image texture from another. A possible solution that has been suggested is to use the Lacunarity analysis, which considers multiple textural characteristics for a single fractal object (Mandelbrot, 1983).

The efficiency of fractal measurement for detecting the scaling properties of remote sensing images due to the resampling process was also evaluated with the resampled images. For a single image, their FD values changed with different pixel aggregation levels, implying real remotely sensed images are not true fractals. Overall, it was found that the resampling process tended to complicate the image texture with increasing pixel size. At some pixel aggregation levels, the resampling process also appeared to have stronger impact than other levels on changing image spatial properties. Specifically, for the images collected by the MSS, ETM+, and ASTER sensors, the textural complexity of the original images was less dissimilar to each other once they were resampled by the 480 m pixel size. When only the resampled images were considered, the resampling process could generate more distinct images at 480 m than other aggregation levels for the three image types. The results from image subscenes acquired by the ETM+, ASTER, and IKONOS sensors show that the textural complexity of both of the residential and downtown landscapes recorded in the original images was more comparable after being resampled to the 120 m aggregation level than other scales. Besides, the resampling process could produce image texture that is most different from other resampled images at 120 m for the ETM+ images regardless of the dominant landscape, 60 m for both ASTER and IKONOS residential images, and 30 m for both ASTER and IKONOS residential images. Overall, these pixel aggregation levels may suggest critical scale ranges where the intricacy of the underlying urban landscape recorded in the original image would reach a threshold that may be due to the interoperation of various natural (e.g., geological and ecological) and anthropogenic processes involved in urbanization. These aggregation scales may potentially link to critical thresholds or phenomenological breakpoints that can be used to uncover the specific process having the greatest impact on the urban structure in the real world. However, we should be aware that the use of FD values from the TP algorithm alone may be insufficient in providing enough insights into this topic. The TP algorithm was designed to measure spatial complexity by a single FD across scales, which is inadequate, especially when one realizes that many natural and anthropogenic processes do not have the same influence on urban landscape at different scales in the real world. The application of multifractal models is thus recommended to better address this issue (De Cola, 1993).

REFERENCES

Brown, S. R. 1995. Measuring the dimension of self-affine fractals: Examples of rough surfaces. In C. C. Barton and P. R. LaPointe (Eds.), *Fractals in the Earth Sciences.* New York: Plenum, pp. 77–87.

Clarke, K. C. 1986. Computation of the fractal dimension of topographic surfaces using the triangular prism surface area method. *Computers and Geosciences* 12:713–722.

De Cola, L. 1989. Fractal analysis of a classified Landsat scene. *Photogrammetric Engineering and Remote Sensing* 55:601–610.

De Cola, L. 1993. Multifractals in image processing and process imagine. In N. S. N. Lam and L. De Cola (Eds.), *Fractals in Geography*. Englewood Cliffs, NJ: Prentice Hall; pp. 280–304.

De Jong, S. M., and Burrough, P. A. 1995. A fractal approach to the classification of Mediterranean vegetation types in remotely sensed images. *Photogrammetric Engineering and Remote Sensing* 61:1041–1053.

Dell'Acqua, F., and Gamba, P. 2006. Discriminating urban environments using multiscale texture and multiple SAR images. *International Journal of Remote Sensing* 27:3797–3812.

Emerson, C. W., Lam, N. S. N., and Quattrochi, D. A. 1999. Multiscale fractal analysis of image texture and pattern. *Photogrammetric Engineering and Remote Sensing* 65:51–61.

Emerson, C. W., Lam, N. S. N., and Quattrochi, D. A. 2005. A comparison of local variance, fractal dimension, and Moran's I as aids to multispectral image classification. *International Journal of Remote Sensing* 26:1575–1588.

Emerson, C. W., Quattrochi, D. A., and Lam, N. S. N. 2002. Spatial metadata for global change investigations using remote sensing. The GIScience 2002 2nd International Conference on Geographic Information Science, Boulder, CO.

Frohn, R. C. 1998. *Remote Sensing for Landscape Ecology: New Metric Indicators for Monitoring, Modeling, and Assessment of Ecosystems*. Boca Raton, FL: Lewis Publishers, CRS Press.

Jaggi, S., Quattrochi, D. A., and Lam, N. S. N. 1993. Implementation and operation of three fractal measurement algorithms for analysis of remote-sensing data. *Computers and Geosciences* 19:745–767.

Lam, N. S. N. 1990. Description and measurement of Landsat TM images using fractals. *Photogrammetric Engineering and Remote Sensing* 56:187–195.

Lam, N. S. N., and De Cola, L. 1993. *Fractals in Geography*. Englewood Cliffs, NJ: Prentice Hall.

Lam, N. S. N., Qiu, H.-L., and Quattrochi, D. A. 1997. An evaluation of fractal surface measurement methods using ICAMS (Image Characterization and Modeling System). ACSM/ASPRS Annual Convention, Seattle, WA. ASPRS and ACSM, pp. 377–386.

Lam, N. S. N., Qiu, H.-L., Quattrochi, D. A., and Emerson, C. W. 2002. An evaluation of fractal methods for characterizing image complexity. *Cartography and Geographic Information Science* 29:25–35.

Lam, N. S. N., and Quattrochi, D. A. 1992. On the issues of scale, resolution, and fractal analysis in the mapping sciences. *Professional Geographer* 44:88–98.

Lam, N. S. N., Quattrochi, D. A., Qiu, H.-L., and Zhao, W. 1998. Environmental assessment and monitoring with Image Characterization and Modeling System using multiscale remote sensing data. *Applied Geographic Studies* 2:77–93.

Liang, B., and Weng, Q. 2013. An evaluation of fractal characteristics of urban landscape in Indianapolis, USA, using multi-sensor satellite images. *International Journal of Remote Sensing* 34:804–823.

Mandelbrot, B. B. 1977. *Fractals: Form, Chance and Dimension*. San Francisco: WH Freeman.

Mandelbrot, B. B. 1983. *The Fractal Geometry of Nature*. San Francisco: WH Freeman.

Myint, S. W. 2003. Fractal approaches in texture analysis and classification of remotely sensed data: Comparisons with spatial autocorrelation techniques and simple descriptive statistics. *International Journal of Remote Sensing* 24:1925–1987.

Myint, S. W. 2007. Urban mapping with geospatial algorithms. In Q. Weng and D. Quattrochi (Eds.), *Urban Remote Sensing*. Boca Raton, FL: Taylor and Francis Group, CRC Press, pp. 109–136.

Nunn, S. 1995. Role of local infrastructure policies and economic development incentives in metropolitan interjurisdictional cooperation. *Journal of Urban Planning and Development* 121:41–56.

Qiu, H.-L., Lam, N. S. N., Quattrochi, D. A., and Gamon, J. A. 1999. Fractal characterization of hyperspectral imagery. *Photogrammetric Engineering and Remote Sensing* 65:63–71.

Read, J. M., and Lam, N. S. N. 2002. Spatial methods for characterizing land cover and detecting land-cover changes for the tropics. *International Journal of Remote Sensing* 23:2457–2474.

Soille, P., and Rivest, J.-F. 1996. On the validity of fractal dimension measurements in image analysis. *Journal of Visual Communication and Image Representation* 7:217.

Turner, M. G., and Garnder, R. H. 1991. *Quantitative Methods in Landscape Ecology*. New York, NY: Springer.

United States Census Bureau. 2012. http://www.census.gov/. Assessed December 3.

Weng, Q. 2003. Fractal analysis of satellite-detected urban heat island effect. *Photogrammetric Engineering and Remote Sensing* 69:555–566.

Zhao, W. 2001. Multiscale Analysis for Characterization of Remotely Sensed Images. Ph.D. Dissertation, Louisiana State University.

13

SPATIOTEMPORAL SCALES OF REMOTE SENSING PRECIPITATION

Yang Hong and Yu Zhang

13.1 BACKGROUND ON PRECIPITATION SCALE

Precipitation variability, both in space and time, plays a key role in global climate and water cycle, regional water resources management, and local flash flood warning. However, currently in situ precipitation measurements are limited due to discrete-point observations; such sparse data cannot well represent precipitation high spatial and temporal variability across scales. Remote sensing techniques such as radar and satellite have advanced rainfall measurements at relatively high resolutions to a new era for numerous applications. In most of the studies, remote sensing precipitation data are usually resampled into certain spatial and temporal scales for particular application purposes, and consequently this often causes scale-related uncertainty. Today it is well known that the error of remote sensing precipitation estimation is nonlinearly related to the scaling (e.g., Anagnostou and Krajewski, 1998; Gourley et al., 2010; Hong et al., 2006; Jordan et al., 2000; Seo and Krajewski, 2010; Smith et al., 2004, 2005; Villarini et al., 2008). Hong et al. (2006) describe that the satellite rainfall estimation error is a function of spatial and temporal scales where higher spatial and temporal resolution is subject to larger uncertainty. On the other hand, from the end-user perspective, precipitation is a key forcing in hydrological modeling and natural hazard forecasting. Wood et al. (1990) pointed out that in hydrological modeling major factors of heterogeneity leading to spatial variability in runoff are precipitation, topography, and soil types. The latter two factors are considered as static through time; thus, spatiotemporal variability of precipitation dominates the uncertainty of hydrological modeling at a range of scales.

Scale Issues in Remote Sensing, First Edition. Edited by Qihao Weng.
© 2014 John Wiley & Sons, Inc. Published 2014 by John Wiley & Sons, Inc.

This chapter overviews the precipitation measurement methods and quantifies its uncertainty as a function of spatiotemporal scales and precipitation intensity in Section 13.2, with a case study on its scale-based uncertainty impact on and error propagation into hydrological prediction in Section 13.2, followed by a conclusion in Section 13.4.

13.2 SPATIOTEMPORAL SCALING FOR PRECIPITATION

13.2.1 Overview of Precipitation Measurements

13.2.1.1 In Situ Precipitation Measurements Liquid precipitation is traditionally measured by rain gauges, which is a point-based measurement but usually considered as "ground truth." The basic idea of the rain gauge is to collect rainfall into a cylindrical container of a fixed diameter during storms. There are different types of rain gauges: weighing gauges, tipping-bucket (TB) gauges, capacitance gauges, optical gauges, disdrometers, and so on. Among all those types of rain gauges, the TB gauge is commonly used by agencies such as the National Weather Service (NWS) and United States Geological Survey (USGS) (Habib et al., 2001). Before the era of radar and satellite precipitation missions, rain gauges were applied for operational as well as calibration purposes [e.g., calibrate radar precipitation estimation algorithm (Anagnostou and Krajewski, 1998)].

Although rain gauges provide surface rainfall measurements at relatively high accuracy compared to remote sensing precipitation estimations at a specific point, in most cases, rain gauge instruments are so sparsely distributed that they are constrained from accurately characterizing spatial and temporal variability (Villarini et al., 2008).

13.2.1.2 Remote Sensing Precipitation Measurements Unlike gauges that measure precipitation at point scales, recent developments of radar and satellite precipitation estimation techniques have provided much broader coverage beyond ground in situ observations. Ground radar measures the electromagnetic backscatter power return from raindrops, also expressed as the reflectivity factor Z (mm^6/mm^3). Usually an empirical Z–R relationship (where R indicates the precipitation) is applied for converting measured reflectivity to rainfall intensity estimation. Although ground radar provides extended precipitation measurements over rain gauge stations, it is still limited for remote areas, mountain regions, and vast oceanic surface. The recent development of satellite-based precipitation retrieval techniques has provided extended precipitation coverage beyond both in situ data and ground radar networks. Spaceborne radar on the satellite platforms, together with passive radiometers and infrared sensors, are combined to derive global precipitation information.

13.2.2 Spatiotemporal Scales of Precipitation

With the recent advances in remote sensing techniques, remotely sensed precipitation can provide extended rainfall information at relatively high temporal and spatial resolution. Current spatial resolution of ground radar can reach as high as around 250 m from the NWS WSR-88 radar network in the Continental United States while the spatial resolution of satellite precipitation is around 5–25 km from the Tropical Rainfall Measuring Mission (TRMM) at near global coverage. The combined regional and global precipitation products can be synergistically developed into 5-min products from radar networks and every 3-h products from satellites, which enables users to accumulate to longer time scales such as 6 h, 12 h, daily, monthly, annually, and up to decade scales for a variety of researches and applications.

13.3   PRECIPITATION UNCERTAINTY QUANTIFICATION AND ERROR PROPAGATION INTO HYDROLOGICAL PREDICTION

As a key forcing variable to the land surface hydrological process, it is vital to better understand precipitation uncertainty quantification and error propagation into hydrological prediction as a function of scales. This section is mainly based on Hong et al. (2006), which developed an end-to-end uncertainty propagation analysis framework to first quantify the scale-dependent satellite rainfall uncertainty and further assess its error propagation into hydrological modeling and stream flow predictions.

13.3.1   Uncertainty Quantification Framework

Uncertainty quantification is defined as a function of the space–time–intensity of precipitation as shown in Figure 13.1. To quantify the influence of estimation uncertainty on hydrological responses by ensemble simulation would further place an appropriate degree of confidence from data users' perspectives. The advantage of this error formula, for example, is that the various users of the precipitation products can approximate the error by directly implementing the function at their specified space–time scales and rain intensities and quantifying the error propagation into a conceptual rainfall–runoff hydrological model through Monte Carlo simulation. By flowing the framework as shown in Figure 13.1, an ensemble of satellite precipitation

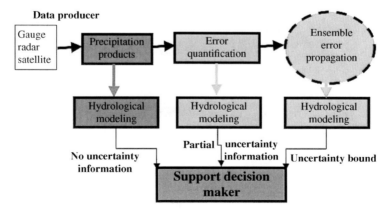

FIGURE 13.1   A conceptual "end-to-end" error analysis framework for precipitation data producers and users.

forcing data is generated and then their hydrological prediction uncertainty is quantified at different confidence internal levels.

13.3.2   Uncertainty Quantification of Satellite Precipitation Estimation

In this study, the error of precipitation σ_E is considered as a function of its spatial resolution L, time integration T, sampling frequency Δt, and the space–time average of rain rate $(\hat{R})$, as shown in the equation

$$\sigma_E = f\left(\frac{1}{L}, \frac{\Delta t}{T}, \hat{R}\right) = a\left(\frac{1}{L}\right)^b \left(\frac{\Delta t}{T}\right)^c (\hat{R})^d \tag{13.1}$$

The error of precipitation σ_E refers to the standard deviation between the reference rainfall data (i.e., the NWS WSR-88D stage IV data) and the estimated satellite rainfall data [i.e., PERSIANN-CCS: Precipitation Estimation from Remote Sensed Information Using Artificial Neural Network–Cloud Classification System (Hong et al., 2004) in this study]. The NWS WSR-88D stage IV data, with manual quality control, serve as ground reference rainfall data at 4 km × 4 km spatial and hourly temporal resolution. Stage IV data with high spatial and temporal resolution rainfall analysis provide data useful for testing of satellite rainfall estimation algorithms. Additional information about the National Centers for Environmental Prediction (NCEP) stage IV analysis can be found at http://wwwt.emc.ncep.noaa.gov/mmb/ylin/pcpanl/stage4/. The PERSIANN-CCS is a segment-based cloud classification and rainfall estimation system based on coregistered passive microwave and infrared images from low Earth-orbiting and geostationary satellites by using computer image processing and pattern recognition techniques. We operate this system (http://hydis8.eng.uci.edu/GCCS/) with the goal of producing data at spatial–temporal resolution (reaching $4 \times 4\,\text{km}^2$, 30 min) suitable for basin scale hydrological research and applications. PERSIANN-CCS has been generating precipitation estimates at resolution $0.04° \times 0.04°$ scale and 30-min time interval since 2000.

The study area is shown as a shaded area in Figure 13.2. Given their processed basic resolution ($0.04° \times 0.04°$ and hourly), both the satellite and radar data sets were

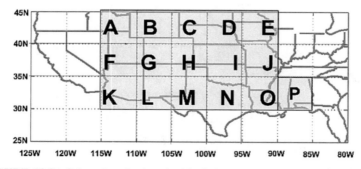

FIGURE 13.2   Selected study domain (shaded area), each box is a $5° \times 5°$ grid.

aggregated into a range of discrete temporal (1, 3, 6, 12, and 24 h) and spatial (0.04°, 0.12°, 0.24°, 0.48°, and 0.96°) scales. Therefore each box is divided into subgrids with side lengths of 0.04°, 0.12°, 0.24°, 0.48°, and 0.96° latitude–longitude; each subgrid holds rainfall intensities at time resolutions of 1, 3, 6, 12, and 24 h for this analysis. The error analysis is conducted for each box separately.

The satellite rainfall estimates were incrementally divided into five bins according to their intensities; the reference errors are then calculated for each rain rate bin at various spatial and temporal scales. The reference error (σ_E) distribution at the discrete temporal and spatial scales and increasing rain rates for box I in August 2003 are displayed in Figure 13.3 (left). The same reference error is shown in Figure 13.3 (right), but as a percentage of rainfall estimates $\sigma_E/\hat{R}(\%)$. From Figure 13.3, the satellite rainfall estimation reference error is a function of spatial and temporal scales where higher spatial and temporal resolution is subject to larger reference error.

For each box shown in Figure 13.2, we divide an entire year data set at various spatial–temporal scales and 10 bins of incremental rain intensities. By defining the minimum satellite sampling frequency as 1 h ($\Delta t = 1$), the parameters a, b, c, and d of Equation (13.1) were calibrated using the root mean-square error (RMSE) as the calibration criterion. Using the calibrated optimal parameter set, Figure 13.4 plots the reference error estimates (σ_E) with respect to spatial–temporal scales and rainfall intensities; the same reference error estimates but as a percentage of rainfall intensity ($\sigma_E/\hat{R}$) are plotted in Figure 13.5. Note that in the current study we arguably assume that the effect of sampling frequency is negligible because the PERSIANN-CCS has relatively high sampling frequency (30 min), one of the advantages of rainfall estimates using multiple satellites.

13.3.3 Uncertainty Propagation from Precipitation Data to Hydrological Prediction

As the key forcing variable of hydrological processes, the precipitation is largely responsible for the uncertainty in model outputs. Clearly, evaluation of the error associated with precipitation products into model behavior is an indispensable element of improving hydrological modeling and data assimilation. In the current study, the influence of the error of input-forcing data, that is, precipitation, through a conceptual rainfall–runoff hydrological model to output forecasting uncertainty is evaluated by propagating the approximated PERSIANN-CCS rainfall error estimates with a Monte Carlo simulation approach. This approach generates an ensemble of precipitation data as forcing input to fit to the conceptual rainfall–runoff model, and the resulting uncertainty in the forecasted stream flow is then quantified. The applicability and usefulness of this procedure are demonstrated in the case of the Leaf River basin, located north of Collins, Mississippi. The size of the Leaf River basin is about 1949 km^2. A map of the Leaf River basin is shown in Figure 13.6. We have used a daily time step for precipitation input. This may be larger than desirable to capture the hydrological response. See, for example, Burges (2003) for a perspective on this issue.

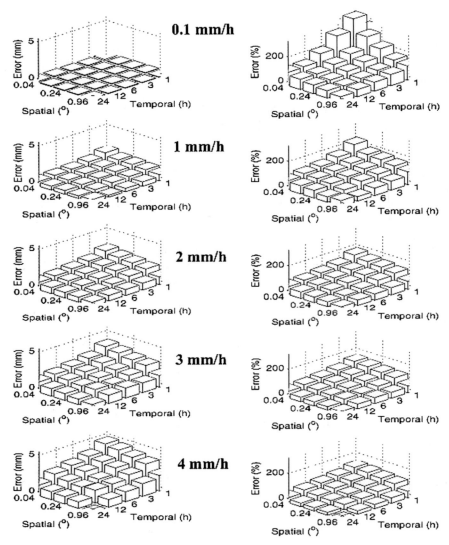

FIGURE 13.3 Error distribution as a function of spatial and temporal scales (*left*) at increasing rain rates (*right*) Percentage of error to rain rate.

To demonstrate the rainfall error propagation through a hydrological model, we employed a parsimonious conceptual hydrological model (HyMOD) as described by Boyle et al. (2001) (Figure 13.7) in the model simulation.

HyMOD stream flow simulation proceeded under two scenarios: (1) the error as a variable ratio of rain rates based on Equation (13.1) and (2) the error as a fixed ratio of rain rates. In each error scenario, 100 ensemble members of the HyMOD simulation were used to derive the confidence interval of stream flow prediction by using the Monte Carol method. The simulated runoff output and its 95% uncertainty bound

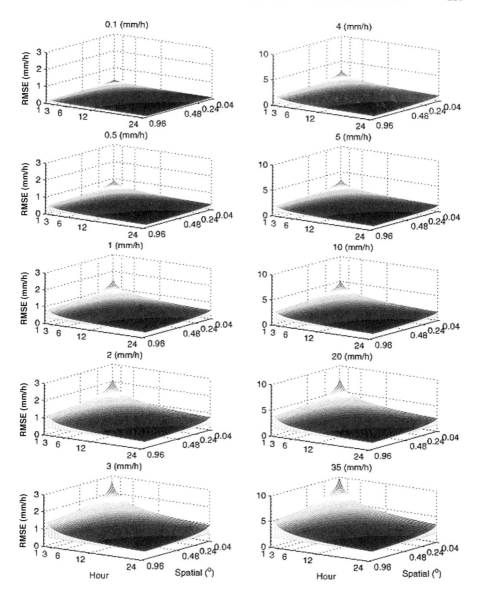

FIGURE 13.4   Plot of the error distribution with respect to space scale L, time scale T, and rain intensity R using the optimal parameters. (See the color version of this figure in Color Plates section.)

derived from the second scenario are displayed in Figure 13.8a; the results from the first scenario are given in Figure 13.8b. The red dots represent the runoff time series generated from the satellite rainfall estimates. The results show that the reliability of estimated stream flow is highly related to the intensity of the input-forcing data. For both error propagation cases, the uncertainty bound is significantly higher during

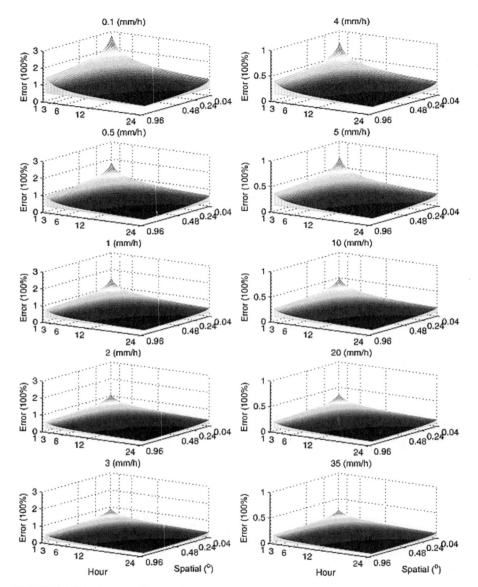

FIGURE 13.5   Same as Figure 13.3, expect the error is expressed as percentage (%) of rain rates. (See the color version of this figure in Color Plates section.)

high-flow periods and is lower in low-flow regions. However, the 95% uncertainty bound of the simulated runoff as shown in Figure 13.8*b* is lower than the one generated from the fixed error ratio case pictured in Figure 13.8*a*. The first scenario, scale-dependent error propagation, offers more realistic uncertainty assessment of

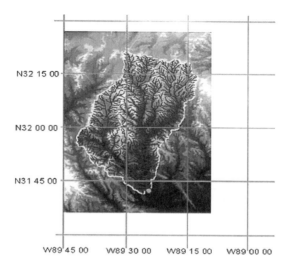

Leaf River Near Collins, Mississippi
USGS # 02472000
Basin Area : 1949 km²
~ 44 x 44 km²

FIGURE 13.6   Study area for streamflow uncertainty estimation associated with rainfall uncertainty. (See the color version of this figure in Color Plates section.)

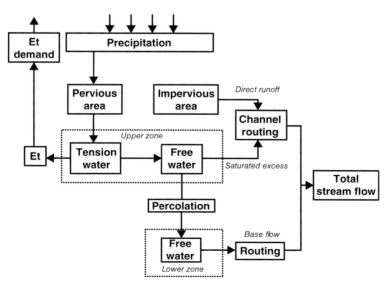

FIGURE 13.7   Hydrological model (HyMOD) after [Boyle *et al.*, 2001].

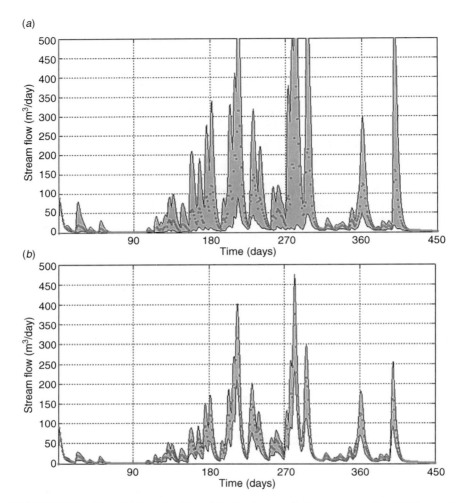

FIGURE 13.8   Uncertainty estimation of stream flow prediction corresponding to 95 percentile confidence using rainfall error propagation: (a) fixed rainfall error 50% and (b) scale-dependent rainfall error according to Equation (13.1). The red dots are mean of the 100 ensemble streamflow sequences simulated from satellite rainfall estimates. (See the color version of this figure in Color Plates section.)

stream flow prediction than the fixed ratio error propagation, particularly during high flow, that is, heavy rainfall periods.

13.4   CONCLUSIONS

As satellite-based precipitation estimates are increasingly applied to atmospheric and hydrological studies at various space–time scales, the quantification of scale-based uncertainties and the effect of how the uncertainties propagate from precipitation data through models need to be evaluated as a whole so that the users have a certain degree of confidence while making decisions (Figure 13.1). Additionally, evaluation of uncertainties associated with precipitation products and its propagation into model behavior is an indispensable element of evaluating data quality and improving hydrological simulation techniques. As an evaluation result, the uncertainty of satellite-based precipitation estimation error is a function of several factors, including spatial and temporal resolution of estimates, the estimated rain rates, and sampling frequency. Furthermore, when compared with the conventional error propagation procedure, that is, fixed ratio error estimate, this strategy not only provides more realistic quantification of precipitation estimation error but also offers improved uncertainty assessment of the error propagation from the precipitation input into the hydrological models (Figure 13.8).

The proposed uncertainty analysis framework in this case study not only provides a general framework of scale-based satellite precipitation estimation error quantification procedure but also can assess its influence on the uncertainty of hydrological prediction through Monte Carlo simulation of the error propagation. As a first attempt to quantify uncertainty of one of the major satellite-based precipitation data sets (PERSIANN-CCS) at the fine scale ($0.04° \times 0.04°$ and hourly), the satellite rainfall error model [Equation (13.1)] does not consider rainfall detection and false-alarm probabilities. In our continuing effort to quantify the uncertainty of high-resolution satellite-based rainfall estimates, the error property of spatial–time integrated precipitation related to the undetected rainfall and false-alarm scenarios will be investigated and embedded in the end-to-end uncertainty analysis framework report here.

REFERENCES

Anagnostou, E. N., and Krajewski, W. F. 1998. Calibration of the WSR-88D precipitation processing subsystem. *Weather Forecast* 13:396–406.

Boyle, D. P., Gupta, H. V., Sorooshian, S., Koren, V., Zhang, Z. Y., and Smith, M. 2001. Toward improved streamflow forecasts: Value of semidistributed modeling. *Water Resources Research* 37:2749–2759.

Burges, S. J. 2003. Process representation, measurements, data quality, and criteria for parameter estimation of watershed models, in Calibration of Watershed Models. *Water Science and Application* 6:283–299.

Gourley, J. J., Hong, Y., Flamig, Z. L., Li, L., and Wang, J. H. 2010. Intercomparison of rainfall estimates from radar, satellite, gauge, and combinations for a season of record rainfall. *Journal of Applied Meteorology and Climatology* 49:437–452.

Habib, E., Krajewski, W. F., and Kruger, A. 2001. Sampling errors of tipping-bucket rain gauge measurements. *Journal of Hydrologic Engineering* 6:159–166.

Hong, Y., Hsu, K.-L., Moradkhani, H., and Sorooshian, S. 2006. Uncertainty quantification of satellite precipitation estimation and Monte Carlo assessment of the error propagation into hydrologic response. *Water Resources Research* 42:W08421.

Hong, Y., Hsu, K. L., Sorooshian, S., and Gao, X. G. 2004. Precipitation estimation from remotely sensed imagery using an artificial neural network cloud classification system. *Journal of Applied Meteorology* 43:1834–1852.

Jordan, P., Seed, A., and Austin, G. 2000. Sampling errors in radar estimates of rainfall. *Journal of Geophysical Research Atmospheres* 105:2247–2257.

Seo, B. C., and Krajewski, W. F. 2010. Scale dependence of radar rainfall uncertainty: Initial evaluation of NEXRAD's new super-resolution data for hydrologic applications. *Journal of Hydrometeorology* 11:1191–1198.

Smith, D. F., Gasiewski, A. J., Jackson, D. L., and Wick, G. A. 2005. Spatial scales of tropical precipitation inferred from TRMM microwave imager data. *IEEE Transactions on Geoscience and Remote Sensing* 43:1542–1551.

Smith, M. B., Koren, V. I., Zhang, Z., Reed, S. M., Pan, J. J., and Moreda, F. 2004. Runoff response to spatial variability in precipitation: An analysis of observed data. *Journal of Hydrology* 298:267–286.

Villarini, G., Mandapaka, P. V., Krajewski, W. F., and Moore, R. J. 2008. Rainfall and sampling uncertainties: A rain gauge perspective. *Journal of Geophysical Research* 113: D11102.

Wood, E. F., Sivapalan, M., and Beven, K. 1990. Similarity and scale in catchment storm response. *Reviews of Geophysics* 28:1–18.

PART V

NEW FRONTIERS IN EARTH OBSERVATION TECHNOLOGY

14

MULTISCALE APPROACH FOR GROUND FILTERING FROM LIDAR ALTIMETRY MEASUREMENTS

José L. Silvan-Cárdenas and Le Wang

14.1 INTRODUCTION

The increase in spatial resolution of sensors has enabled human interpreters to *see* more details of the landscape and, at the same time, has challenged remote sensing scientists to develop fully automated or semiautomated algorithms for efficiently extracting desired features from such massive data sets. Examples of desired features are digital terrain models (DTMs), which constitute a basic source of information to many applications, such as watershed analysis, hydrological modeling, river dynamics studies, coastal erosion estimation, environmental resource evaluation, and urban growth projections, to list just a few.

The small-footprint, discrete-return light detection and ranging (Lidar) system has become one of the most important means to produce high-resolution DTM data. This has been due in part to a number of advantages over competing aerial photogrammetric techniques such as independence of sunlight, higher vertical accuracy, less missing data by occlusion, low redundancy and because it does not rely on the existence of textured surfaces and discontinuities for a successful point matching (Pfeifer and Mandlburger, 2009). The small-footprint, discrete-return Lidar system is based on the accurate measurement of the elapsed time between emitted and backscattered laser pulses. The emitted pulse is typically short time and unimodal, whereas the backscatter may spread over longer times exhibiting multiple modes called returns. Returns are associated with distinct layers with which the laser interacted.

Scale Issues in Remote Sensing, First Edition. Edited by Qihao Weng.

For each return, the three-dimensional (3D) position is determined and its intensity recorded. Hence the raw Lidar data consist of a dense cloud of 3D points with associated return intensity. Each point provides the location where the laser hit Earth's surface during the scanning process, whereas the intensity is a digital representation of the fraction of pulse energy reflected at that location.

The z coordinate of points corresponds to terrain elevation with respect to a horizontal datum, typically the mean sea level, plus the height of nonterrain features in some instances. In order to produce a DTM by interpolation of ground points, a discrimination of ground from nonground points must be carried out first. This discrimination process is referred to as ground filtering and is generally considered a preprocessing for generating not only the DTM but also the height of nonground components (Axelsson, 1999). Methods to ground filtering can be grouped into two major categories, point-based and raster-based methods. Methods in the first category classify directly the point cloud whereas methods in the second category first interpolate the point cloud onto a regular grid surface, namely, the digital surface model (DSM). Each approach has advantages and disadvantages. In general, rasterizing the data first allows to take advantage of digital image processing algorithms which run much faster than point-based operations, whereas point-based processing tends to be more accurate (Axelsson, 1999). Point-based methods tend to include techniques such as clustering analysis (Roggero, 2001), local surface fitting, discrimination by slope/terrain difference/surface curvature (Sithole, 2001), active contours, and adaptive triangulated irregular networks (Axelsson, 2000), whereas in raster-based methods approaches such image segmentation, edge detection (Brovelli, 2002), mathematical morphology (Zhang et al., 2003), repetitive interpolations, and analysis–synthesis framework are more common. The method presented in this chapter falls in the last category.

Filtering methods vary in complexity, accuracy, and sensitivity to changes in parameters. A recent review of methods and critical issues of the ground filtering problem have been revised by Meng et al. (2010). The International Society for Photogrammetry and Remote Sensing (ISPRS) Working Group III/3 has conducted a test to determine the performance of several filters developed over the past years to extract bare-Earth points from point clouds. Among other things, the study concluded that, in general, filters that estimate local surfaces are found to perform best (Sithole and Vosselman, 2004). Other comparative studies have found that using several window sizes in a progressive filtering tends to be more accurate as it removes features of different sizes and tends to be less sensitive to parameter selection (Zhang et al., 2003; Zhang and Whitman, 2005). Interestingly, the need for processing at multiple scales/ resolutions has been increasingly recognized by several studies. For instance, the so-called progressive morphological filters (Chen et al., 2007), which apply morphological operators with structuring elements of increasing sizes, have been shown to outperform implementations with a single-size operator. Other methods have taken advantage of the progressive smoothing in a scale-space representation for detecting high surface curvatures that occur at ground/nonground transitions (Evans and Hudak, 2007). These notions are more formally articulated in multiresolution image decompositions, which other studies have also explored in the context of ground filtering; these include the multiresolution wavelet transform (Thuy and Tokunaga, 2004) and the multiscale Hermite transform (Silván-Cárdenas and Wang, 2006).

The latter method is based on the so-called multiscale erosion operator defined in the multiscale Hermite transform (MHT) domain. This method progressively removes the contribution of above-ground objects to the transform coefficients, so that when the inverse transform is applied, an approximation of the DTM results. This chapter had the objectives of revising the theoretical basis of this method, discussing the practical aspects of its operation, and presenting an extension of the original method by revising the erosion operation and testing its performance with ISPRS data sets. The chapter starts with the background section, where the scale-space representation is first discussed as the theoretical foundation of the MHT. This includes a description of the MHT of discrete 2D signals. In this part the local spatial rotation in the transform domain is also introduced. The extension of the single-scale case to multiple scales is presented next by building upon previous definitions. The following section presents the ground filtering method with emphasis on new features added to the previous method. In this part, it is shown how the multiscale erosion operation, which is the main mechanism to estimate the MHT coefficients of the bare terrain from the MHT expansion of the DSM, can be generalized to higher order coefficients through a scale-space shifting operation. This section also includes the parameter selection and other processing steps needed by the ground filtering method. The last part of the chapter presents some results from a few tests and relevant discussion.

14.2 BACKGROUND

14.2.1 Scale-Space Representations

The Gaussian scale-space theory was first developed by Iijima in the late 1950s (Iijima, 1959), but it was not adopted by the computer vision and image processing community until further developments by Witkin and Koenderink in the 1980s (Witkin, 1984; Koenderink, 1984). For the latter, the theory served as a means to generalize existing notions of Gaussian pyramids and as a well-founded way to perform multiscale analysis.

The original axiomatics of scale-space theory postulated the scale space to (1) be linear, (2) be shift invariant, (3) be scale and rotation invariant, (4) preserve positivity, and (5) fullfil the semigroup property. The latter means that the scale-space representation is equipped with an associative binary operator, the convolution operator. These axioms led Iijima and others to conclude the Gaussian kernel was the only possibility for the linear scale space, so that "Gaussian scale space" was also a term to refer to "linear scale space." However, a deeper look into the matter led Felsberg and Sommer (2004) to discover another non-Gaussian linear scale space based on a Poisson kernel. They concluded that for the Gaussian kernel to be the unique kernel of the linear scale space a sixth axiom needed to be added to the original list, one that requires the frequency response of the scale space kernel to be continuosly differentiable at the origin.

The main idea of the scale-space representation of a measured signal, such as the DSM, is to embed the signal into a one-parameter family through Gaussian smoothing, where the parameter controlling the width of the Gaussian kernel is termed the *scale*. Furthermore, if scale is equated to time, the evolution of the scale-space family can also be described by the diffusion equation, a differential equation that governs heat diffusion in a homogeneous medium. In this sense, the scale-space representation is the result of letting an initial heat distribution, given by the original signal, to evolve over time, so that fine-scale features in the signal will disappear monotonically with increasing time or scale. The diffusion equation description was also convenient because the Gaussian kernel and all its derivatives are solutions to this equation, which had the important implication that Gaussian derivatives are natural operators of the linear scale space (Koenderink and Van Doorn, 1992).

In light of the scale-space theory, decomposing an input signal in terms of Gaussian derivatives at multiple scales is not only desirable but also convenient because these basis functions exhibit a wide range of orientation and scale character-istics that make them more efficient to detect primitive structures, such as ridges, edges, and lines. Moreover, psychophysical and biophysical evidence has showed that early processing of the visual signal by the human visual system performs similar operations. At this point, it should be noted that a DSM can be treated as an image because many editing processes may require visualizing and interpreting the DSM as an image. Furthermore, digital derivatives of the DSM have more natural meaning than for images. For instance, the derivative along the steepest ascent direction cor-responds to the slope around the derivation point. Also, level contour lines have a close relation with the gradient field.

14.2.2   Multiscale Hermite Transform

Several decompositions based on Gaussian derivatives have been developed (Martens, 1990; Reed and Bloom, 1996; Silván-Cárdenas and Escalante-Ramírez, 2006); however, most of them differ in the selection of scaling constants and implementation. One such decomposition is the multiscale Hermite transform, which is an overcomplete signal decomposition based on the difference of Gaussian filters that are further decomposed in scale and rotated Gaussian derivatives (Silván-Cárdenas and Escalante-Ramírez, 2006). Its development had an inspiration on models of the receptive fields of ganglion cells in the human visual system. It can be implemented in a pyramidal fashion, similar to a wavelet transform, with the additional property of being *steerable*, that is, it can be computed in a rotated coordinate system as a linear combination of the unrotated version.

The MHT is defined for one, two, and higher dimensions and for both continuous and discrete domains. Of particular interest is the case for 2D discrete signals, which is relevant for the ground filtering problem. For that particular case the theoretical development has been revised in several studies, including for ground filtering (Silván-Cárdenas and Wang, 2006) and for building detection (Silván-Cárdenas and Wang, 2011) from Lidar measurements.

The single-scale discrete Hermite transform (DHT) of a two-dimensional signal $z(x, y)$ defined on a discrete domain corresponds to a set $\{z_{n,m}(p,q)\}$ for $n, m = 0, \ldots, N$ of surfaces that approximates the partial derivatives of a Gaussian-smoothed version of the input signal at locations (p, q). Each surface $z_{n,m}(p, q)$ is referred to as a transform coefficient of a given order (n, m) that indicates the derivation order with respect to the spatial coordinates. From a frequency filtering point of view, the zero-order coefficient $z_{0,0}(p, q)$ corresponds to the low-pass filtered version of the input, whereas higher order coefficients correspond to bandbass filtered versions of the input surface.

For the ground filtering problem the signal corresponds to a DSM in raster format. The transform is implemented as a convolution of the input signal with a bank of 1D filters of compact support along each dimension followed by a subsampling with a rate factor of 2. Because of the subsampling, the DHT coefficients are approximately one-quarter the size of the input surface. The bank of filters corresponds to the binomial family, which is the discrete counterpart of the continuous family formed by Gaussian derivatives. In the case of the Gaussian family, the width of the Gaussian defines the scale, whereas in the case of the binomial family, the filter length defines the scale. In either case, the scale controls the degree of the smoothness of the transform coefficients.

The full set of transform coefficients allows recovering the input signal without any loss of information. However, the DHT expansion compacts most of the signal information in the first few coefficients so that a near-perfect reconstruction can be obtained with a truncated expansion. The inversion procedure is implemented through an upsampling of the coefficients by a factor of 2 and followed by their convolution with corresponding interpolation filters, which are nothing but scaled and reflected versions of the binomial filters. The transform is symmetric in the sense that the same operations are performed both in the forward (analysis) and inverse (synthesis) directions.

The multiscale discrete Hermite transform (MDHT) is an extension of the single-scale case that simulates multiple filter lengths and hence multiple scales. It simulates increasing filter lengths by recursively replacing the zero-order coefficient of previous level by its DHT expansion with fixed filter length ($N = 6$) times a scaling factor, thus yielding a waveletlike pyramidal decomposition. In this case, the replaced zero-order coefficients are not part of the MDHT expansion but are only used for the computation of the next coarser level. In contrast, the coarsest low-pass coefficient is part of the MDHT expansion, as it is not replaced by its expansion. The result is a set of coefficients $\{z_{n,m}^{(k)}(p,q)\}$ also indexed by a scale level k for $k = 1, \ldots, K$. The coefficients in the kth level still correspond to partial derivatives of the order indicated by n and m, but their size or support is reduced by a factor of 2^k, and the degree of smoothness is defined by a filter length $N_k = 2 \times 4^k$. In other words, the representation is equivalent to filtering the original signal with filters of length $N_k = 2 \times 4^k$ and then resampling the output by a factor 2^k. Conversely, the reconstruction of the original signal is carried out through successive reconstruction of low-pass coefficients, starting with the coarsest resolution layer. Figure 14.1 illustrates the decomposition–reconstruction process for a raster DSM.

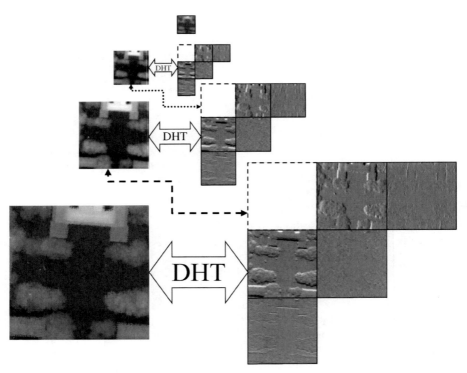

FIGURE 14.1 Multiscale disccrete Hermite transform (MDHT). Labeled arrows indicate signal decomposition through single-scale DHT. Dashed line indicates replacement of low-pass coeffcient by its DHT expansion. Modified from Silván-Cárdenas and Wang (2006).

14.2.3 Local Spatial Rotation

The local spatial rotation is a convenient way for spatial orientation analysis as well as for further compaction of the information through aligning the coordinate axis along the strongest signal variation. This process consists of expressing the transform coefficients in a coordinate system (u, v) that has been rotated by an angle θ with respect to the original coordinate system (x, y). Furthermore, since different orientation angles are chosen for different sampling cells (p, q), this operation is referred to a local spatial rotation. Figure 14.2 shows an example of original (left image) versus rotated (right image) coefficients when displayed as gray-level images.

While the local spatial rotation is a powerful analysis, the operation is compu-tationally efficient and easy to implement because the rotated coefficients are related with the original coefficients through linear combinations. The notation for the MDHT coefficients in a rotated framework is now extended to $z_{n,m}^{(k,\theta)}$, which includes also the rotation angle θ. For further details on how the rotated coefficients are computed, see Silván-Cárdenas and Escalante-Ramírez, 2001 (2006). In the

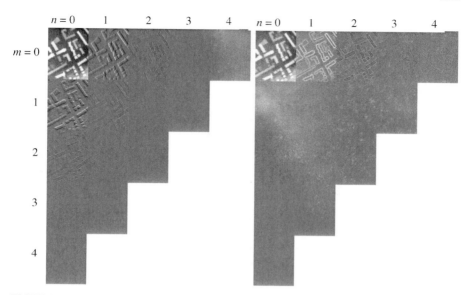

FIGURE 14.2 Example of DHT coefficients (subimages) up to fourth order (left) and corresponding rotated coefficients (right). In both cases, order of derivation with respect to *x* and *y* coordinate is indicated by *n* and *m*, respectively. Transform coefficients were linearly stretched to a common range so that the actual variability is proportional to the display contrast shown.

case presented here, the rotation angle was set to the orientation of the local gradient, which can be directly computed from the first-order coefficients as expressed in Equation (14.1):

$$\theta = \arctan\left(\frac{z_{0,1}^{(k)}}{z_{1,0}^{(k)}}\right) \tag{14.1}$$

$$z_{0,1}^{(k,\theta)} = 0 \qquad z_{1,0}^{(k,\theta)} = \sqrt{\left(z_{1,0}^{(k)}\right)^2 + \left(z_{0,1}^{(k)}\right)^2} \tag{14.2}$$

This selection of the angle yields the rotated coefficients as given by Equations (14.2). The nonzero coefficient corresponds (up to a constant factor) to the magnitude of the local gradient. These coefficients are also illustrated in the left panel of Figure 14.2. The former is in the intersection of the second row and the first column of the image array, whereas the latter is in the intersection of the second column and first row. Other angles may be computed from higher order coefficients; however, the gradient angle generally achieves higher energy compaction of the signal along the first coordinate. This is the case if the input signal embeds strongly oriented features, such as the edges of buildings in a DSM (see Figure 14.2). Such energy compaction has been shown useful for building detection from the digital height model (DHM).

14.3   PROBLEM FORMULATION

Variations in elevation can be due to natural terrain relief, height variation in vegetation, and human-made features as well as to transitions between ground and nonground surfaces. From the ground filtering problem point of view, the latter are the most important elevation variations. These variations can occur at various scales, that is, within a few meters or over long distances. Linear operators such as those employed in the MDHT are effective tools for detecting elevation variations that occur at different scales.

A terrain signal $t(x, y)$ is assumed embedded in the DSM surface $z(x, y)$ through addition with the nonground feature height $h(x, y)$ and a vertical noise $\varepsilon(x, y)$ from the acquisition process, or

$$z(x, y) = t(x, y) + h(x, y) + \varepsilon(x, y) \tag{14.3}$$

Furthermore, in the MDHT domain the surface model above can be written as

$$z_{n,m}(p, q) = t_{n,m}(p, q) + h_{n,m}(p, q) \tag{14.4}$$

where the noise component is assumed negligible and the scale and angle indices (k, θ) have been omitted to simplify notation. Hence, the filtering problem is reduced to a source separation problem with two components, namely the height of objects above the ground and the terrain elevation. Evidently, even in the case of a negligible noise component, the problem is still ill-conditioned as there are more unknown variables than equations. Therefore, no unique solution exists, and different solutions will differ in the way unknown variables are further constrained.

14.4   FILTERING METHOD

The ground filtering method described here is based on the MDHT expansion of the DSM. The method was first introduced by Silván-Cárdenas and Wang (2006), so only the major additions are detailed here. The major processing steps of the ground filtering method are depicted in Figure 14.3. The process takes as main input the original point cloud and produces as output a labeled set of points. Some intermediate products are the eroded DSM and the ground/nonground mask. The original method performed a single decomposition of the DSM. However, a new feature has been added, which includes the possibility for repeated application of the multiscale erosion operator to an estimated DTM. This part is indicated with dashed lines in Figure 14.3. The following sections describe the processing steps indicated in the flowcharts.

14.4.1   Point-to-Raster Conversion

Point-to-raster conversion is a necessary step for processing in the raster mode. The conversion is carried out in two steps. In the first step, a grid covering the same area of

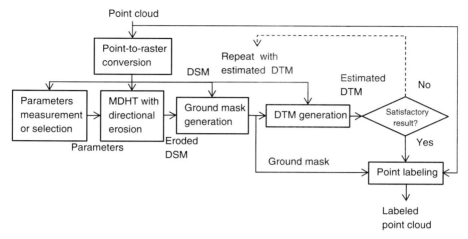

FIGURE 14.3   Flow diagram of ground filtering method. Dashed line indicates optional loop for repeated processing.

the point cloud is created and each cell of the grid is assigned the maximum elevation value of all points within the cell. In a second step, the cells not assigned in the first step are interpolated using the nearest neighbor interpolation method. It is recommended that the cell size of the grid is chosen in the order of the average point spacing or larger, so that holes are not too large.

In some data sets, it is necessary to handle outlayers and gaps prior to the point-to-raster conversion process above. Outliers are isolated points with very high or very low elevation values which can influence the filtering process, especially because the smoothing mechanism of the MHT decomposition will tend to diffuse such extreme values to other cells. These extreme values may be produced by flying birds, high antennas, and sensor faultier, among others. On the other hand, gaps are areas where no recording of points was made. Gaps can be due to low or no energy detected by the sensor, such as in water bodies where both high-energy absorption and specular reflection may occur. The interpolation on gaps may cast artifacts on the final DSM, especially in complex terrain areas. One way to deal with this problem is to use a higher order interpolation method for gaps, such as the cubic convolution interpolation method. This should lessen the diffusive mechanism of the MHT, which tends to propagate errors from interpolated gaps.

14.4.2   MDHT with Directional Erosion

The MDHT with directional erosion is a decomposition that adaptively processes the transform coefficients in such a way that the inverse transform of the processed coefficients is a surface with above-ground features removed. This process consists in detecting ground/nonground transition cells and in applying an erosion operator to the coefficients on detected cells. The eroded zero-order coefficient is then used for generating the next coarser layer in the decomposition and the operations are repeated for all subsequent layers, so that contributions from above-ground features are gradually removed from the coefficients.

The surface gradient along transitions between ground and nonground points is assumed higher than the maximum terrain slope. This is generally the case for features of sizes much larger than the average sampling distance of Lidar data. In contrast, small trees, cars, or other low features embedded in data sets with average point distance greater than 1 m tend to produce smaller gradient values than some terrain features. Ground/nonground transitions are detected by comparing the local gradient $z_{1,0}^{(\theta,k)}$ with a threshold T_k defined for each layer k as given by Equation (14.5). This threshold has been derived from a 1D surface model parameterized in terms of maximum terrain slope m_{max} and the maximum elevation difference between two contiguous terrain terraces Δ_{max}. Cells where the gradient exceeds the threshold are detected as ground/nonground transitions and then an erosion process is applied:

$$T_k = \frac{2^k m_{max}}{\sqrt{2 + 2\pi \left(\frac{2^k m_{max}}{\Delta_{max}}\right)^2}} \tag{14.5}$$

As shown below, the erosion operation can be seen as a local spatial shifting of the signal so that a portion of the surface in the vicinity of a ground/nonground transition is replaced by a portion of the signal located along neighboring ground points. The local spatial shifting is based on a property of the binomial filters according to which linear combinations of members of a binomial filter family can reconstruct the members of a binomial filter family at a shifted location and decreased filter length. Furthermore, since the DHT is a linear transform, a similar relation is satisfied by the DHT coefficients. For 2D and higher dimensional signals, one can combine the shifting operation with the rotation operation, so that a directional spatial shifting results. In particular, for 2D signals the spatial shifting along a direction defined by an angle θ is expressed through Equation (14.6), where the filter length is included in the notation to make explicit the shifting in scale as well. It should be noted that the filter length (N) rather than a scale index (k) is used in this equation:

$$\frac{z_{n,l}^{(\theta,N-M)}(p-M/2,q)}{\sqrt{C_{N-M}^n}} = \sum_{m=0}^{M}(-1)^m C_M^m \frac{z_{n+m,l}^{(\theta,N)}(p,q)}{\sqrt{C_N^{n+m}}} \quad n=0,\ldots,N-M \quad l=0,\ldots,N \tag{14.6}$$

$$z_{0,0}^{(\theta,6)}(p-1,q) = z_{0,0}^{(\theta,8)}(p,q) - \frac{z_{1,0}^{(\theta,8)}(p,q)}{\sqrt{2}} + \frac{z_{2,0}^{(\theta,8)}(p,q)}{\sqrt{28}} \tag{14.7}$$

The particular case for $N=8, M=2$, and $n=l=0$ is written in Equation (14.7), which corresponds to the erosion operator developed in previous work (Silván-Cárdenas and Wang, 2006) if the scale shifting and the third term on the left-hand side are neglected. In that study, such an erosion operator was derived from an explicit model of a 1D surface in the continuous domain and then extrapolated for the 2D discrete case. In contrast, here we show that the erosion operator is an approximation of the local shifting operator expressed in Equation (14.6). Hence, this represents a generalization of the erosion operator that allows shifting the higher order coefficients as well, instead of simply setting them to zeros as in the previous work. Furthermore, in the original formulation, the terrain elevation underneath nonground features was assumed flat because the erosion operator was essentially a local spatial shifting. This assumption is relaxed here.

Since the local spatial shifting is essentially a zero-order interpolation, a natural generalization can be carried out through a truncated Taylor expansion with higher order terms. In particular, here we use an expansion up to first order, which could account for sloppy terrain underneath nonground features. This implementation uses a Taylor expansion up to first-order terms which, after taking advantage of the relation between the transform coefficients and the surface derivatives, can be expressed as in Equation (14.8). Hence the erosion operator is implemented by computing the right term of Equation (14.8) using the local spatial shifting of Equation (14.6):

$$z_{n,m}^{(\theta,N)}(p,q) \approx z_{n,m}^{(\theta,N)}\left(p - \frac{M}{2}, q\right) - \frac{M}{4}\sqrt{n+3}z_{n+1,m}^{(\theta,N)}\left(p - \frac{M}{2}, q\right) \qquad (14.8)$$

Furthermore, one of the problems in representing local support signals with the MDHT occurs along boundary cells. As with any neighborhood operators, the local derivative operators used in MDHT are influenced by neighboring surface values. Hence neighbours outside the support or coverage of the surface need to be extrapolated. The way these cell values are created is called the boundary condition. The boundary condition used previously was the symmetric extension, which consists in reflecting the surface values horizontally around boundary cells. In this study an alternative was adopted, which consists in reflecting both horizontally (x–y plane) and vertically (z coordinate) the surface values around the boundary cells. This yields a boundary condition called the antisymmetric reflection, which naturally extrapolates preserving the surface slope beyond boundary cells.

14.4.3   Parameter Selection

There are many parameters that need to be properly defined for the filtering method to work. First, the number of pyramid layers in the MDHT expansion must be selected according to the largest above-ground feature that is to be removed from the input surface. Since the erosion operator of the pyramid layer k erodes the features of size 2^k, then we can compute the minimum number of layers that will warrant the effective erosion of all features up to a given size. Equation (14.9) is used to compute the number of pyramid layers in terms of the width W of largest above-ground features and the cell size δ of the input surface. The factor of 2 in the formula accounts for the fact that erosion is performed from two opposite sides. Hence the argument of the logarithm in the formula is the largest semiwidth in number of cells. The maximum

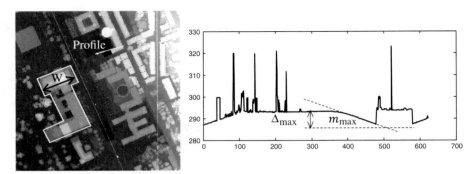

FIGURE 14.4 Interactive measurements of filter parameters. Left: DSM displayed as gray-level image. Right: plot of terrain profile indicated in DSM image. Measurements of the nonground maximum width (W), maximum terrain elevation difference (Δ_{max}), and maximum terrain slope (m_{max}) are also indicated.

width can be measured using an oriented rectangle that encloses the largest nonground feature or group of features clumped together. Figure 14.4 (left) illustrates the measurement of this parameter from the DSM:

$$K = \left\lceil \frac{\log_2 W}{2\delta} \right\rceil \tag{14.9}$$

Second, the maximum terrain slope and the maximum elevation difference required for detecting ground/nontransition cells in the multiscale pyramid can be measured directly from 1D surface profiles. This is illustrated in Figure 14.4 (right). Alternatively, since both parameters are associated to a simple 1D surface model parameterized in terms of slope and elevation differences, a 2D histogram of the surface elevation–slope may also reveal these parameters. This is illustrated in Figure 14.5.

Third, the shifting parameter M must be properly selected so that artifacts from approximations made are reduced. Figure 14.6 illustrates the effect of the shifting parameter on a reconstructed profile from its shifted transform coefficients. The shifting in scale can be ignored only if the surface is smooth enough (bottom panel). However, for sharp transitions the inverse transform requires to account for the scale shifting (middle panel), or otherwise signal distortions may occur (upper panel) and this is accentuated with increasing values of M. Based on preliminary tests, a value of $M = 2$ was observed to perform well. This value of the shifting parameter ensures a shifting of 2^k pixels at the kth resolution level while maintaining a minimal scale shifting. This is particularly important for an adaptive shifting scheme, as the inverse transform uses a constant filter length. In other words, even when the shifting is performed on some pixels while others are left unshifted, the inverse transform can still have the same form for all the pixels without introducing important distortions in the reconstructed terrain surface. Also, since most information from the surface variations is contained within the first few-order coefficients, the erosion operator is

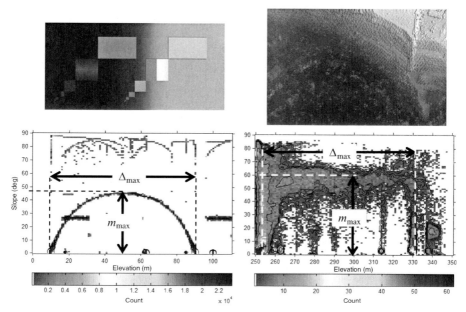

FIGURE 14.5 Measurement of terrain parameters from visualizations of elevation–slope histograms. Top: input DSM; bottom; elevation–slope histograms as images. (See the color version of this figure in Color Plates section.)

applied only to the low-pass and first-order coefficients ($z_{0,0}^{(\theta)}$ and $z_{1,0}^{(\theta)}$), while higher order coefficients along detected ground/nonground transitions are set to zero.

14.4.4 Ground Mask and DTM Generation

Once the MDHT coefficients of the terrain component have been estimated through the erosion operation, the inverse MDHT transform is computed and the resynthesized surface used for generating a ground/nonground mask. A cell is labeled as ground in the ground mask if its elevation value falls below the estimated surface plus a tolerance value; otherwise it is labeled nonground. The suggested value for the tolerance value corresponds to the vertical accuracy of the sensor, which is typically around 15 cm (Habib, 2009).

In order to determine if the result is satisfactory, a preliminary DTM is produced by masking the nonground features in the original DSM and filling the holes. Hole filling in the raster mode can be efficiently accomplished through the nearest-neighbor interpolation method. However, since evaluation is generally visual, a more pleasant surface can be generated using a higher order interpolation. The approach adopted here was to select the maximum elevation value of randomly selected neighboring cells, where the probability for selecting a cell was set proportional to the inverse distance. This method has the advantage of running

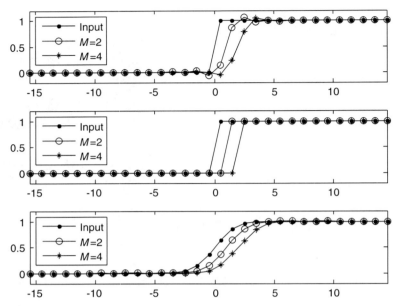

FIGURE 14.6 Plots illustrating effect of shifting parameter (M) in reconstructed surface profiles (circles and stars). Input profile (dots) corresponds to a sharp step for the first and second cases and to a smooth step for third case. The DHT of the input was computed with $N = 8$ and the transform coefficients were processed with a scale-space shifting of Equation (14.6) for $M = 2$ and $M = 4$. Then, the profiles were reconstructed from processed coefficients with the inverse DHT using the same filter length ($N = 8$) for the first and third cases but a decrease filter length ($N–M$) for the second case.

relatively fast while producing surfaces with little artifacts. The maximum value was used because such values would be associated with remaining small nonground features, if any, and which could be filtered in a second pass.

14.4.5   Repeat Pass and Point Labeling

Some data sets, where complex terrain is present, may require additional passes of the multiscale erosion operator to refine the ground/nonground mask. If this is the case, the input DSM is replaced by the estimated DTM and the process repeated. At this point the only parameter that should be readjusted is the maximum width of nonground features. A rule of thumb is to decrease this value by a factor of 4 in every pass, since remaining nonground features are much smaller in every pass. The repetition of the filtering stops if the number of cells that changed in the ground mask between consecutive passes falls below a threshold, typically one percent of the total. Alternatively, the process stops if a maximum number of iterations is reached, which should be set to a small number. Here we use three. Once the ground/nonground mask has been refined, the labeling of points is performed with a simple overlay operation in which the point is labeled as ground if it falls within a ground cell.

14.5   FILTERING TESTS

Both the previously developed multiscale erosion and the new local shifting method were applied to the ISPRS data sets for two sites (www.commission3.isprs.org/wg3/). The first site corresponds to an urban area (CSite2) exhibiting large and irregular shaped buildings as well as a road with a bridge and small tunnel. The second site corresponds to a rural area (FSite5) with vegetation on steep slopes, quarry, vegetation on river banks, and gaps. The original point cloud included elevation and intensity from the first and last returns; however, only the first return was used in the tests as it was generally the cleanest measurement due to multiple bounces of the laser signal. The parameters used for the filtering of each data set are provided in Table 14.1. Cell sizes for point-to-raster conversions were roughly equivalent to the original average point spacing. Filter parameters were determined using the visualization of the elevation–slope histogram as illustrated in Figure 14.5. In both methods the MDHT decomposition used extended boundary cells through antisymmetric reflections of boundary cells. According to this condition, the extension of a sloppy terrain near the edge will maintain its slope rather than change it with a symmetric reflection.

The accuracy assessment of the filtering results was carried out for up to four sample insets that were available for each test area. The ground/nonground masks obtained with the new method for the entire test area are displayed in Figure 14.7 with the limits of validation samples overlaid. The quantitative accuracy revealed a slight to moderate improvement between the two methods. Table 14.2 summarizes the overall accuracy, that is, the percent of correctly classified points for each sample set and method. In both cases, most commission errors (data not shown) were located along the eastern side due to the edge effect of the multiscale filtering, yet this was less significant for the new method. This is verified in the greater accuracy obtained for the samp23 and samp24. The reason for this was the sloppy terrain in that area, which was better represented with the first-order Taylor expansion of the new method. For the forested site, the major problem was the omission errors (data not shown) along the southern side, where a sharp terrain shape is present. The improvement was observed only in samp52 and samp53, which are precisely located along an area of complex terrain.

TABLE 14.1   Selected Filter Parameters for Each Test Site

Parameter	CSite2	Fsite5
Maximum feature width	119	55
Maximum terrain elevation difference	26	110
Maximum terrain slope	28	57
Maximum tolerance	0.15	0.15
Cell size	1	2
Iterations	2	1

TABLE 14.2   Overall Accuracy of Filtering Result for Various Samples Indicated in Figure 14.7

	CSite2		Fsite5	
	Old Method	New Method	Old Method	New Method
samp*1	69	69	75	76
samp*2	83	83	68	81
samp*3	82	85	59	78
samp*4	68	75	87	86
Mean	75	78	72	80

Note: Asterisk (*) means insert 2 or 5 depending on site number.

 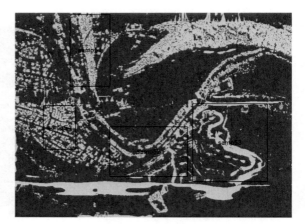

FIGURE 14.7   Filtering result for urban and forest sites. Dark gray indicates ground and light gray indicates detected nonground features. Boxes indicate insets used for accuracy assessment.

14.6   DISCUSSION

Ground filtering is a processing step of the Lidar altimetry data that consists in separating ground from nonground points. Since both terrain and off-ground features happen at somewhat overlapping range of scales, separation of these two components becomes tremendously difficult. Several studies have proven that the size of a two-dimensional neighborhood is critical for ground filtering performance (Zhang et al., 2003; Zhang and Whitman, 2005). Others have shown that the multiscale/multiresolution representations can be used for discriminating nonground points (Thuy and Tokunaga, 2004; Evans and Hudak, 2007). In general, these methods tend to be less sensitive to parameter selection and are able to detect features of various sizes.

This chapter presented a filtering method based on a multiscale signal decomposition termed the multiscale Hermite transform, which has been formulated in the context of the scale-space theory for signal processing. The original method had been ranked among the top three most accurate methods of a number of previously developed methods (Silván-Cárdenas and Wang, 2006). In this work, we further improved the method by incorporating several novel features, including (1) the generalization of the erosion operator for shifting not only the zero-order coefficient but also the first-order coefficients in a rotated framework, (2) relaxing the assumption on maximally flat terrain underneath nonground features by allowing linear prediction of coefficients at these points through a truncated Taylor expansion, (3) the replacement of the symmetric condition to an antisymmetric condition when computing the MDHT coefficients along boundary cells, and (4) a mechanism to refine the DTM through repeated application of the multiscale erosion operator. In addition, practical considerations on parameter selection were provided.

The new method was compared with the original method and results showed improvement along areas where terrain is particularly complex or sloppy. The method had been shown to be robust on low-resolution data sets and able to effectively remove large buildings of complex size. However, further testing is necessary to automatize the parameter selection and to determine in advance an appropriate number of interactions. Also, it has been claimed that the raster-based approach adopted here is more efficient than the point-based approach; however, a comparative study on time complexity of ground filtering algorithms is still lacking.

REFERENCES

Axelsson, P. 1999. Processing of laser scanner data—algorithms and applications. *ISPRS Journal of Photogrammetry & Remote Sensing* 54:138–147.

Axelsson, P. 2000. DEM generation from laser scanner data using adaptive TIN models. *International Archives of the Photogrammetry, Remote Sensing and Spatial Information Sciences* 33:110–117.

Brovelli, M. A., Cannata, M., and Longoni, U. M. 2002. Managing and processing LIDAR data within GRASS. *GRASS Users Conference.* University of Trento, Trento, Italy: 29 pp.

Chen, Q., Gong, P., Baldocchi, D., and Xie, G. 2007. Filtering airborne laser scanning data with morphological methods. *Photogrammetric Engineering & Remote Sensing* 73(2):175–185.

Evans, J. S., and Hudak, A. T. 2007. A multiscale curvature algorithm for classifying discrete return lidar in forested environments. *IEEE Transactions on Geoscience and Remote Sensing* 45(4):1029–1038.

Felsberg, M., and Sommer, G. 2004. The monogenic scale-space: A unifying approach to phase-based image processing in scale-space. *Journal of Mathematical Imaging and Vision* 21(1):5–26.

Habib, A. 2009. Accuracy, quality assurance, and quality control of LiDAR data. In J. Shan and C. K. Toth (Eds.) *Topographic Laser Ranging and Scanning*, Boca Raton, FL: CRC Press, pp. 269–294.

Iijima, T. 1959. Basic theory of pattern observation. *Papers of Technical Group on Automata and Automatic Control.* Japan: IECE.

Koenderink, J. J. 1984. The structure of images. *Biological Cybernetics* 50:363–370.

Koenderink, J. J., and Van Doorn, A. J. 1992. Generic neighboorhood operators. *IEEE Transaction on Pattern Analysis and Machine Intelligence* 14(6):597–605.

Martens, J. B. 1990. The hermite transform-theory. *IEEE Transactions on Acoustics, Speech and Signal Processing* 38(9):1595–1606.

Meng, X., Currit, N., and Zhao, K. 2010. Ground filtering algorithms for airborne LiDAR data: A review of critical issues. *Remote Sensing* 2(3):833–860.

Pfeifer, N., and Mandlburger, G. 2009. LiDAR data filtering and DTM generation. In J. Shan and C. K. Toth (Eds.), *Topographic Laser Ranging and Scanning*, Boca Raton, FL: CRC Press, pp. 307–333.

Reed, J. A., and Bloom, T. R. 1996. A gaussian derivative-based transform. *IEEE Transactions on Image Processing* 5(3):551–553.

Roggero, M. 2001. Airborne laser scanning: Clustering in raw data. *International Archives of Photogrammetry, Remote Sensing and Spatial Information Sciences* 34:227–232.

Silván-Cárdenas, J. L., and Escalante-Ramírez, B. 2001. Image coding with a directional-oriented discrete Hermite transform on a hexagonal sampling lattice. In A. Tescher (Ed.), *Applications of Digital Image Processing XXIV*, San Diego: SPIE. pp. 528–536.

Silván-Cárdenas, J. L., and Escalante-Ramírez, B. 2006. The multiscale Hermite transform for local orientation analysis. *IEEE Transaction on Image Processing* 15(5):1236–1236.

Silván-Cárdenas, J. L., and Wang, L. 2006. A multi-resolution approach for filtering LiDAR altimetry data. *ISPRS Journal of Photogrammetry and Remote Sensing* 61(1):11–22.

Silván-Cárdenas, J. L., and Wang, L. 2011. Extraction of buildings footprint from LiDAR altimetry data with the Hermite transform. *Pattern Recognition—MCPR'11 Proceedings of the Third Mexican Conference on Pattern Recognition*. Cancun, Mexico, Berlin: Springer, pp. 314–321.

Sithole, G. 2001. Filtering of laser altimetry data using a slope adaptive filter. *International Archives of the Photogrammetry, Remote Sensing and Spatial Information Sciences* 34:203–210.

Sithole, G., and Vosselman, G. 2004. Experimental comparison of filter algorithms for bare-Earth extraction from airborne laser scanning point clouds. *ISPRS Journal of Photogrammetry and Remote Sensing* 59(1):85–101.

Thuy, T., and Tokunaga, M. 2004. Filtering airborne laser scanner data: A wavelet-based clustering method. *Photogrammetric Engineering and Remote Sensing* 70(11):1267–1274.

Witkin, A. 1984. Scale-space filtering: A new approach to multiscale description. *Image Understanding* 3:79–95.

Zhang, K., Chen, S., Whitman, D., and Shyu, M. 2003. A progressive morphological filter for removing nonground measurements from airborne LiDAR data. *IEEE Transactions on Geoscience Remote Sensing* 41:872–882.

Zhang, K., and Whitman, D. 2005. Comparison of three algorithms for filtering airborne lidar data. *Photogrammetric Engineering & Remote Sensing* 71(3):313–324.

15

HYPERSPECTRAL REMOTE SENSING WITH EMPHASIS ON LAND COVER MAPPING: FROM GROUND TO SATELLITE OBSERVATIONS

GEORGE P. PETROPOULOS, KIRIL MANEVSKI, AND TOBY N. CARLSON

15.1 INTRODUCTION

Land cover is a fundamental variable of Earth's system intimately connected with many human activities and the physical environment (Styers et al., 2009; Otukei and Blascke, 2010). At local and regional scales, knowledge of land cover forms a basic dimension of recourses available to any political unit (Kavtzoglou and Colkesen, 2009). At a larger scale, land cover information is of key importance in delineating the broad patterns of climate and vegetation that form the environmental context for human activities. Information on land cover distribution is useful in policy decision making, such as concerning environmentally or ecologically protected areas or native habitat mapping and restoration (Fassnacht et al., 2006; Sanchez-Hernandez et al., 2007). Last but not least, thematic maps of land cover are also linked to the monitoring of desertification and land degradation, key environmental parameters pronounced in many areas on Earth (Castillejo-Gonzalez et al., 2009).

Field spectroradiometry investigates the interrelationships between the spectral characteristics of a target, its biophysical attributes, and the field environment in a spectrally contiguous manner of the reflective wavelengths (Milton et al., 2009). As such, field spectroradiometry "mimics" hidden spectral information from higher scales such as air- or spaceborne platforms. More detailed spectral characteristics are represented by the reflectance–absorption–transmission properties of the target

Scale Issues in Remote Sensing, First Edition. Edited by Qihao Weng.
© 2014 John Wiley & Sons, Inc. Published 2014 by John Wiley & Sons, Inc.

recorded on the ground. Specifically, field spectroradiometry deals with analysis of the position of specific reflectance features, shape of the spectrum, spectral variability, and similarity (Rao et al., 2007). These properties have naturally high degree of variation for most of the land covers on Earth. As a result, field spectroradiometry is capable of producing a "spectral fingerprint" for such variations of many land cover types (McCoy, 2004).

On the other hand, remote sensing has been an attractive source in the determination of land cover spatial distribution, providing valuable information for delineating the extent of land cover classes as well as for performing temporal land cover change analysis at various scales (Kavtzoglou and Colkesen, 2009). The general circumstances that make it attractive for this purpose include its capability to provide inexpensive and repetitive data over large regions, even for inaccessible locations, and at a wide range of spatial and temporal scales. Remote sensing data in land cover classification mapping started to be routinely applied from the late 1960s (Petropoulos et al., 2012a). Since then, a wide range of spaceborne multispectral systems have been placed in orbit. Multispectral imagery from either airborne or satellite sensors has been widely used to map broad vegetation groups or land cover classes. Nevertheless, those systems, although able to measure at very high spatial resolution, record information in a small number of distinct spectral bands, thereby providing obviously limited spectral information content on characteristics related to Earth's surface objects, including land cover.

Recent advances in remote sensor technology have led to the launch of hyperspectral systems. Similar to field spectroradiometry, those are able to record reflected light from land surface objects in numerous narrow, virtually continuous spectral bands from the visible to the short-wave infrared parts of the electromagnetic spectrum, thereby acquiring vast amounts of spectral information observed from higher altitudes. The potential of hyperspectral remote sensing imagery to improve discrimination among similar land cover classes, especially in diversely covered and landscape-fragmented regions on Earth, in comparison with traditional multispectral images, has already been highlighted by many investigators (e.g., Rao et al., 2007; Cho et al., 2009; Zomer et al., 2009; Petropoulos et al., 2011, 2012a,b; Elatawneh et al., 2012). Furthermore, as the challenge in remote sensing is concerned with identifying complex surface features from aircraft and spaceborne imagery, field spectroradiometry emerges as a highly suitable means for such investigations. Indeed, recent studies have clearly demonstrated the contribution of field spectroradiometry and in situ hyperspectral libraries as a means for achieving accurate discrimination of vegetation cover on a species level. The use of hyperspectral data on both ground and space scales has already been presented as a relative success in mapping vegetation cover at species level (Cho et al., 2009; Mathur et al., 2002; Price, 1994; Thenkabail et al., 2000, 2004a; Zomer et al., 2009).

An identical radiometric principle is followed in recording data from spaceborne and field scale observations. However, spaceborne-level studies that investigate land cover with the support of field spectroradiometry require careful consideration of the factors arising from the different scales. At the spaceborne scale, the radiance received at the sensor accounts for the effects of absorption and scattering of Earth's atmosphere. The physical interaction of the sunlight with the gases and particles in the atmosphere with the land cover surface and its transmission along a different path upward through the atmosphere affect the data recorded by the sensor in Earth's orbit (Adler-Golden et al., 1999; Gao et al., 2009; Song et al., 2001). On the other, the instrumentation set-up, the local environment, and the target properties relative to the sun's illumination significantly interfere with the energy measured by the ground sensor. In order to normalize the remote measurement discrepancy, these factors have to be accounted for, suppressed, or standardized prior further land cover analysis. This is of special importance considering the fact that many published remote sensing studies strongly emphasize the image processing techniques while little attention is given to the methods used for collecting field spectral data and information on the field campaign.

The present chapter aims at discussing the contribution of field spectroradiometry and hyperspectral remote sensing in extracting information related to land cover mapping. The chapter starts by providing an overview on the use of field spectroradiometry in examining the spectral discrimination between different land cover targets. Within this framework, the main factors affecting field spectroradiometric measurements are discussed. Subsequently, the recent developments in spectral libraries from field spectroradiometric measurements are touched upon. Following this, an overview on the main statistical approaches employed in spectral separation of different land cover targets is provided, linked with the most important scale factors. In this framework, a critical review and some examples of recent related studies and spectral libraries are furnished as well. The second part of the chapter is focused on the use of hyperspectral remote sensing imagery from airborne or satellite platforms for obtaining regional estimates of land use/cover. An overview of the different techniques employed in mapping land cover types from such data is first presented with selective examples from case studies. Next, the importance of observation scale in land cover extraction using remote sensing data is discussed. Finally, the main conclusions are drawn and the challenges toward a more precise estimation of land use/cover from field spectroradiometry and hyperspectral remote sensing with respect to spectral information acquired at different spatial scales are highlighted.

15.2   FIELD SPECTRORADIOMETRY

The science of spectroradiometry began in the nineteenth century with the use of spectrometry (the study of human vision only) and later with the interpretation of aerial photography (Schaepman, 2007). Yet, the real scientific involvement of spectroradiometry started with the development of airborne multispectral sensors in the middle of the twentieth century (Short, 2009). Ever since, field spectroradiometry is an integral part of the remote sensing science since both use the sun's radiation as the primary light source. The key variable in spectroradiometry is the spectral reflectance. This is emphasized as opposed to imaging spectroscopy, which primarily analyzes the causes and exhibition of spectral absorbance from imagery and scattering processes that occur when light strikes the target (Milton et al., 2009). The spectral reflectance is not a natural characteristic of the materials, but it is rather derived as "spectral reflectance factor" $R(\lambda)$, a ratio of the radiance reflected by a surface to that reflected into the same beam geometry by an ideal (lossless) and diffuse (isotropic, equally in all directions) standard surface irradiated under the same conditions (Milton et al., 2009). It is dimensionless and ranges from 0 to 1 but may reach values beyond 1, especially for highly reflective surfaces, for instance, snow. In addition, the spectral reflectance depends on the wavelength of the incident radiation, the properties of the target, and the incident radiation angle.

The bulk solar radiation is concentrated at wavelengths of 100–2500 nm, with high irradiance intensity in the visible spectrum (VIS) from 400 to 700 nm, which declines toward the short-wave infrared (SWIR) from 1300 to 2500 nm. That is because of the complex absorption and light scattering reactions between the solar energy and the atmosphere constituent gases, primarily H_2O, CO_2, O_2, O_3, CH_4, and N_2O (Avery and Berlin, 1992). Therefore, atmospheric gases make certain wavelengths more selective transmission bands (called atmospheric windows) and absorption bands (called atmospheric blinds) to the incoming solar radiation. The water vapor forms the major atmospheric blinds (Price, 1998) by strong energy absorption in the near infrared (NIR) from 700 to 1300 nm through the SWIR from 1300 to 2500 nm, with broad regions of complete absorption near 1450, 1900, and 2400 nm. The other atmospheric gases display less strong absorption bands.

Compared to land cover types, such as water or soil, vegetation reflectance is the most inconsistent in shape and generally looks similar, irrespective of vegetation type and health status (Hadjimitsis et al., 2009). The typical spectral signature of vegetation in the VIS spectrum is composed of the maximum reflectance (minimum absorption) at 550 nm and increased absorption in the red and blue portions of the VIS due to the selective reflection/absorption properties of chlorophyll and the other auxiliary pigments such as carotenoids in the leaves. Further, the points of inflection at about 600 and 630 nm and minimum reflectance (maximum absorption) in the neighborhood of 680 nm are due to the presence of specific accessory pigments, while beyond 680 nm of the electromagnetic spectrum vegetation becomes highly reflective. The red edge (680–800 nm) is the region of rapid change in reflectance before the NIR. Maximum reflectance is reached in the NIR plateau from 800 to 1300 nm due to the multiple scattered reflectances in the air spaces—in both the mesophyll cells of the leaves and the canopy. The SWIR reflectance is characterized by two major water vapor absorption zones: 1450–1530 nm and 1900–2000 nm.

Finally, the natural anisotropy (direction dependency) of the reflectance introduces the bidirectional reflectance distribution function (BRDF) of each target (Schaaf, 2009). It is a four-dimensional function with unit-inverse steradian (sr^{-1}) which describes the change of the reflectance with the solar position geometry (Brennan and Bandeen, 1970). The underlying concept for characterization of the BRDF and the directional issues of radiation and reflectance are discussed in Nicodemus et al. (1977). The consideration of BRDF in the field spectroradiometry is discussed later in this section.

Field spectroradiometry has progressively evolved in both design and mobility aspects. Modern measurement devices—called spectroradiometers—are nowadays available, measuring spectra in a wide spectral range (300–2500 nm) with high precision and accuracy and are portable and easy to use (Milton et al., 2009). In addition, mobile platforms such as aerial lift trucks that carry spectroradiometers above a tree canopy are exploited, for example, MUFSPEM@MED from the Mediterranean Agronomic Institute of Chania (Manevski et al., 2011, 2012), preceded by MUFSPEM from the Technical University of Munich, or FIGOS from the Remote Sensing Laboratories at the University of Zurich (Schopfer et al., 2008). Such technologies provide repeatable spatial sampling with negligible vegetation disturbance. Nevertheless, at the field scale, reaching the top of high land cover targets such as trees is a challenging task, as not just the instrument has to be suspended and operated on a height, but factors that influence the field measurements also have to be simultaneously taken into consideration. Moreover, the definition of the spectral signatures has proven difficult due to the variable nature of both the material and the local environment (e.g., atmosphere, moisture content, or illumination) (Cochrane, 2000; Pfitzner et al., 2006).

The rest of this section is focused on field spectroradiometry. First, the main factors affecting field spectroradiometric measurements are discussed. Then, the development of spectral libraries and the main types of those libraries depending on the spectra preprocessing performed are outlined. Finally, an overview of the statistical approaches employed for vegetation species discrimination from field spectroradiometry data is provided, including relevant case studies.

15.2.1 Factors Affecting Field Spectroradiometric Measurements

15.2.1.1 Instrumentation Factors Consideration of the instrumentation factors consists of the evaluation of the measurement equipment which includes processes describing the spectroradiometric behavior with respect to the environment and its calibration. However, the inherent assumption of the illumination characteristics of the calibration and the target will always introduce certain error. Therefore, calibration is used to offset as much as possible from the target the changes in reflectance due to variations in the sun's natural illumination, the atmospheric influences, and the scattering effects between the measurements. It is usually done by the use of a surface with a known reflectance of almost 100% (white panel) which has diffuse reflectance properties, thus maintaining constant contrast over a wide range of lighting conditions. The most common is the Spectralon® material, a white reference reflector with a nominal 99% reflectance manufactured by Labsphere (North Sutton, NH). Periodical reflectance measurements of the white panel on the field can give direct indication of how "perfect" a diffuse reflector is during a field campaign. Conditions of high stability and similarity in its spectral responses as well as low variation of the absolute spectra of the measured targets during a field campaign allow the influence of the fluctuations caused by the clear-sky weather to be assumed as negligible (Manevski et al., 2012).

Before collecting spectra, the instrument has to be warmed up for at least 30 min so as to reduce the sensitivity drift (often called a "step"). The latter is a sudden change in the reflectance curve of a target in a wavelength where the strands in the fiber-optic cable of the spectroradiometer overlap. The sensitivity drift can be corrected directly on the field by holding the fiber-optic cable away from the target and allowing the field of views (FoVs) of the strands to overlap [Analytical Spectral Devices (ASD), 2009]. However, sensitivity drift is often saved in the spectral data and it has to be corrected (Beal and Eamon, 1999; Salisbury, 1998).

Implementation of the spectrum averaging additionally ensures sufficient quantity of usable information obtained expressed as the relative amount of noise contained in the signal (i.e., signal-to-noise ratio, SNR). The higher the number of spectra averaged, the higher the SNR value, and consequently the higher the amount of usable information (ASD, 2009; Milton et al., 2009).

15.2.1.2 Environmental Factors   The environmental factors that affect the reflectance measurements can be divided into atmospheric gases and illumination. As outlined earlier, the complex reactions and absorption of the solar energy by the atmosphere constituent gases, such as H_2O, O_2, O_3, CH_4, and NO, cause rapid energy drop at longer wavelengths, especially in the SWIR spectrum. The spectroradiometer production companies compensate for this natural effect by reduction of the spectral resolution of the spectroradiometer at NIR and SWIR wavelengths (Salisbury, 1998). Most remote sensing instruments operate in atmospheric windows by tuning the detectors to specific frequencies that pass through the atmosphere with minimum disturbances, that is, random noise. Such noise is in particular induced by the atmospheric water vapor which exhibits strong absorption of the incident and reflected radiation by water molecules. The optical spectrum is thus commonly partitioned into four distinct wavelength regions: VIS, NIR, and SWIR-1 (1300–1900 nm) and SWIR-2 (1900–2500 nm) (Figure 15.1).

In order to decrease the complexity that arises from the sun angle, the sun azimuth-view angle relationship (the BRDF), spectral measurements on a field scale are often assumed as isotropic, and the smallest nadir position of the sensor above the target with the sun in its zenith is commonly used (Manakos et al., 2010; Pfitzner et al., 2006). The parameters in the BRDF equation that describe the relative position of the sensor to the sun are omitted. Thus, the sensor view on the ground FoV and the orientation of the sun azimuth relative to any preferred orientations of the target are the only significant geometrics. Since the FoV of high-altitude sensors is usually at nadir, with a predetermined flight path, there is very little that can be done to accommodate for the vegetation land cover variability that might affect the BRDF response on the field. Spectral measurements are performed in close temporal proximity between solar noon at high sun angle between 11:00 AM and 2:00 PM, with sun height variation of less than 30°. At higher latitudes, directional measurements with off-nadir angles and wider FoV are implemented (Eklundh et al., 2011). Measurements should be performed in such a way in terms of sensor position that the BRDF variations are assumed as negligible. Additional minor variations can be further minimized with spectra averaging (Manevski et al., 2012).

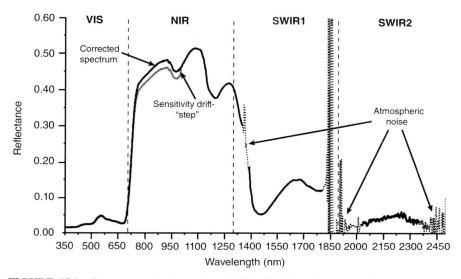

FIGURE 15.1 Four spectral regions from VIS to SWIR and preprocessing of vegetation reflectance spectrum. The sensitivity drift effect from 350 to 1000 nm (gray line) is corrected (full line). The spectral regions around 1400, 1950, and 2400 nm (dashed line) are removed due to excessive atmospheric noise.

It should be emphasized that, even if solar elevation angle and azimuth do not vary during routine data collection on the field, they should be taken into consideration when data acquired at other times are used, especially when used on satellite data for calibration and/or validation. Atmospheric corrections for BRDF are necessary for highly elevated sensors that view very wide (e.g., 100-km) areas on the ground FoV from a given altitude, as the direction-dependent effects are more expressed, rather than for small FoV sensors or field spectral measurements.

Other illumination factors such as skylight and background radiance from the surroundings, compared to direct solar illumination and skylight, are typically small and insignificant sources of error, except if a large background object or something close enough to the target covers a large portion of the solid angle viewed by the target. Clouds, shades, and wind attenuate the solar irradiance and the target's reflectance (Chang et al., 2005). Therefore, clouds should be avoided as much as possible, while the wind affects the spectral signature of the target as it moves the canopy and changes its shadow. It is nevertheless minimized by the fast scanning time (100 ms for a complete 350–2500-nm spectrum) of the spectroradiometers nowadays (ASD, 2009). In addition, replicate measurements are essential to find the average reflectance value for a specific target in the field in order to minimize both the wind background radiance synergetic effect and instrumentation factors described previously (Hemmer and Westphal, 2000).

15.2.1.3 Measurement Errors

15.2.1.3 Measurement Errors   Measurement errors are typically present in every field work and should be accounted for in the spectral characterization procedure. The most important process includes obtaining as pure as possible signals from the target, which can contain a mixture of scene elements if proper height adjustments of the FoV are not performed. The FoV should be broad enough to cover a homogeneous species composition. At the same time, the FoV should minimize the contribution of the underlying soil to the spectral response of the target or should represent a pixel on the ground (Ben-Dor et al., 2009). Multiple relative reflectance readings on different positions in close proximity over the canopy can also be recorded, following a white panel reading, in order to account for the variability contained in a single pixel of the currently offered image products (Manakos et al., 2010; Manevski et al., 2012).

Emphasis is also placed on the collection of metadata that describe all the instrumentation and environmental factors. Such data provide an evaluation of the suitability of the collected spectral data sets for potentially new applications in the future (Hueni et al., 2009). Digital photos of the vegetation offer benefits in the interpretation stage, such as screening the vegetation state and composition (Milton et al., 2006, 2009; Zomer et al., 1999). Details regarding the measurement process also have to be documented in specific measurement field logs (Salisbury, 1998). If platforms are reflective, a common practice is the application of low reflective materials to the wider spectral range possible on it to suppress the scattered light (Zomer et al., 2009). The self-shadow effect is usually prevented by positioning the platform and the sensor toward the south for the North Hemisphere and vice versa for the South Hemisphere.

15.2.2 Developing Field Spectral Library

At the field scale, fundamental to the spectral discrimination between land cover is the success in extracting pure spectra of the targets. That can be achieved if parameters that affect the measurement environment are considered, including the time of the measurement which reflects the phenological status of vegetation land cover. Such information, organized and stored together with the spectral signatures, is called a spectral library (Hueni et al., 2009). The field scale spectral signatures are related to different geo–bio–chemo–physical parameters of the land cover analyzed, such as the soil or understory vegetation cover in the background surface, the phenology of the vegetation, shades, pigments, and water content (Williams, 1991). A special difficulty is that land cover parameters vary in time and space and are difficult to be controlled on the field (Chang et al., 2005; Schaaf, 2009). That increases the variability of the field spectra reflectance data and additionally contributes to the difficulties in transferability of the spectral libraries and the spectral discrimination of different land cover. Therefore, it is understandable that a commonly accepted and clear methodology for construction of a spectral library is a challenging task and of crucial importance for ensuring adequate data quality and relevance to the application considered each time (Milton et al., 2009; Salvaggio et al., 2005). Most efforts are concentrated on obtaining unaltered field reflectance spectra in order to better characterize the spectral signatures of diferent land cover. The most common types of spectral libraries recorded at the field scale are presented next, providing also reference to examples of spectral libraries available today.

15.2.2.1 Unaltered Reflectance Spectral Library In case no quantitative manip-
ulations of the estimates of the field-sensed data have been performed, the spectral
reflectance is unaltered, that is, "instrument-produced" reflectance. Unaltered reflectance of vegetation land cover has been widely studied because it is commonly viewed
as an indicator of the properties of the landscape and the interaction between the local
climate, rocks and soil, landforms, fauna, water and humans, and landscape factors
integrated within digital imagery (Schmidt and Skidmore, 2003). This implies that the
value of the unaltered spectral libraries is taken into account in the above-mentioned
variation sources inherent in the pixel data sensed by the air- and spaceborne scale
remote sensing systems (Manevski et al., 2012). Therefore, unaltered spectral libraries
are most often used as sets of reference spectra to delineate different land covers and
mixed communities within a certain spatial extent. Apart from the effect of canopy
structure and soil background from the canopy spectra, other factors that can cause
vegetation to show variability in unaltered spectral reflectance from one target to
another include the leaf shape and size, the canopy architecture and density, its
internal water content, as well as the soil type beneath the vegetation cover
(Price, 1994).

15.2.2.2 Continuum-Removed Spectral Library Continuum removal is a mathematical manipulation of the unaltered spectral reflectance with the use of a continuum
line fitted over the top of the spectral signature local maxima that are connected using
straight-line segments according to the equation

$$R_{(\lambda)cr} = \frac{R_{(\lambda)}}{C_{(\lambda)}} \tag{15.1}$$

where

$R_{(\lambda)cr}$ = continuum-removed spectral reflectance
$R_{(\lambda)}$ = unaltered (raw) spectral reflectance
$C_{(\lambda)}$ = continuum line spectral value

Since the local maxima spectral data form the continuum line segments and are
thus on the continuum line, their values in the output continuum-removed data are
equal to 1.0 (Figure 15.2). The continuum-removed spectral libraries are used
primarily in geological spectral subsets to isolate absorption pits and falls for further
spectral analysis, which would otherwise be difficult to detect on unaltered reflectance. The continuum removal applied on vegetation spectra eliminates the canopy
variations induced by soil surface, moisture content, and canopy structure. This
technique has been therefore used as a method to standardize vegetation land cover
field spectral libraries obtained at different measurement setups at the field scale and to
test if the spectral discrimination between different vegetation is improved (Psomas
et al., 2005; Manevski et al., 2011). It should be borne in mind that variations induced
by specific canopy characteristics such as different amounts of dry matter or
mesophyll structure can be important in the spectral discrimination of different
vegetation covers. Moreover, applying the continuum removal on a wide spectral

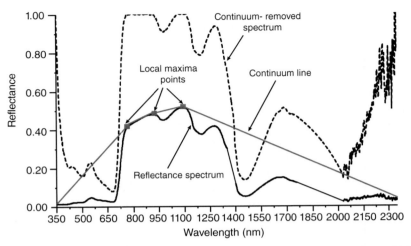

FIGURE 15.2 Vegetation reflectance spectrum (full line), its continuum (gray line) with local maxima (gray triangles), and the continuum-removed reflectance spectrum (dashed line).

range such as the one commonly used in remote sensing from 350 to 2500 nm might not provide optimal vegetation classification. This is because slightly low local reflectance maxima do not fall on the continuum curve and thus potential spectral absorption differences are avoided (Figure 15.2). In addition, placing a peak in a more noisy or poorly defined spectral signature on the continuum curve can result in artificial representation of the absorption features. Some algorithms (Exelis, 2009) are able to define local maxima in order to increase the likelihood of identifying real absorption features.

15.2.2.3 First- and Second-Derivative Spectral Library

Overall similar to the continuum removal, derivatives are applied on an unaltered spectral library to eliminate the interference from changes in illumination or soil background in the remote sensing of vegetation or for resolving complex spectra of several target species within individual pixels on digital imagery (Asner et al., 2000). Two derivatives of the unaltered spectral reflectance are commonly used. The first derivative is calculated for a wavelength intermediate between two wavelengths, while the second-derivative determination involves three closely spaced wavelength reflectance values. First and second derivatives emphasize sudden spectrum changes, for example, from a negative slope to a positive slope in the case of the red edge feature of vegetation reflectance spectrum (Figure 15.3). Such distinguishable features are especially useful for discriminating between peaks of overlapping bands (Morrey, 1968). In addition, derivative indices are recently being used for analyzing absorption of different vegetation at the leaf scale or for monitoring vegetation status over different soil backgrounds (Frank and Menz, 2003). Derivative spectral indices turn out to better describe vegetation cover status such as chlorosis, compared to the conventional broad-band indices such as the near-infrared/red reflectance ratio, due to their

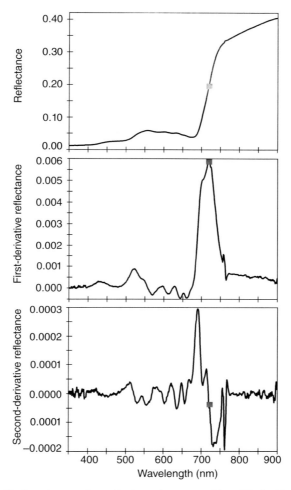

FIGURE 15.3   Unaltered vegetation reflectance (upper graph) and its first (middle graph) and second (lower graph) derivatives in 350–900-nm spectrum. The gray edge has an inflection point (point where where the slope is maximum) at 730 nm. This corresponds to the maximum in its first derivative and to local minimum in the second derivative (gray squares).

sensitivity to leaf cover and color (Kochubey and Kazantsev, 2007). Well-described details on derivatives of spectral reflectance can be found in, for example, Morrey (1968).

15.2.2.4   Spectral Libraries Available from Field Spectra Data Sets   A spectral library is designed to hold pure spectra information of various land cover types in order to assist spectral discrimination at the field scale and further large-scale imagery mapping. In addition, the possibility of addressing the issue of mixed pixels, encountered in many medium- to high-resolution sensors, is one of the main tasks of the spectral libraries. Efforts are therefore being made toward the creation of extensive spectral libraries of land cover for different environments (wetlands, deserts, etc.) and land cover types (vegetation, soils, minerals) and across a broad range of bioclimatic, edaphic, and disturbance conditions (Hueni et al., 2009; Milton et al., 2009).

A few publicly available spectral libraries exist and are available today, such as the recent SPECCHIO online spectral database system from the Remote Sensing Laboratories (RSL) of the University of Zurich (http://www.specchio.ch/). The latter includes rich metadata sets enclosed within the reference spectra and spectral campaign data in order to ensure longevity and shareablity of spectral data between research groups (Hueni et al., 2009).

The U.S. Geological Survey (USGS) offers a comprehensive spectral library (http://speclab.cr.usgs.gov/) and integrates the ASTER spectral library, the Johns Hopkins University (JHU) Spectral Library, and the Jet Propulsion Laboratory (JPL) Spectral Library (Baldridge et al., 2009). This library offers a comprehensive choice of land cover such as vegetation and man-made materials, measured with different spectrometers, such as the Beckman 5270 (200–3000 nm), the ASD portable field spectrometer (350–2500 nm), the Nicolet Fourier transform infrared (FTIR) interferometer spectrometer (1300–15000 nm), and the NASA Airborne Visible/Infra-Red Imaging Spectrometer (AVIRIS) (400–2500 nm).

Other small-scale spectral libraries are also developed, such as the Vegetation Spectral Library (VSL) of the Systems Ecology Laboratory at the University of Texas at El Paso (UTEP) in cooperation with colleagues at the University of Alberta (http://spectrallibrary.utep.edu/). Metadata regarding type of target and measurement site and images are also enclosed. The spectral library is open to contributions of new users to upload their spectral libraries toward building up an extensive knowledge base worldwide.

15.2.3   Statistical Approaches in Field Spectroradiometry for Vegetation Discrimination

15.2.3.1 Basic Assumptions   The natural variation of vegetation reflectance results in its distribution, which usually does not cluster around the mean in a bell shape if graphically presented; that is, it does not have a normal, or Gaussian, distribution (Manevski et al., 2012). Other studies report normality in the spectral reflectance data distribution (Mutanga and Skidmore, 2007; Vaiphasa et al., 2005). As a consequence, statistical techniques coupled with such data can be grouped into parametric and nonparametric tests. Parametric tests assume normal data distribution and homogeneous variance as a measure of the amount of variation in the data being compared. They are considered robust because violations of the assumptions, especially of the data distribution, do not significantly affect the tests (Robson, 1994). On the other hand, when the distribution is obviously not bell shaped, neither is the variance homogeneous, alternative approaches, such as nonparametric techniques, are recommended. In both cases, the resulting p-value demonstrates the probability of making an error in conluding a difference between the groups being compared when none really exists. Methods have their prerequisites, strengths, and weaknesses, but when coupled on a certain spectral range, they should reduce the number of wavelengths to the most relevant for spectral discrimination of land covers compared without the loss of important information related to the study objectives (Ben-Dor et al., 2009; Thenkabail et al., 2004a,b). Decreasing the dimensionality of both field and satellite hyperspectral data is of special importance and it still remains a challenge. Statistical approaches include analyses of variance (Adam and Mutanga, 2009; Manevski et al., 2011, 2012; Vaiphasa et al., 2005), correlation analysis (Mariotti et al., 1996), linear discriminant analysis (Abdel-Rahman et al., 2010; Clark et al., 2005), and canonical discriminant analysis (van Aardt and Wynne, 2001). Based on the number of variables used in the explanation of certain phenomena, the statistical methods can be grouped as univariate and multivariate techniques. Those are elaborated on next.

15.2.3.2 Univariate Statistical Techniques
Statistical techniques that utilize only one dependent variable, that is, factor (e.g., vegetation type or plant species), in explanations of one or more independent variables (e.g., reflectance) are referred to as univariate statistical techniques. The most common is the single-factor analysis of variance (commonly known as one-way ANOVA), a parametric test that works well even if the distribution is approximately Gaussian, especially when large samples (usually more than 20) are used. The concept of ANOVA in the determination of spectral regions where different land covers are most likely to be discriminated lies in maximizing the spectral variability between the covers, at the same time minimizing the spectral variability within them, for every wavelength in a certain spectral domain. ANOVA explores the variability between the groups as a deviation of each group's mean from the "grand mean"— the mean of the means of all groups. Details on the computation of variance and ANOVA can be found in traditional statistics textbooks (Robson, 1994).

Similar univariate parametric statistical vegetation discrimination methods based on the parametric ANOVA include t-tests (Jacobsen et al., 1995; Vaiphasa, 2006) or Duncan's multiple-range test (Jurado-Exposito et al., 2003). Even though statistical tests such as ANOVA are practical data exploration tools used to find spectral windows for statistical discrimination of different vegetation types, the results may still not be independently interpreted without additional data analyses. This is because the possibility of making a type I error, that is, rejection of the null hypothesis when it is actually true, is increased due to the high number of variables, that is, wavelengths (Vaiphasa et al., 2005).

An alternative approach that makes no assumptions about population distribution is a nonparametric test. For example, one or more reflectance values are off scale, that is, too high or too low, which is often true. This could cause variance homogeneity— the most important requirement in parametric ANOVA—to not be achieved. Although not completely free from the assumption of homogeneity of variance, nonparametric analysis is an alternative which ranks the reflectance values from low to high and the variance analysis is based on the distribution of the ranks (Artigas and Yang, 2006). To this end, the Kruskal–Wallis nonparametric ANOVA compares the medians of reflectance spectra from the land cover sensed in a similar manner to the parametric one in order to determine the similarity and their usefulness in potential mapping. The Mann–Whitney U-test is a common nonparametric test used for further pairwise comparisons in detailed discriminative spectral analysis.

15.2.3.3 Multivariate Statistical Techniques

In contrast to the univariate, the multivariate statistical techniques provide an optimal linear combination of more dependent variables that satisfy specific statistical criteria as an explanation of more independent variables. Consequently, such techniques are used as both dimension reduction and explanatory tools.

Principal-component analysis (PCA) is a nonparametric multivariate data analysis technique that takes a cloud of data points analyzed and rotates it such that the maximum variability is visible. The outputs are uncorrelated explanatory variables, as linear combinations of the original variables, called principal components, and usually the first principal component accounts for the maximum variability in the data, and each following component accounts for as much of the remaining variability as possible (Castro-Esau et al., 2004).

Canonical discriminant analysis is another dimension–reduction multivariate technique which explains which variables (here wavelengths) discriminate the best between classes (vegetation groups). It summarizes the between-class variation in much the same way that PCA summarizes the total variation (Dimitrakopoulos, 2001). Cluster analysis, or clustering, is the procedure of an assignment of a set of observations into subsets (called clusters) so that observations in the same cluster are similar in some sense, that is, in a way that the degree of association between two objects is maximal if they belong to the same group and minimal otherwise. As given above, cluster analysis can be used to determine how the data are organized without providing an explanation or interpretation, that is, it is an unsupervised approach (Holden and LeDrew, 1998).

Other statistical techniques and quantitative methods coupled on the spectral data obtained at the field scale attempt not only to determine spectral regions where different covers are most likely to be statistically discriminated but also to measure how well different covers can be separated. Such separability measures look at either the distance between the class means (e.g., the Euclidean distance) or both the differences between the class means and the distribution of the values around those means (e.g., Jeffreys–Matusita or Bhattacharyya distance) at one or at more wavelengths at a time.

15.2.3.4 Field Spectroradiometry Applications Related to Land Cover Mapping

Hyperspectral discrimination of different land covers at the field scale is a topic that has been the subject of research over the last two decades (Clark et al., 2005; Cochrane, 2000; Lewis, 2001; Manevski et al., 2011, 2012; van Aardt and Wynne, 2001). On the one hand, laboratory spectroradiometry omits the natural field conditions, such as the variation of the sun's energy, or the effect of canopy architecture, just to mention few, thus resulting in investigation of prerequisites, rather than applicability to the future investment of hyperspectral sensors onboard (Vaiphasa et al., 2005). On the other hand, field spectroradiometry studies deal with the time–space variations of the factors affecting the measurements as detected naturally on the field. In addition, even though variability is high, the level of accuracy of the methods in collecting spectral data of different land covers on the field is at an acceptable level (Van Aardt and Wynne, 2007). Statistical techniques and quantitative methods applied on the field hyperspectral data prior to further digital image analysis aim to solve the variability between different land cover targets, which, in the case of vegetation cover, is often very low due to the similar canopy architecture, color, or physicochemical properties (Williams, 1991; Zang et al., 2008).

One of the main applications of field spectroradiometry is in investigating the potential means of mapping vegetation physiognomic types along different environmental or phenological gradients. For instance, Artigas and Yang (2006) analyzed dominant saltmarsh species in New Jersey meadows in the fall using the VNIR spectrum from 350 to 950 nm with the Mann–Whitney U-test at $p < 0.01$ and suggested the orange and red segments of the VIS and few bands in the NIR as unique for discrimination. Psomas et al. (2005) have used the same statistical test for comparison between the spectral signatures of four dry-mesic grassland types in Switzerland. The statistically significant seasonal variation between the spectral reflectance of the vegetation demonstrated the potential of using spectral information for discriminating different grassland type changes during one growing period on satellite imagery. Manevski et al. (2012), using the Mann–Whitney U-test at $p < 0.01$, have recently found the distinctive phenology of three Mediterranean trees and shrubs and the seasonal effect of rapid flowering in spring as important factors determining the spectral discrimination at the field scale. That vegetation cover spectral variability is generally not parametric deserves more attention. Even if normality is assumed when a large number of sampled spectra are available (the central limit theorem), such analysis in addition requires the assumption of equal reflectance variances to be met between the targets being compared. This is something which should be considered (Figure 15.4) but is often not discussed or it is ignored.

Field spectroradiometry has also been used for estimating and mapping vegetation quality in terms of macroelements such as nitrogen. For example, Mutanga et al. (2003) investigated the discrimination of the tropical pasture grass *Cenchurus ciliaris* under different nitrogen treatments at the canopy level. The results from the parametric ANOVA tests at $p < 0.001$ with a 95% confidence level (CL) coupled on both unaltered and continuum-removed spectra showed significant differences between the nitrogen treatments, demonstrating the possibility to map variation in pasture quality using hyperspectral remote sensing.

It is well known that the high-dimensional complexity of hyperspectral data imposes some issues in terms of image processing algorithms, extensive field campaigns and high cost. Therefore, identification of the optimal bands is required for discriminating and mapping vegetation cover without losing important information. For that purpose, Adam and Mutanga (2009) have proposed a hierarchical method which initially involves one-way ANOVA as an attempt to spectrally discriminate papyrus (*Cyperus papyrus* L.) from three other species within the 300–2500-nm spectral region, measured in summer in swamp wetlands in South Africa. Their results demonstrated that papyrus was significantly different then its

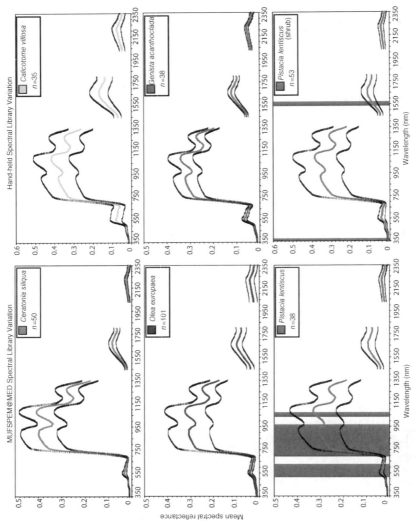

FIGURE 15.4 Mean reflectance spectrum of common Mediterranean vegetation obtained with aeral lift mobile platform MUFSPEM@MED (left) and hand held (right). Spectra are flanked by standard deviation of mean (black lines) at 95% confidence level. Wavelengths where the variance between three plants compared in each spectral library was equal are shaded gray on the lower graphs. Spectral regions around 1400, 1940, and 2400 nm were removed due to atmospheric noise. Figure adopted from Manevski et al. (2012). (See the color version of this figure in Color Plates section.)

coexisting species in 412 wavelengths at $p < 0.05$ (95% CL), with maximized discrimination in the red edge (741–746 nm) and the NIR (982–1297 nm). Similarly, Lacar et al., (2001) explored the differences between four grape vine (*Vitis vinifera*) varieties, that is, cabernet sauvignon, merlot, semillon, and shiraz, and their pairs, in a southern Australian vineyard. The authors found ANOVA coupled with the Tukey posthoc test [National Institute of Standards and Technology (NIST), 2003] powerful enough to identify the red edge ($\sim$720 nm) followed by the green reflectance peak and its wings in the VIS where the field reflectance spectra showed the greatest statistically significant difference.

Post hoc tests are designed for situations in which significant test results have already been obtained and additional exploration is needed on which means are significantly different from each other. A post hoc test would control the type of error for all possible comparisons between the groups being analyzed. Manevski et al. (2011) have used the Bonferroni post hoc test (Abdi, 2007) at both $p < 0.05$ and $p < 0.01$ to study the spectral discrimination of common Mediterranean tree and shrub land cover at the field scale. The Bonferroni test is conservative and strict when the number of comparisons exceeds the number of degrees of freedom between groups. However, the type II error rates are high for individual tests, that is, the Bonferroni overcorrects for type I error. Though significant spectral differences were concluded, significant data reduction was not achieved. This might be due to the use of the continuum-removed reflectance, which enhanced absorption pits and reflection peaks, yielding significantly high number of wavelengths where the reflectances of the plants are statistically different.

While originally used in spectroscopy of rocks and minerals, the continuum removal has been recently used in field spectroradiometry of vegetation cover. Schmidt and Skidmore (2003) attempted to discriminate 27 different Dutch saltmarsh vegetation types measured in the 400–2500-nm spectral region in summer using the Mann–Whitney U-test at $p < 0.01$. Despite the finding that for some of the vegetation species there is no single band that is significantly different to all other vegetation communities, still the authors found the approach robust enough to determine the NIR plateau around 771 nm as an important spectral region for vegetation discrimination. The red edge at 707 nm was reported as having the lowest frequency of statistically significant different medians of reflectance of vegetation types.

Finally, the synergistic use of field spectroradiometry and hyperspectral remote sensing imagery for land cover characterization has not been as extensively explored. Methods such as spectral distance analysis or spectral feature selection algorithms have been applied in some studies for selection of a single band or the "best" band combination for vegetation species discrimination. Yet, one can argue that such methods of dimension reduction objectively evaluate the intrinsic redundancy of the hyperspectral data, taking into consideration the biochemical and physical processes such data mimic in vegetation land cover. Moreover, most of those field scale studies have not been tested on hyperspectral data acquired from airborne sensors, with only few previous studies bearing explicit discrimination (van Aardt and Wynne, 2001, 2007). Continuation of such work has special importance in demonstrating a limited number of wavelength bands that can be used for accurate land cover classification and inventory.

15.3   HYPERSPECTRAL REMOTE SENSING IN LAND COVER EXTRACTION

15.3.1   Introduction

Remote sensing data have been an attractive source in the determination of land cover thematic mapping, providing valuable information in delineating the extent of land cover classes, as well as in performing temporal land cover change analysis at various scales (Kavtzoglou and Colkesen, 2009). Multispectral remote sensing data have been widely used for land use/cover mapping at diverse spatial scales (Cihlar, 2000; Carrao et al., 2008). The multispectral remote sensors produce images in few relatively broad spectral bands for every unit—such as pixel—on digital imagery. The repetitive availability, relatively low cost per unit area, and high spatial resolution (e.g., less than 1 m from IKONOS, QuickBird, and WorldView-2 sensors) make the multispectral information of potentially high interest to scientists and decision makers, thus being exploited for land cover mapping and enhancing change detection applications (Xie et al., 2008). However, the coarse spectral resolution is an important limitation of such data for detailed land cover analysis. Even though the spatial resolution of some multispectral sensor systems is very high, spectral differences of land cover such as vegetation recorded with multispectral sensors are often very small (Avery and Berlin, 1992). Vegetation cover has naturally high spectral variability driven by the bio-chemical complexity, diversity, and phenology. At low canopy cover, where spectral dominance of soil background occurs, analysis of vegetated landscapes becomes even more difficult (Curran, 2001). Therefore, vegetation cover analysis on multispectral imagery is performed at a community level or on land cover classes of broader physical meaning (e.g., deciduous vs. evergreen vegetation) (Thenkabail et al., 2004b).

Unlike the multispectral sensors, hyperspectral remote sensors have been available for about two decades or so. They provide spectral data for each measurement unit in numerous continuous spectral bands. As the very narrow spectral bands characterize the target reflectance better, this abundant spectral information can be used to more effectively detect and identify the variability of different land covers on the basis of the spectral signatures than is possible with broader band multispectral sensors (Abdel-Rahman et al., 2010; Xie et al., 2008). Differences in the spectral, spatial, and temporal characteristics between several multi- and hyperspectral imaging systems are illustrated in Table 15.1.

The remainder of this section provides an overview of the techniques employed in deriving information on the spatial distribution of land cover from hyperspectral remote sensing imagery, providing results from selected case studies conducted recently. In this framework, the importance of the observation scale in land cover extraction using remote sensing data is also discussed. It should be noted that apart from the techniques developed for performing land cover mapping from hyperspectral

TABLE 15.1 Comparison of Spectral Properties and Land Cover Applications of Several Multi- and Hyperspectral Sensors

Sensor	Number of Bands	Range (Spectral Resolution)	Spatial Resolution and Features	Land Cover (LC) Mapping Applications
Multispectral Sensors				
Landsat ETM+	7	Blue: 0.45–0.52 μm Green: 0.52–0.60 μm Red: 0.63–0.69 μm Near IR: 0.76–0.90 μm Mid IR: 1.55–1.75 μm Thermal IR: 10.4–12.5 μm Mid IR: 2.08–2.35 μm	15 m panchromatic; 60 m thermal infrared; 30 m multispectral; single scene is 185 × 185 km; temporal resolution is 16 days; launched 1999 (Landsat 1 in 1972)	Regional scale vegetation mapping, usually at community level. The blue shows the green reflectance of healthy vegetation. Red is important for discriminating among different kinds of vegetation. Near IR is especially responsive to the amount of vegetation biomass present in a scene.
SPOT 5	5	Pan: 0.48–0.71 μm Green: 0.50–0.59 μm Red: 0.61–0.68 μm Near IR: 0.78–0.89 μm Shortwave IR: 1.58–1.75 μm	10 m; single scene is 60 × 60 km; launched 2002 (first SPOT in 1986)	Regional scale usually capable of mapping vegetation at community level or species level or global/national/regional scale mapping LC types (i.e., urban area, classes of vegetation, water area, etc.).
Ikonos	4	Pan 0.45–0.90 μm Blue 0.44–0.51 μm Green 0.51–0.59 μm Red 0.63–0.69 μm Near IR 0.75–0.85 μm	0.82 m panchromatic; 3.2 m multispectral; single scene is 11 × 11 km; temporal resolution is 3–5 days; launched 2000	Local to regional scale LC mapping at higher level (e.g. vegetation species) or used to validate other classification results.

(continued)

Table 15.1 *(Continued)*

Sensor	Number of Bands	Range (Spectral Resolution)	Spatial Resolution and Features	Land Cover (LC) Mapping Applications
Hyperspectral Sensors				
AVIRIS	224	0.4–2.5 μm (10 nm)	Airborne sensor since 1998; depending on satellite platforms and latitude of data collected, spatial resolution ranges from a few to dozens of meters; single scene from several to dozens of kilometers	At local to regional scale usually capable of mapping vegetation at community level or species level. As images are carried out as one-time operations, data are not readily available as they are obtained on an "as needed" basis.
Hyperion	220	0.4–2.5 μm (10 nm)	30 m; data available since 2003	At regional scale capable of mapping vegetation at community level or species level.
CHRIS Proba	up to 63	0.415–1.05 μm (1.25 nm)	15–30 m; data available since 2001	Applications in imaging spectrometry and production of laboratory-like reflectance spectra for each pixel in an image.

a According to Xie et al., 2008.

imagery, various hyperspectral vegetation indices have been proposed for studying specific properties of different targets covering Earth's surface, including soil, water, and vegetation. However, this topic is not covered herein as it can form the topic of another chapter by itself and lies outside the objectives of this chapter. Interested readers are referred to a detailed recent review of hyperspectral indices as regards particular vegetation properties by Thenkabail et al. (2000).

15.3.2 Land Cover Mapping from Hyperspectral Remote Sensing Imagery

The ability of hyperspectral sensors to better discriminate ground cover classes than traditional multispectral sensors has been demonstrated elsewhere (Zhang and Ma, 2009). Nowadays, hyperspectral remote sensing imagery is regarded as one of the most significant Earth observation data sources (Du et al., 2010) and is being used for various applications, including land cover classification (Li and Liu, 2010). Producing land use/cover mapping thematic maps using hyperspectral remote sensing data is commonly performed by digital image classification (Chintan et al., 2004). A comprehensive review of the variety of classification approaches applied to remote sensing data, including hyperspectral sensors, was made available recently by Lu and Weng (2007). Generally, a widely used categorization of classification techniques includes three main groups of approaches: pixel-, subpixel-, and object-based classification. These are briefly reviewed next.

15.3.2.1 *Pixel-Based Techniques* Techniques that belong to this group perform classification by assigning pixels to land cover classes using either supervised or unsupervised classifiers. Unsupervised classifiers group pixels with similar spectral values into unique clusters according to some statistically predefined criteria where the classifier combines and reassigns the spectral clusters into information classes. Supervised classifiers use samples of known identity for each land cover class, known as "training sites," to classify image pixels of unknown identity (Campbell, 1996). Supervised classifiers are also commonly divided into parametric and nonparametric. In comparison to the nonparametric [such as artificial neural networks (ANNs) or support vector machines (SVMs) (Vapnik, 1995)], parametric pixel-based classifiers [e.g., the maximum-likelihood (ML)] (Harris, 1998) require prior knowledge/assumptions regarding the statistical distribution of the data to be classified for the different classes used, information often difficult to attain in practice. A pixel-based classification technique that has gained popularity, particularly when implemented with hyperspectral imagery, is the spectral angle mapper (SAM) (Kruse et al., 1993). This is a supervised classification technique based on the computation of spectral angle similarity between a reference source and the target spectra. The popularity of SAM is due to its simplicity and rapid implementation for the examination of the spectral similarity of image spectra to reference spectra. It is also a very powerful classification method because it suppresses the influence of shading effects to accentuate the target reflectance characteristics (De Carvalho and Meneses, 2000). Despite the potential of pixel-based methods, some of these techniques require making assumptions regarding

the probability distribution of the training data sets (e.g., in ML), which might not always coincide with reality. Other classifiers such as ANNs may require a significant amount of effort in terms of calibration and fine tuning of neural nets before obtaining a satisfactory level of classification accuracy. It is also generally argued that such classification approaches do not make use of the spatial concept of the imagery, such as textural or contextual information present in a remotely sensed imagery (Yan et al., 2006). Finally, an important advantage of both classification techniques implemented herein is that for both their implementation is not confined by the so-called Hughes phenomenon known also as the "curse of dimensionality" (Hughes, 1968). This occurs when the number of training samples is limited compared with the number of input features (and thus of classifier parameters) and results in an accuracy loss as the function of the data dimensionality increases (Dalponte et al., 2009; Zhang and Ma, 2009). Although this challenge does not exist in the classification approaches examined in our study, this phenomenon can be important when parametric classifiers (such as ML) are implemented with hyperspectral imagery.

15.3.2.2 Object-Based Classification   Object-based methods were introduced in the early 1970s (de Kok et al., 1999). They are based on the concept that information necessary to interpret an image is represented not in single pixels but in meaningful image objects. The first step in object-based classification is image segmentation based on which remote sensing imagery is divided into regions where each is homogeneous and no two adjustment regions are homogeneous (Pal and Pal, 1993). In the next step, the segmented image is used along with textural and contextual information as well as the spectral information to produce a thematic map of land use/cover. A hierarchy of levels of segmentation, ranging from a few large objects to many smaller objects, can be obtained with each object belonging to a superobject at a higher level of segmentation and smaller objects at a lower level of segmentation. The spectral characteristics of each object include not only an average value for each band from the pixels participating in the object but also statistical values such as minimum, maximum, and standard deviation regarding each band. In addition, the objects are described by shape, size, tone, texture, compactness, and other characteristics describing the spatial features of the object (Bock et al., 2005). All of those variables can be used in the classification process to assist in the discrimination of the objects and their correct assignment to the land use/cover classes. The characteristic of inclusion of spectral as well as spatial information of image objects constitutes the main advantage of the object-based classifiers. Most importantly, consideration of object attributes (e.g., shape, heterogeneity) results in the reduction of the "salt-and-pepper" effect and "edge" usually seen in pixel-based classification. Also, it should be noted that in comparison to pixel-based classifiers, object-based classification generally requires higher user expertise to be implemented. Also, generally, the accuracy of segmentation directly affects the performance of the object-based classification, and several studies have shown that only good segmentation results can lead to object-oriented image classification outperforming pixel-based classification (e.g., Yan et al., 2006).

15.3.2.3 *Spectral Unmixing*   This is a very different classification approach that divides the pixel into proportions of different spectral components called end members, and further knowledge of their proportion to the overall spectral response of the pixel is required (Hostert et al., 2003; Okin et al., 2001; Zhang et al., 2006). In general, the number of end members selected must be less than the number of sensor spectral bands in order for this technique to be applied. Also, all of the end members present in the image should be employed in order to get reliable results. The output of spectral unmixing is one image for each end member, with pixel values between zero and one, representing the fraction of the original image attributed to the particular end member. Subpixel classification approaches are generally divided into linear and nonlinear unmixing, depending on whether it is assumed that the reflectance at each pixel of the image is a linear or a nonlinear combination of the reflectance of each material present within the pixel (Small, 2001; Plaza et al., 2009). Hyperspectral data are appropriate for linear mixture model since it has much more spectral bands. The mixture model is also good for small object detection or early warning of invasive species. A good summary of the linear mixture model and different constraints can be found in Miao et al. (2006).

15.3.2.4 *Land Cover Mapping Accuracy Assessment*   Classification accuracy of the land cover maps derived from hyperspectral images is generally evaluated based on the computation of overall accuracy (OA), user's (UA) and producer's (PA) accuracy, and the kappa (K_c) statistic (Congalton and Green, 1999). The OA expresses as a percentage the probability that a pixel is classified correctly by the thematic map and is a measure of the overall classification accuracy. The statistic K_c measures the actual agreement between reference data and the classifier used to perform the classification versus the chance of agreement between the reference data and a random classifier. The PA for a certain class expresses what percentage of a category on the ground is correctly classified by the analyst and can define a measure of pixels omitted from its reference class (omission error). Likewise, UA expresses the percentage of pixels of a category that do not "truly" belong to the reference class but are committed to other ground truth classes (commission error). In mathematical terms, these parameters are expressed as follows (Congalton and Green, 1999; Liu et al., 2007):

$$OA = \frac{1}{N} \sum_{i=1}^{r} n_{ii} \tag{15.2}$$

$$PA = \frac{n_{ii}}{n_{i,\text{col}}} \tag{15.3}$$

$$UA = \frac{n_{ii}}{n_{i,\text{row}}} \tag{15.4}$$

$$K_c = N \sum_{i=1}^{r} n_{ii} - \sum_{i=1}^{r} \frac{n_{i,\text{col}} n_{i,\text{row}}}{N^2} - \sum_{i=1}^{r} n_{i,\text{col}} n_{i,\text{row}} \tag{15.5}$$

where n_{ii} is the number of pixels correctly classified in a category, N is the total number of pixels in the confusion matrix, r is the number of rows, and $n_{i,col}$ and $n_{i,row}$ are the column (reference data) and row (predicted classes) total, respectively.

In computing the above statistical measures, independent validation points (i.e., pixels) from each classification class need to be selected. Those validation points are generally selected in homogeneous regions of each land cover class included in the classification scheme and away from the locations where the training points are collected, ensuring nonoverlap of pixels between the training data and validation sites.

15.3.3   Hyperspectral Remote Sensing of Land Cover Mapping: Case Studies

During the last decades, a number of airborne and satellite hyperspectral sensing system sensors have been launched. The recent availability of such data has also encouraged the development of several techniques for analyzing the rich information content provided by hyperspectral imagery. Hyperion is a satellite hyperspectral sensor onboard the Earth Observer-1 (EO-1) satellite, launched under NASA's New Millennium Program near the end of 2000. Hyperion acquires images at 30 m spatial resolution and at about 10 nm spectral resolution in 242 spectral bands in total, 70 of which are found in the VNIR and 172 bands in the short-wave infrared (SWIR). This sensor is regarded as the first "real" spaceborne hyperspectral remote sensing instrument in orbit. The availability of Hyperion data has created unique opportunities for remote sensing studies to be conducted exploring their potential use in land use/cover thematic mapping extraction.

Various studies have examined the combined use of Hyperion with different pixel-based classification techniques for land use/cover mapping (Goodenough et al., 2003; Walsh et al., 2008a,b; Pignatti et al., 2009). Furthermore, both linear and nonlinear unmixing classification combined with Hyperion for land classification has also been investigated in a few studies (e.g., Walsh et al., 2008a,b; Pignatti et al., 2009). Others have also explored the use of object-based classification with Hyperion imagery analysis for performing land use/cover mapping (Walsh et al., 2008; Wang et al., 2010). For example, Walsh et al. (2008) compared the performances of three classifiers, namely Spectral Angle Mapper (SAM), spectral unmixing, and object-based classification combined with Hyperion imagery for mapping invasive plant species in Ecuador, and reported that the object-based classification outperformed the other two techniques. Wang et al. (2010) applied an object-based classification to a Hyperion imagery acquired for a test region in China and showed an overall accuracy ranging from 72 to 88%, depending on the number of classes delineated.

Yet, from a review of the literature it can be observed that a few studies reporting comparative analysis of the performance of different classification approaches with hyperspectral imagery for land use/cover mapping have been conducted so far (Petropoulos et al., 2012a). For example, Pal and Mather (2005) compared different pixel-based classifiers, including SVMs and ANNs from the multispectral LANDSAT ETM+ and Digital Airborne Imaging Spectrometer (DAIS) airborne hyperspectral sensor respectively, for two test sites in the United Kingdom and Spain. Other authors also reported higher classification accuracy from SVMs in comparison to the other two classifiers, with a difference in overall accuracy of ~2–5%. Karimi et al. (2006) evaluated the usefulness of SVMs and ANN for hyperspectral imagery classification over an agricultural area using data from an airborne hyperspectral sensing system flown over a region in Canada. Authors also found the SVMs to outperform the ANN classifier by approximately 15% and 0.114 in overall accuracy and kappa coefficient, respectively. Pal (2006) examined the combined use of SVMs with hyperspectral imagery from the DAIS airborne sensor for a test region in Spain and reported overall classification accuracy for SVMs higher that 91%. Koetz et al. (2008) examined the combined use of hyperspectral and lidar data with SVMs for fuel types mapping for a region in France. Other authors reported an overall accuracy and kappa coefficient of 69.15% and 0.645, respectively, when the SVMs was applied only with the hyperspectral imagery, which increased by 6.3% and 0.115, respectively, when the two data sets were combined. More recently, Petropoulos et al. (2012a) evaluated the combined use of Hyperion imagery with SVMs and ANN classifiers for a heterogeneous region in Greece. Their results showed a close classification accuracy between the two classifiers (higher than 85% in both cases), with the SVMs somehow outperforming the ANN by 3.31% overall accuracy and by 0.038 kappa coefficient (Figure 15.5). In another study, the same authors also compared SVMs and object-based classification for another region in Greece representative of typical Mediterranean conditions. Findings from their work showed that both classifiers were able to produce comparatively accurate land cover maps of the studied area. Overall accuracy and kappa for object-based classification were 81% and 0.779, respectively, whereas for SVMs were 76% and 0.719 respectively.

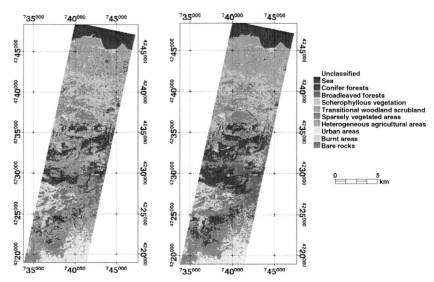

FIGURE 15.5 Hyperion pixel-based classification using SVM classifier (left) and ANNs (right). Adopted from Petropoulos et al. (2012a). (See the color version of this figure in Color Plates section.)

15.3.4 Importance of Observation Scale in Land Cover Extraction from Remote Sensing

The most stable and reliable source of land cover information that can be obtained on the ground and be linked between remote sensing studies obtained at different scales is the spectral response of the cover (McCoy, 2004). Field spectroradiometry and spectral library data have demonstrated already that there is a lot to offer. For example, Rao et al. (2007) showed the application of field spectroradiometry data of agricultural crops in India, at the canopy and pixel scales, for classification on Hyperion imagery. Promising results of 86.5 and 88.8% overall classification accuracy for both scales respectively were achieved. Nevertheless, the similarity and variability of the spectral properties of land cover inevitably render high mapping uncertainty in pixel-based analysis of heterogeneous and cover fragmented landscapes, present in many geographic regions (Xie et al., 2008).

The spectral signature of vegetation is primarily driven by the different bio-chemical content and pigment quantity, plant architecture, and growing factors. A certain degree of variation is thus naturally expected to occur in the reflectance values between different plants. Field spectroradiometry takes the benefit of the ground scale in detecting such variation because it is naturally contained in the spectral information recorded on the field and can indeed be used in enhancing the discrimination between different covers. It also has potential for identification of spectral bands that are correlated with land cover properties other than type, such as vegetation health status or leaf chemical compounds, with high accuracy, though it remains challenging (Abdel-Rahman et al., 2009; Thenkabail et al., 2004b). Furthermore, discriminant analysis of field spectral data offers an important understanding of the cover discrimination potential prior to image classification (McCoy, 2004). If the differences of the reflectance between the land covers being compared in certain wavelengths are significant in statistical means, the results can then provide a sound basis for future remote sensing mapping. In addition, field hyperspectral data contain information about the land cover internal properties in its natural environment as observed from the highly elevated remote sensors. Such data can be additionally used in digital imagery for proper selection of training pixels for land cover mapping or for pixel unmixing in more detailed analysis of a subscene (Borengasser, 2007). Therefore, the advances in sensor technology in higher spatial and spectral resolution over broader spectral regions impose the need for more accurate and representative spectral signatures for solving the land cover variability on imagery data (Salisbury, 1998). Both field discrimination and image classification face the need for distinct spectral signatures obtained on the field scale, especially for economically and ecologically important land covers such as vegetation.

Of special importance in land cover information extraction and analysis on the ground scale is the process of planning and executing of fieldwork in remote sensing studies. The spectral reflectance is an inherent property of an object, independent of time, location, illumination intensity, and atmospheric conditions (Peddle et al., 2001). However, this property is strongly affected by the local environment and the behavior of the field equipment. The benefit of a systematic approach to planning and properly executing field spectroradiometric work in remote sensing lies in identifying pitfalls and problems and selecting appropriate solutions in advance (McCoy, 2004). The variability of the ground measurements factors, together with the natural variability in land cover such as vegetation, constrain the results from hyperspectral image vegetation analysis studies. Support of ground-collected spectral libraries are to be interpreted only relative to their specific collection phenomenology/methodology and study area (Salvaggio et al., 2005). That makes the linkage and comparison of in situ data with other remote sensing data complex. If in situ spectra measurements are significantly linked with the image end members, only then should the field data serve to validate the identity of the vegetation type mapped.

15.4 FINAL REMARKS

The rapid development of remote sensing sensor technology makes it increasingly and steadily more and more feasible to derive land cover information from hyperspectral Earth observation data acquired at different observational scales (Ben-Dor et al., 2009; Herold, 2009; Underwood et al., 2007).

Although a novel technology, field spectroradiometry land cover study, especially for vegetation, has led to remarkable achievements and has proven the significant capability in land cover analysis domain over the past 15 years. Both multi- and univariate statistics proved their greatest ability to discriminate vegetation land cover reflectance, as determined by the reflectance analysis of variance, in bands of the VIS and especially in the NIR spectrum, including the red edge between 680 and 730 nm. Single bands yielded by field scale dimension–reduction studies for discrimination between targets of a certain land cover such as vegetation species are reported within the NIR, but findings remain to be confirmed on satellite imagery in future investigations. In addition, experiences gained by vegetation discriminant analyses at the field scale so far indicate an increasing importance of the SWIR spectrum, considering its relative ability to characterize water content in vegetation cover.

Many studies have been performed in highly heterogeneous or vegetation fragmented regions, aiming to derive information on land use/cover distribution. It is thus expected land cover to cluster on satellite image to satisfactory classification accuracy values if the imagery scale is smaller than or equal to the FoV used on the ground. The full potential of field spectroradiometry is yet to come.

Indeed, various studies conducted in dissimilar ecosystems and implementation settings have already demonstrated that very accurate maps of land cover can be derived from hyperspectral systems mounted in both airborne and satellite platforms. Different methods combined with such data via the computation of standardized approaches based on error matrix statistics have generally reported, high overall accuracies and kappa coefficients, reaching or exceeding in some cases 85% and 0.800. These are generally regarded as very satisfactory and accurate estimates for many practical applications on which regional maps of land cover might be required.

However, field scale remote sensing such as field spectroradiometry to hyperspectral imaging remote sensing analysis of land cover has not yet been fully implemented in practice. Many of these studies suggest that, prior to any land cover analysis from remotely sensed air- and spaceborne imagery at the species level, having knowledge about species spectral separability is vital. Yet this is not a trivial task. This is because comparison of spectral signatures obtained from the field to satellite hyperspectral remote sensing sensors is hindered by complications involved in accounting for various factors, such as the physical setup of the sensors and the measurement environment, the latter being especially variable in time and space on Earth's surface. Also, scale-related factors have to be taken into account to minimize the remote measurement discrepancy between what is actually at ground level and what is perceived from remotely sensed imagery before data can be of use.

All in all, to improve estimation of land cover from hyperspectral remote sensing data, it is important to acquire an integrated knowledge of the spectral properties of the land cover targets coupled with an understanding of the factors that affect the variations of their spectral signature at given spectral and spatial scales. In this framework, the synergy of contemporary image processing techniques combined with sophisticated hyperspectral imagery available nowadays from a range of highly sophisticated hyperspectral sensing systems supported by hyperspectral data collection acquired at different observational scales should be further investigated at different ecosystem settings.

ACKNOWLEDGMENTS

Authors wish to thank the anonymous reviewers for useful comments on the manuscript. Dr. Petropoulos wishes to thank INFOCOSMOS E.E. (http://www.infocosmos.eu/rsgis/index.html) for the support and encouragement provided in completing the present work.

REFERENCES

Abdel-Rahman, E. M., Ahmed, F. B., and van den Berg, M. 2010. Estimation of sugarcane leaf nitrogen concentration using in situ spectroscopy. *International Journal of Applied Earth Observation and Geoinformation* 12:S52–S57.

Abdi, H. 2007. *Bonferroni and Sidak Corrections for Multiple Comparisons*. Thousand Oaks, CA: Sage.

Adam, E., and Mutanga, O. 2009. Spectral discrimination of papyrus vegetation (*Cyperus papyrus L.*) in swamp wetlands using field spectrometry. *ISPRS Journal of Photogrammetry and Remote Sensing* 64(6):612–620.

Adler-Golden, S. M., Matthew, M. W., Bernstein, L. S., Levine, R. Y., Berk, A., Richtsmeier, S. C., and Burke, H. H. 1999. Atmospheric correction for short-wave spectral imagery based on MODTRAN4. *Imaging Spectrometry V* 3753:61–69.

Artigas, F. J., and Yang, J. S. 2006. Spectral discrimination of marsh vegetation types in the New Jersey Meadowlands. *USA Wetlands* 26(1):271–277.

Analytical Spectral Devices (ASD), A. S. D. I. 2009. *Technical Guide* (available upon request to ASD Inc.). In D. C. Hatchell (Ed.) (Vol. 4th edition), Boulder, CO.

Asner, G. P., Wessman, C. A., Bateson, C. A., and Privette, J. L. 2000. Impact of tissue, canopy, and landscape factors on the hyperspectral reflectance variability of arid ecosystems. *Remote Sensing of Environment* 74(1):69–84.

Avery, T. E., and Berlin, G. L. 1992. *Fundamentals of Remote Sensing and Airphoto Interpretation,* 5th ed. New York: MacMillan.

Baldridge, A. M., Hook, S. J., Grove, C. I., and Rivera, G. 2009. The ASTER spectral library version 2.0. *Remote Sensing of Environment* 113(4):711–715.

Beal, D., and Eamon, M. 1999. Preliminary Results of Testing and a Proposal for Radiometric Error Correction Using Dynamic, Parabolic Linear Transformations of "Stepped" Data: PCORRECT.EXE (available upon request to ASD Inc.).

Ben-Dor, E., Chabrillat, S., Dematte, J. A. M., Taylor, G. R., Hill, J., Whiting, M. L., and Sommer, S. 2009. Using imaging spectroscopy to study soil properties. *Remote Sensing of Environment* 113:S38–S55.

Benz, U. C., Hofmann, P., Willhauck, G., Lingenfelder, I., and Heynen, M. 2004. Multi-resolution, object-oriented fuzzy analysis of remote sensing data for GIS-ready information. *ISPRS Journal of Photogrammetry and Remote Sensing* 58:239–258.

Bock, M., Xofis, P., Mitchley, J., Rossner, G., and Wissen, M. 2005. Object-oriented method for habitat mapping at multiple scales—case studies from northern Germany and Wye Downs, UK. *Journal for Nature Conservation* 13:75–89.

Borengasser, M. 2007. *Hyperspectral Remote Sensing and the Atmosphere Hyperspectral Remote Sensing.* Boca Raton, FL: CRC Press, pp. 31–38.

Borengasser, M., Hungate, W. S., and Watkins, R. 2007. *Imaging Spectrometers Hyperspectral Remote Sensing.* Boca Raton, FL: CRC Press, pp. 17–29.

Brennan, B., and Bandeen, W. R. 1970. Anisotropic reflectance characteristics of natural earth surfaces. *Applied Optics* 9(2):405.

Campbell, J. B. 1996. *Introduction to Remote Sensing,* 2nd ed. London: Taylor & Francis.

Carrao, H., Goncalves, P., and Caetano, M. 2008. Contribution of multispectral and multi-temporal information from MODIS images to land cover classification. *Remote Sensing of Environment* 112:986–997.

Castillejo-González, I. L., López-Granados, F., García-Ferrer, A., Peña-Barragán, J. M., Jurado-Expósito, M., Sánchez de la Orden, M., and González-Audicana, M. 2009. Object- and pixel-based analysis for mapping crops and their agro-environmental associated measures using QuickBird imagery. *Computers and Electronics in Agriculture* 68:207–215.

Castro-Esau, K. L., Sanchez-Azofeifa, G. A., and Caelli, T. 2004. Discrimination of lianas and trees with leaf-level hyperspectral data. *Remote Sensing of Environment* 90(3):353–372.

Chang, J., Clay, S. A., and Clay, D. E. 2005. Clouds influence precision and accuracy of ground-based spectroradiometers. *Communications in Soil Science and Plant Analysis* 36 (13–14):1799–1807.

Chintan, A. S., Arora, M. K., and Pramod, K. V. 2004. Unsupervised classification of hyperspectral data: An ICA mixture model based approach. *International Journal of Remote Sensing* 25:481–487.

Cho, M. A., Skidmore, A. K., and Sobhan, I. 2009. Mapping beech (*Fagus sylvatica L.*) forest structure with airborne hyperspectral imagery. *International Journal of Applied Earth Observation and Geoinformation* 11(3):201–211.

Cihlar, J. 2000. Land cover mapping of large areas from satellites: Status and research priorities. *International Journal of Remote Sensing* 21:1093–1114.

Clark, M. L., Roberts, D. A., and Clark, D. B. 2005. Hyperspectral discrimination of tropical rain forest tree species at leaf to crown scales. *Remote Sensing of Environment* 96(3–4):375–398.

Cochrane, M. A. 2000. Using vegetation reflectance variability for species level classification of hyperspectral data. *International Journal of Remote Sensing* 21(10):2075–2087.

Congalton, R., and Green, K. 1999. *Assessing the Accuracy of Remotely Sensed Data: Principles and Practices.* Boca Raton, FL: CRC/Lewis Press.

Council Directive 92/43/EEC. 1992. Of the Conservation of Natural Habitats and Wild Flora and Fauna, the Council of the European Communities.

Curran, P. J. 2001. Imaging spectrometry for ecological applications. *International Journal of Applied Earth Observation and Geoinformation* 3(4):305–312.

Dalponte, M., Bruzzone, L., Vescovo, L., and Gianelle, D. 2009. The role of spectral resolution and classifiers complexity in the analysis of hyperspectral images of forest areas. *Remote Sensing of Environment* 113(11):2345–2355.

De Carvalho, O. A., and Meneses, P. R. 2000. Spectral Correlation Mapper (SCM): An Improvement on the Spectral Angle Mapper (SAM). Summaries of the 9th JPL Airborne Earth Science Workshop, JPL Publication 00-18.

De Kok, R., Schneider, T., and Ammer, U. 1999. Object-based classification and applications in the alpine forest environment. In fusion of sensor data, knowledge sources and algorithms: Proceedings of the joint ISPRS/EARSEL workshop, 3–4 June 1999, Valladolid, Spain. *International of Archives Photogrammetry and Remote Sensing* 32:7-4-3 W6.

Dimitrakopoulos, A. P. 2001. A statistical classification of Mediterranean species based on their flammability components. *International Journal of Wildland Fire* 10(2):113–118.

Du, P., Tan, K., and Xing, X. 2010. Wavelet SVM in reproducing kernel Hilbert space for hyperspectral remote sensing image classification. *Optics Communications* 283 (24):4978–4984.

Elatawneh, A., Kalaitzidis, C., Petropoulos, G. P., and Schneider, T. 2012. Evaluation of diverse classification approaches for land use/cover mapping in a Mediterranean region utilizing Hyperion data. *International Journal of Digital Earth* 1–23.

Exelis. 2009. ENVI User's Guide: ENVI Software, Exelis Visual Information Solutions, www.exlivis.com.

Fassnacht, K. S., Cohen, W. B., and Spies, T. A. 2006. Key issues in making and using satellite-based maps in ecology: A primer. *Forest Ecology and Management* 222:167–181.

Frank, M., and Menz, G. 2003. Detecting Seasonal Changes in a Semi-arid Environment Using Hyperspectral Vegetation Indices. Paper presented at the Proceedings of the 3rd EARSel Workshop on Imaging Spectrometry.

Galvao, L. S., Roberts, D. A., Formaggio, A. R., Numata, I., and Breunig, F. M. 2009. View angle effects on the discrimination of soybean varieties and on the relationships between vegetation indices and yield using off-nadir Hyperion data. *Remote Sensing of Environment* 113:846–856.

Gao, B. C., Montes, M. J., Davis, C. O., and Goetz, A. F. H. 2009. Atmospheric correction algorithms for hyperspectral remote sensing data of land and ocean. *Remote Sensing of Environment* 113:S17–S24.

Goodenough, D. G., Dyk, A., Niemann, O., Pearlman, J. S., Chen, H., Han, T., Murdoch, M., and West, C. 2003. Processing HYPERION and ALI for Forest Classification. *IEEE Trans. Geosci. Remote Sensing* 41(2):1321–1331.

Hadjimitsis, D. G., Padavid, G., and Agapiou, A. 2009. Surface reflectance retrieval from Landsat TM/ETM+ images for monitoring irrigation demand in cyprus. Paper presented at the EARSeL Symposium "Imagine Europe," Chania, Greece.

Harris, J. W., and Stocker, H. 1998. Maximum likelihood method. §21. 10. 4 in *Handbook of Mathematics and Computational Science*. New York: Springer, p. 824.

Hemmer, T. H., and Westphal, T. L. 2000. Lessons learned in the post-processing of field spectroradiometric data covering the 0.4 to 2.5 im wavelength region. In S. S. Shen and M. R. Descour (Eds.) *Algorithms for Multispectral, Hyperspectral, and Ultraspectral Imagery VI, Proceedings of SPIE*, Vol. 4049, p. 249.

Henkabail, P. S., Smith, R. B., and Pauw, E. D. 2000. Hyperspectral vegetation indices and their relationships with agricultural crop characteristics. *Remote Sensing Environment* 71:158–182.

Herold, M. 2009. Assessment of the status of the development of the standards for the Terrestrial Essential Climate Variables. Rome, Italy: The Global Terrestrial Observing System-GTOS.

Holden, H., and LeDrew, E. 1998. Spectral discrimination of healthy and non-healthy corals based on cluster analysis, principal components analysis, and derivative spectroscopy. *Remote Sensing of Environment* 65(2):217–224.

Hostert, P., Roder, A., and Hill, J. 2003. Coupling spectral unmixing and trend analysis for monitoring of long-term vegetation dynamics in Mediterranean rangelands. *Remote Sensing of Environment* 87(2–3):183–197.

Hueni, A., Nieke, J., Schopfer, J., Kneubuhler, M., and Itten, K. I. 2009. The spectral database SPECCHIO for improved long-term usability and data sharing. *Computers & Geosciences* 35(3):557–565.

Hughes, G. F. 1968. On the mean accuracy of statistical pattern recognition. *IEEE Transactions on Information Theory IT* 14:55–63.

Jacobsen, A., Broge, N. H., and Hansen, B. U. 1995. Monitoring wheat fields and grasslands using spectral reflectance data. Paper presented at the 79 International Symposium on Spectral Sensing Research (ISSSR), Victoria, Australia.

Jurado-Exposito, M., Lopez-Granados, F., Atenciano, S., Garcia-Torres, L., and Gonzalez-Andujar, J. L. 2003. Discrimination of weed seedlings, wheat (*Triticum aestivum*) stubble and sunflower (*Helianthus annuus*) by near-infrared reflectance spectroscopy (NIRS). *Crop Protection* 22(10):1177–1180.

Karimi, Y., Prasher, S. O., Patel, R. M., and Kim, S. H., 2006. Application of support vector machines technology for weed and nitrogen stress detection in corn. *Computers and Electronics in Agriculture* 51:99–109.

Kavzoglu, T., and Colkesen, I. 2009. A kernel functions analysis for support vector machines for land cover classification. *International Journal of Applied Earth Observation and Geoinformation* 11:352–359.

Kochubey, S. M., and Kazantsev, T. A. 2007. Changes in the first derivatives of leaf reflectance spectra of various plants induced by variations of chlorophyll content. *Journal of Plant Physiology* 164(12):1648–1655.

Koetz, B., Morsdorf, F., Linden, S., van der Curt, T., and Allgower, B. 2008. Multi-source land cover classification for forest fire management based on imagine spectrometry and LiDAR data. *Forest Ecology and Management* 256(3):263–271.

Kruse, F. A., Lefkoff, A. B., Boardman, J. W., Heidebrecht, K. B., Shapiro, A. T., Barloon, P. J., and Goetz, A. F. H. 1993. The Spectral Image Processing System (SIPS)—Interactive visualization and analysis of imaging spectrometer data. *Remote Sensing of Environment* 44:145–163.

Lacar, F. M., Lewis, M. M., and Grierson, I. T. 2001. Use of hyperspectral reflectance for discrimination between grape varieties. Paper presented at the Geoscience and Remote Sensing Symposium, 2001. IGARSS '01. IEEE 2001 International.

Lewis, M. M. 2001. Discriminating vegetation with hyperspectral imagery—what is possible? Paper presented at the Geoscience and Remote Sensing Symposium, 2001. IGARSS '01. IEEE 2001 International.

Li, D-C., and Liu, C-W. 2010. A class possibility based kernel to increase classification accuracy for small data sets using support vector machines. *Expert Systems with Applications* 37:3104–3110.

Liu, C., Frazier, P., and Kumar, L. 2007. Comparative assessment of the measures of thematic classification accuracy. *Remote Sensing of Environment* 107:606–616.

Lu, D., and Weng, Q. 2007. A survey of image classification methods and techniques for improving classification performance. *International Journal of Remote Sensing* 28(5):823–870.

Manakos, I., Manevski, K., Petropoulos, G. P., Elhag, M., and Kalaitzidis, C. 2010. Development of a spectral library for Mediterranean land cover types. Paper presented at the 30th EARSeL Symp.: Remote Sensing for Science, Education and Natural and Cultural Heritage, Chania, Greece.

Manevski, K., Manakos, I., Petropoulos, G. P., and Kalaitzidis, C. 2011. Discrimination of common Mediterranean plant species using field spectroradiometry. *International Journal of Applied Earth Observation and Geoinformation* 13(6):922–933.

Manevski, K., Manakos, I., Petropoulos, G. P., and Kalaitzidis, C. 2012. Spectral discrimination of Mediterranean maquis and phrygana vegetation: Results from a case study in greece. *IEEE Journal of Selected Topics in Applied Earth Observations and Remote Sensing* 5(2):604–616.

Mariotti, M., Ercoli, L., and Masoni, A. 1996. Spectral properties of iron-deficient corn and sunflower leaves. *Remote Sensing of Environment* 58(3):282–288.

Mathur, A., Bruce, L. M., and Byrd, J. 2002. Discrimination of subtly different vegetative species via hyperspectral data. Igarss 2002: IEEE International Geoscience and Remote Sensing Symposium and 24th Canadian Symposium on Remote Sensing, Vols I–Vi, Proceedings, pp. 805–807.

McCoy, R. M. 2004. *Field Methods in Remote Sensing*. Guilford Publications. New York, NY.

Miao, X., Gong, P., Swope, S., Pu, R., Carruthers, R., Anderson, G., Heaton, J., and Tracy, C. R. 2006. Estimation of yellow starthistle abundance through CASI-2 hyperspectral imagery using linear spectral mixture models. *Remote Sensing of Environment* 101:329–341.

Milton, E. J., Fox, N. P., and Schaepman, M. E. 2006. Progress in field spectroscopy. *IEEE International Geoscience and Remote Sensing Symposium*, Vols. 1–8, pp. 1966–1968.

Milton, E. J., Schaepman, M. E., Anderson, K., Kneubuhler, M., and Fox, N. 2009. Progress in field spectroscopy. *Remote Sensing of Environment* 113:S92–S109.

Morrey, J. R. 1968. On determining spectral peak positions from composite spectra with a digital computer. *Analytical Chemistry* 40(6):905–914.

Mutanga, O., Skidmore, A. K., and van Wieren, S. 2003. Discriminating tropical grass (*Cenchrus ciliaris*) canopies grown under different nitrogen treatments using spectroradiometry. *ISPRS Journal of Photogrammetry and Remote Sensing* 57(4):263–272.

National Institute of Standards and Technology (NIST) 2003. *E-Handbook of Statistical Methods: Tukey's Method.* Retrieved June, 2010, from http://www.itl.nist.gov/div898/handbook/prc/section4/prc471.htm.

Nicodemus, F. F., Richmond, J. C., Hsia, J. J., Ginsberg, I. W., and Limperis, T. L. 1977. *Geometrical Considerations and Nomenclature for Reflectance* (Vol. Monograph 160) Washington, DC: National Bureau of Standards.

Okin, G. S., Roberts, D. A., Murray, B., and Okin, W. J. 2001. Practical limits on hyperspectral vegetation discrimination in arid and semiarid environments. *Remote Sensing of Environment* 77(2):212–225.

Otukei, J. R., and Blaschke, T. 2010. Land cover change assessment using decision trees, support vector machines and maximum likelihood classification algorithms. *International Journal of Applied Earth Observation and Geoinformation* S12:S27–S31.

Pal, M. 2006. Support vector machine-based feature selection for land cover classification: A case study with DAIS hyperspectral data. *International Journal of Remote Sensing* 27(14):2877–2894.

Pal, M., and Mather, P. M. 2005. Some issues in the classification of DAIS hyperspectral data. *International Journal of Remote Sensing* 27:2895–2916.

Pal, N. R., and Pal, S. K. 1993. A review of image segmentation techniques. *Pattern Recognition* 26:1277–1294.

Peddle, D. R., White, H. P., Soffer, R. J., Miller, J. R., and LeDrew, E. F. 2001. Reflectance processing of remote sensing spectroradiometer data. *Computers & Geosciences* 27(2):203–213.

Petropoulos, G. P., Kalaitzidis, C., and Vadrevu, K. P. 2011. Support Vector machines and object-based classification for obtaining land use/cover cartography from hyperion hyperspectral imagery. *Computers and Geosciences* 41:99–107.

Petropoulos, G. P., Arvanitis, K., and Sigrimis, N. 2012a. Hyperion hyperspectral imagery analysis combined with machine learning classifiers for land use/cover mapping. *Expert Systems with Applications* 39:3800–3809.

Petropoulos, G. P., Vadrevu, K. P., and Kalaitzidis, C. 2012b. Spectral Angle mapper and object-based classification combined with hyperspectral remote sensing imagery for obtaining land use/cover mapping in a mediterranean region. *Geocarto International* 28(2):1–16.

Pfitzner, K., Bollh, F. A., and Carr, G. 2006. A standard design for collecting vegetation reference spectra: Implementation and implications for data sharing. *Journal of Spatial Science* 51(2):79–92.

Pignatti, S., Cavalli, R. M., Cuomo, V., Fusilli, V., Pascucci, S., Poscolieri, M., and Santini, F. 2009. Evaluating hyperion capability for land cover mapping in a fragmented ecosystem: Pollino National Park, Italy. *Remote Sensing of Environment* 113:622–634.

Plaza, J., Plaza, A., Perez, R., and Martinez, P. 2009. On the use of small training sets for neural network-based characterisation of mixed pixels in remotely sensed hyperspectral images. *Pattern Recognition* 42:3032–3045.

Price, J. C. 1994. How unique are spectral signatures? *Remote Sensing of Environment* 49(3):181–186.

Price, J. C. 1998. An approach for analysis of reflectance spectra. *Remote Sensing of Environment* 64(3):316–330.

Psomas, A., Zimmermann, N. E., Kneubühler, M., Kellenberger, T., and Itten, K. 2005. Seasonal variability in spectral reflectance for discriminating grasslands along a dry- mesic gradient in Switzerland. Paper presented at the 4th EARSeL Workshop on Imaging Spectroscopy. New Quality in Environmental Studies, 27–30 April, Warsaw, Poland.

Rao, N. R., Garg, P. K., and Ghosh, S. K. 2007. Development of an agricultural crops spectral library and classification of crops at cultivar level using hyperspectral data. *Precision Agriculture* 8(4–5):173–185.

Robson, C. 1994. *Experiment, Design and Statistics in Psychology* (3rd ed.). Penguin Books.

Salisbury, J. W. 1998. *Spectral Measurements Field Guide*. Defense Technical Information Center.

Salvaggio, C., Smith, L. E., and Antoine, E. J. 2005. Spectral signature databases and their application/misapplication to modeling and exploitation of multispectral/hyperspectral data. *Algorithms and Technologies for Multispectral, Hyperspectral, and Ultraspectral Imagery XI, Proceedings of SPIE*, Vol. 5806, 531.

Sanchez-Hernandez, C., Boyd, D. S., and Foody, G. M. 2007. Mapping specific habitats from remotely sensed imagery: Support vector machine and support vector data description based classification of coastal saltmarsh habitats. *Ecological Informatics* 2:83–88.

Schaaf, C. B. 2009. Albedo and reflectance anisotropy: Assessment of the status of the development of the standards for the Terrestrial Essential Climate Variables (Vol. 12). Global Terrestrial Observing System. United National Framework Convention on Climate Change, 30th ed. June 1–10, 2009, Bonn.

Schaepman, M. E. 2007. Spectrodirectional remote sensing: From pixels to processes. *International Journal of Applied Earth Observation and Geoinformation* 9(2): 204–223.

Schmidt, K. S., and Skidmore, A. K. 2003. Spectral discrimination of vegetation types in a coastal wetland. *Remote Sensing of Environment* 85(1):92–108.

Schopfer, J., Dangel, S., Kneubuhler, M., and Itten, K. I. 2008. The improved dual-view field goniometer system FIGOS. *Sensors* 8(8):5120–5140.

Short, N. M. 2009. Remote Sensing Tutorial—Reference Publication 1078 Retrieved June 22, 2012, from http://rst.gsfc.nasa.gov/Front/tofc.html.

Small, C. 2001. Estimation of urban vegetation abundance by spectral mixture analysis. *International Journal of Remote Sensing* 22:1305–1334.

Song, C., Woodcock, C. E., Seto, K. C., Lenney, M. P., and Macomber, S. A. 2001. Classification and change detection using Landsat TM data: When and how to correct atmospheric effects? *Remote Sensing of Environment* 75(2):230–244.

Styers, D. M., Chappelka, A. H., Marzen, L. J., and Somers, G. L. 2009. Developing a land-cover classification to select indicators of forest ecosystem health in a rapidly urbanizing landscape. *Landsc. Urban Plan* 94(3–4):158–165.

Thenkabail, P. S., Smith, R. B., and De Pauw, E. 2000. Hyperspectral vegetation indices and their relationships with agricultural crop characteristics. *Remote Sensing of Environment* 71:158–182.

Thenkabail, P. S., Enclona, E. A., Ashton, M. S., Legg, C., and De Dieu, M. J. 2004a. Hyperion, IKONOS, ALI, and ETM plus sensors in the study of African rainforests. *Remote Sensing of Environment* 90(1):23–43.

Thenkabail, P. S., Smith, R. B., and De Pauw, E. 2000. Hyperspectral vegetation indices and their relationships with agricultural crop characteristics. *Remote Sensing of Environment* 71(2):158–182.

Thenkabail, P. S., Enclona, E. A., Ashton, M. S., and Van der Meer, B. 2004b. Accuracy assessments of hyperspectral waveband performance for vegetation analysis applications. *Remote Sensing of Environment* 91(3–4):354–376.

Underwood, E. C., Ustin, S. L., and Ramirez, C. M. 2007. A comparison of spatial and spectral image resolution for mapping invasive plants in coastal California. *Environmental Management* 39(1):63–83.

Vaiphasa, C. 2006. Consideration of smoothing techniques for hyperspectral remote sensing. *ISPRS Journal of Photogrammetry and Remote Sensing* 60(2):91–99.

Vaiphasa, C., Ongsomwang, S., Vaiphasa, T., and Skidmore, A. K. 2005. Tropical mangrove species discrimination using hyperspectral data: A laboratory study. *Estuarine Coastal and Shelf Science* 65(1–2):371–379.

van Aardt, J. A. N., and Wynne, R. H. 2001. Spectral separability among six southern tree species. *Photogrammetric Engineering and Remote Sensing* 67(12):1367–1375.

Van Aardt, J. A. N., and Wynne, R. H. 2007. Examining pine spectral separability using hyperspectral data from an airborne sensor: An extension of field-based results. *International Journal of Remote Sensing* 28(1–2):431–436.

Vapnik, V. 1995. *The Nature of Statistical Learning Theory.* New York: Springer.

Walsh, S. J., McCleary, A. L., Mena, C. F., Shao, Y., Tuttle, J. P., González, A., and Atkinson, R. 2008a. QuickBird and Hyperion data analysis of an invasive plant species in the Galapagos Islands of Ecuador: Implications for control and land use management. *Remote Sensing of Environment* 112:1927–1941.

Walsh, S. J., McCleary, A. L., Mena, C. F., Shao, Y., Tuttle, J. P., González, A., and Atkinson, R. 2008b. QuickBird and Hyperion data analysis of an invasive plant species in the Galapagos Islands of Ecuador: Implications for control and land use management. *Remote Sensing Environment* 112:1927–1941.

Wang, J., Chen, Y., He, T., Lv, C., and Liu, A. 2010. Application of geographic image cognition approach in land type classification using Hyperion image: A case study in China. *International Journal of Applied Earth Observation and Geoinformaiton* 12S: S212–S222.

Williams, D. L. 1991. A comparison of spectral reflectance properties at the needle, branch, and canopy level for selected Conifer species. *Remote Sensing of Environment* 35(2–3):79–93.

Xie, Y. C., Sha, Z. Y., and Yu, M. 2008. Remote sensing imagery in vegetation mapping: A review. *Journal of Plant Ecology–UK* 1(1):9–23.

Yan, G., et al. 2006. Comparison of pixel-based and object-oriented image classification approaches—a case study in a coal fire area, Wud, Inner Mongolia, China. *International Journal of Remote Sensing* 27(18):4039–4055.

Zang, I., Gao, Z., Armitage, R., and Kent, M. 2008. Spectral characteristics of plant communities from salt marshes: A case-study from Chongming Dongtan, Yangtze estuary China. *Frontiers of Environmental Science and Engineering in China* 2:11.

Zhang, J., Fu, M., Tiao, J., Huang, Y., Hassani, F. P., and Bai, Z. 2010. Response of ecological storage and conservation to land use transformation: A case study of a mining town in China. *Ecological Modelling* 221:1427–1439.

Zhang, J. K., Rivard, B., Sanchez-Azofeifa, A., and Castro-Esau, K. 2006. Intra and inter class spectral variability of tropical tree species at La Selva, Costa Rica: Implications for species identification using HYDICE imagery. *Remote Sensing of Environment* 105(2):129–141.

Zhang, R., and Ma, J. 2009. Feature selection for hyperspectral data based on recursive support vector machines. *International Journal of Remote Sensing* 30(14):3669–3677.

Zomer, R. J., Trabucco, A., and Ustin, S. L. 2009. Building spectral libraries for wetlands land cover classification and hyperspectral remote sensing. *Journal of Environmental Management* 90(7):2170–2177.

Zomer, R. J., Ustin, S. L., and Evans, M. 1999. Ground-truth data collection protocol for hyperspectral remote sensing. Retrieved May, 2012, from http://cstars.ucdavis.edu.

INDEX

Scale Issues in Remote Sensing, First Edition. Edited by Qihao Weng.
© 2014 John Wiley & Sons, Inc. Published 2014 by John Wiley & Sons, Inc.